T0392050

Saving Science from Quantum Mechanics

Saving Science from Quantum Mechanics

The Epistemology of the Measurement Problem

EMILY ADLAM

OXFORD
UNIVERSITY PRESS

Oxford University Press is a department of the University of Oxford. It furthers the University's objective of excellence in research, scholarship, and education by publishing worldwide. Oxford is a registered trade mark of Oxford University Press in the UK and in certain other countries.

Published in the United States of America by Oxford University Press
198 Madison Avenue, New York, NY 10016, United States of America.

CIP data is on file at the Library of Congress.

ISBN 9780197808856

DOI: 10.1093/9780197808887.001.0001

Printed by Marquis Book Printing, Canada

The manufacturer's authorized representative in the EU for product safety is Oxford University Press España S.A., Parque Empresarial San Fernando de Henares, Avenida de Castilla, 2 – 28830 Madrid (www.oup.es/en).

Contents

Acknowledgments

I am very grateful to many wonderful teachers and mentors throughout my life. In particular, thanks to David Wallace, Adrian Kent, and Carlo Rovelli, whose intellectual influence on me is no doubt evident in this book; I very much appreciate their guidance and encouragement.

I would also like to thank a number of people who have helped me continue in the field and thus gave me the opportunity to write this book – particularly Francesca Vidotto, Chris Smeenk, Richard Healey, Cai Waegell, Kelvin McQueen, and other colleagues at the University of Western Ontario and Chapman University.

Thanks to Alyx, Jenn and Cassandra for their support and companionship while I was writing this book. Finally, heartfelt thanks to my mother and Tom for their unwavering support of my academic interests, and special thanks to Tom for introducing me to quantum mechanics.

1
Prologue

More than one hundred years have passed since the initial development of quantum mechanics, and yet to this day a satisfactory interpretation of the theory remains frustratingly elusive. The heart of the difficulty is the 'measurement problem,' which refers to the challenges we have faced in seeking to give a full physical account of the process of quantum measurement.

Now, the measurement problem is often presented as if it is just a problem of ontology, a gap in the theory which we would like to fill one day if we can manage it. But this undersells the importance of the problem. There is indeed a gap in the theory, but it is not just any gap. The 'measurement problem' is named thus for a reason – the problem relates specifically to our understanding of *measurement* – and thus to solve it in a satisfactory way we must be cognizant of the special status of measurement in scientific epistemology. Measurements are physical processes but they are also the essence of scientific enquiry, the point at which our experience makes contact with reality, so they have a complex dual role which must be given its due as we attempt to provide a physical account of their nature. Thus in this book I set out to put the quantum measurement problem in the context of a more general understanding of scientific epistemology and the role that measurement has to play in it.

I will argue that a number of popular approaches to the measurement problem, including the Everett interpretation, the Copenhagen interpretation, perspectival interpretations, and QBism, require us to revise deep and foundational beliefs about reality in ways which potentially undermine the methods of enquiry we have used to arrive at quantum mechanics as well as many of our other scientific beliefs. That said, my purpose in this book is not necessarily to suggest that these approaches are completely unworkable but rather to argue that the epistemic facets of the problem cannot be treated as a sideshow: the necessity for a coherent, self-consistent account of the epistemology of measurement is the very reason why the measurement problem matters so much in the first place, and thus arriving at a sensible epistemology should be a central task for any attempt at solving the problem. Thus for each of the approaches I discuss, I will spend some time considering various ways in which one might attempt to offer a coherent account of measurement and scientific knowledge in the face of the epistemic problems they pose. While I am sceptical that any existing proposal can fully overcome these problems, my hope is that this discussion may inspire further work which will help us better understand the prospects for a workable scientific epistemology in the context of radical conceptual revisions.

It is also important to note that although there exist a number of putative solutions to the measurement problem which don't appear to face such severe epistemic difficulties, these non-unitary approaches aren't currently able to reproduce all the predictions of quantum field theory (QFT), and there are reasons to think that approaches of this kind may never do so, in which case they cannot ultimately be

Saving Science from Quantum Mechanics. Emily Adlam, Oxford University Press. © Oxford University Press (2025).
DOI: 10.1093/9780197808887.003.0001

viable solutions to the measurement problem. And this makes it clear that the measurement problem is not just an issue of underdetermination, as it is sometimes billed – the issue is not that we have too many possible solutions but rather that right now we arguably have *no solution at all* which is both empirically adequate and able to satisfactorily address the epistemic aspects of the problem. This is a major lacuna in our scientific knowledge, and although I will close by making a few suggestions about promising ways forward, ultimately I think the ongoing difficulties we face suggest that progress in this area may require further significant conceptual advances in ways which we are yet to understand.

As we make our way through the landscape of solutions to the measurement problem, we will also have the opportunity to explore its links to a variety of interesting issues in the epistemology of science more generally, including the epistemology of probabilistic theories and multiverse theories, the problem of induction, Bayesianism, naturalness and fine-tuning, and the role of scale and fundamentality in science. Contrasting these issues with difficulties encountered in the context of the measurement problem shows them in a useful new light. And it is of great interest to see a real, concrete scientific problem whose solution depends sensitively on issues of epistemology, on questions about how we know what we know and what it takes for scientific beliefs to be empirically confirmed or justified. Here is an opportunity for a fruitful interplay of ideas between these two domains: the measurement problem can help us gain a deeper understanding of the epistemology of science, and meanwhile a careful consideration of epistemology can help us understand what a successful solution to the measurement problem would really look like, and what the obstacles are to arriving at such a thing.

2
What is the Measurement Problem?

In this chapter I will give a preliminary summary of the measurement problem and its epistemic aspects. I begin here with an introduction to quantum mechanics and the history of the measurement problem intended for those new to the subject; readers who are already familiar with these details might wish to go directly to section 2.3.

2.1 Quantum Mechanics

Quantum mechanics is the physical theory we use to describe the behaviour of matter on atomic and subatomic scales. In fact, the term 'quantum mechanics' actually refers to a broad mathematical framework within which we can formulate specific 'theories' describing various concrete physical systems (Wallace, 2018). This framework can be characterised in simple, general terms, prescribing how to write the states of quantum systems, how these states change over time, and how systems are combined. These ideas are formalised in the following postulates, which define the framework known as 'unitary quantum mechanics' (Nielsen and Chuang, 2011):

- To every physical system we ascribe a Hilbert space, $\mathcal{H}$, which represents the state space of the system, so at any given time the system is completely described by a 'pure state,' i.e. a unit vector $|\psi\rangle$ belonging to H.
- Closed quantum systems evolve by unitary transformations. In particular, a closed quantum system can be associated with an operator H known as the Hamiltonian, and the time evolution of the state of the system is given by the Schrödinger equation $H|\psi(t)\rangle = i\hbar\frac{d|\psi(0)\rangle}{dt}$.
- When we combine two physical systems with state spaces $\mathcal{H}_{\mathcal{X}}, \mathcal{H}_{\mathcal{Y}}$, the state space for the resulting composite system is the tensor product $\mathcal{H}_{\mathcal{X}} \otimes \mathcal{H}_{\mathcal{Y}}$ of the individual state spaces.

The first postulate simply tells us that we should write quantum states as vectors in a particular kind of mathematical space known as a Hilbert space. Such quantum states are also sometimes referred to as 'wavefunctions.' It is often convenient to make use of a basis for a Hilbert space, which refers to a set of vectors in the Hilbert space such that all other vectors in the Hilbert space can be written as linear sums of the vectors belonging to the basis; the number of vectors in such a basis gives the dimension of the corresponding Hilbert space. So, for example, a two-dimensional Hilbert space $\mathcal{H}$ might have a basis $\{|0\rangle, |1\rangle\}$, meaning that we can write any vector in $\mathcal{H}$ in

Saving Science from Quantum Mechanics. Emily Adlam, Oxford University Press. © Oxford University Press (2025).
DOI: 10.1093/9780197808887.003.0002

the form $|\psi\rangle = c_0|0\rangle + c_1|1\rangle$. In particular, if $|\psi\rangle$ is a quantum state then the coefficients c_0 and c_1, which are also known as 'amplitudes,' can be real or imaginary, but we require that the mod-squared amplitudes sum to one, i.e. $|c_0|^2 + |c_1|^2 = 1$. A state written in this form is sometimes referred to as a 'superposition' of the states $|0\rangle$ and $|1\rangle$.

The second postulate tells us how quantum states change over time – they undergo time evolution according to the Schrödinger equation, which means that the evolution of any quantum system is determined by a property of that system known as its Hamiltonian, H. In order to determine the correct Hamiltonian we need to know various physical details about the system in question. Evolution according to the Schrödinger equation is a unitary transformation, and one important hallmark of unitary transformations is that they are linear and reversible; these features play an important role in explaining certain characteristic behaviours of quantum systems.

Finally, the third postulate simply tells us how to put individual quantum systems together to form larger composite systems. In particular, if we start out with two systems in states $|\psi\rangle_0$, $|\psi\rangle_1$ and combine them, then the resulting state before any evolution takes place is $|\psi\rangle_0 \otimes |\psi\rangle_1$, where the tensor product $\otimes$ is a mathematical operation for combining the two vectors $|\psi\rangle_0$ and $|\psi\rangle_1$.

These postulates are simple but also profound. Of course, when we want to use quantum mechanics to describe or predict some real experimental scenario, the basic mathematical postulates must be supplemented with additional physical information about the relevant system – for example, we need to know its Hamiltonian and the dimension of its Hilbert space. So in addition to the abstract framework above, actually applying quantum mechanics requires significant technical knowledge about how to correctly model some particular kind of physical system (Wallace, 2018). But nonetheless, all systems describable in this framework exhibit certain characteristics and sometimes counterintuitive quantum behaviours which follow immediately from the mathematical formalism.

One important consequence of the formalism is as follows. If we put two quantum systems together to create a state space from the tensor product of their individual state spaces, as defined by postulate three, and then we apply a generic unitary operation to the resulting composite system, as defined by postulate two, this will usually produce what is known as an 'entangled' state, meaning that the outcomes of measurements performed on the two individual systems will now be correlated. This leads to what is known as 'non-locality' or 'spooky action at a distance' – it appears that measurements on one system have some kind of instantaneous effect on the other system. I will discuss non-locality in more detail in section 4.0.1.

Another important consequence is that distinct terms in the wavefunction can 'interfere' – that is, a state of the form $|\psi\rangle = c_0|0\rangle + c_1|1\rangle$ sometimes exhibits behaviour reminiscent of a classical wave in which the amplitudes c_0, c_1 attached to the individual terms $|0\rangle$ and $|1\rangle$ can be made to reinforce each other or cancel each other out under certain circumstances. Indeed, this wavelike behaviour is the origin of the term 'wavefunction' and the idea of 'wave-particle duality,' referring to the fact that quantum-mechanical systems sometimes behave like particles but sometimes seem to behave more like waves.

2.1.1 Mixed States and Density Matrices

From the above postulates we can also derive a more general set of quantum states and transformations. For if we consider a composite system $X \otimes Y$ which is as a whole in a pure state ψ, usually the individual parts of the system will not be in pure states – that is, it will not be possible to write their states as unit vectors in a Hilbert space. Rather their states must be represented by mathematical objects known as 'density matrices'. A state vector $|\psi\rangle$ can be turned into a density matrix $|\psi\rangle\langle\psi|$ using a mathematical operation known as the 'outer product', and then a generic density matrix can be written as a weighted sum of such objects: $\rho = \sum_i p_i |i\rangle\langle i|$, where the coefficients p_i must be real and non-negative, and $\sum_i p_i = 1$. States of this kind are referred to as mixed states because they have the form of a probabilistic mixture of pure states. The mixed state ρ_X of system X can be derived from the pure state ψ of system $X \otimes Y$ using a mathematical operation known as the 'trace', $Tr_Y()$, which simply amounts to discarding the information about the state of Y, – so we can write $\rho_X = Tr_Y(\psi)$.

Similarly, if the composite system $X \otimes Y$ now undergoes some unitary evolution, then when we try to understand the impact of this evolution on the systems X and Y individually, we will usually find that they undergo changes which cannot be described as a unitary evolution. Instead the evolution of the individual parts of the system must be represented by a mathematical object called a 'completely positive trace-preserving' (CPTP) map – here, the requirements of complete positivity and trace-preservation simply serve to ensure that applying such a map to a valid quantum state always produces another valid quantum state. Thus when we move from composite systems to subsystems, we must generalise quantum states and evolutions beyond pure states and unitary transformations to include mixed states and CPTP maps.

2.1.2 Measurements

The postulates above tell us how to characterise the states of quantum systems and how those states change over time. But there is something else we need to know if we are to actually apply the theory – we must be able to describe measurements on these systems and predict the outcomes of those measurements. And unfortunately, unitary quantum mechanics as characterised above gives a somewhat mysterious description of the measurement process.

A quantum measurement is represented mathematically by a set of operators, each corresponding to one of the possible outcomes of the measurement. The simplest case is a 'projective' measurement, in which each operator has the form $|0\rangle\langle 0|$, where $|0\rangle$ is a unit vector in the appropriate Hilbert space; so we can think of the corresponding outcome as telling us that the system is in the state $|0\rangle$. So, for example, the measurement $\{|0\rangle\langle 0|, |1\rangle\langle 1|\}$ has two possible outcomes which correspond to the two pure states $|0\rangle, |1\rangle$, and therefore if we perform this measurement on a system S, we can think of it as asking the question 'Is S in the state $|0\rangle$ or the state $|1\rangle$?'

Now, if we perform the measurement $\{|0\rangle\langle 0|, |1\rangle\langle 1|\}$ on a system S in the state $|\psi\rangle = c_0|0\rangle + c_1|1\rangle$ where both c_0 and c_1 are non-zero, then since S is actually in a superposition of both $|0\rangle$ and $|1\rangle$, it is not immediately obvious what response this measurement should return. And, in fact, if we try to model the process of performing this measurement within the framework of unitary quantum mechanics by putting the measuring device and system together and performing unitary time evolution on them, we find that in the course of the interaction the measuring device becomes entangled with the system S, and thereafter the two of them are together in a superposition of all of the possible measurement outcomes:

$$|\psi\rangle_{SD} = c_0|0\rangle_S|0\rangle_D + c_1|1\rangle_S|1\rangle_D$$

Here, $|0\rangle_D$ and $|1\rangle_D$ are states of the measuring device corresponding to 'displaying measurement outcome 0' and 'displaying measurement outcome 1,' respectively.

The problem is clear – there is no mechanism to select out any *one* outcome of the measurement, and thus on the face of it the unitary description doesn't seem to predict that measurements actually have specific outcomes! So in order for quantum mechanics to be any use – indeed, in order for it to qualify as a scientific theory at all – we must find a way of extracting some outcomes out of this description.

Because of this problem, it is common to add to unitary quantum mechanics the postulate of wavefunction collapse (Nielsen and Chuang, 2011), which simply says that when we perform a measurement of the kind described above, the state of the system instantaneously collapses into a state corresponding to just one of the possible outcomes of the measurement, thus explaining why we always see a single definite result for such a measurement. For example, if we perform the measurement described by the set of operators $\{|0\rangle\langle 0|, |1\rangle\langle 1|\}$ on a system in a pure state $|\psi\rangle = c_0|0\rangle + c_1|1\rangle$, then the system will probabilistically collapse into either state $|0\rangle$ or $|1\rangle$, thus determining which outcome we see. A prescription known as the 'Born rule' stipulates that the probabilities for the outcomes of such a measurement are given by the mod-squared amplitude of the corresponding pure state vector in the state of the system – that is, in this case the amplitude associated with $|0\rangle$ is c_0, so the probability of obtaining the outcome $|0\rangle\langle 0|$ is $|c_0|^2$, and similarly the probability of obtaining the outcome $|1\rangle\langle 1|$ is $|c_1|^2$. Thus the wavefunction collapse prescription allows us to extract some actual measurement outcomes from the framework of quantum mechanics and to predict probabilities for those measurement outcomes, thus ensuring that the formalism is empirically meaningful and predictively useful.

As before, we can also generalise quantum measurements beyond the pure state case. The most general kind of quantum measurement is a POVM (positive operator-valued measure), which again is represented by a set of operators $\{O_1, O_2...O_n\}$, although these may not necessarily have the projective form described above. In the more general case, the Born rule simply stipulates that when we perform a measurement on a system whose state is described by the density matrix ρ, we obtain the outcome associated with the operator O_m with a probability $Tr(O_m\rho)$, where $Tr(...)$ is a specific mathematical operation on a matrix. Again, measurements of this kind are commonly interpreted as representing the quantum state collapsing probabilistically into some specific final state with probabilities as prescribed by the Born rule, thus producing a single determinate outcome for the measurement. For general POVMs it is not always possible to determine the specific state that the system will be in after the measurement – this will depend on the specific details of how the measurement is implemented. However, the basic idea that a measurement causes a probabilistic collapse which determines the result of the measurement still applies, and thus this remains a common way of thinking about measurement in quantum mechanics.

2.2 History

It is generally agreed that the unitary part of quantum mechanics – that is, both the Hilbert space state vector formalism described in the three postulates above and the density matrix formalism that can be derived from it – is the most satisfactory and securely-founded part of the theory. By contrast, the collapse postulate is more difficult to understand. In particular, it is not a natural fit with the other postulates of quantum mechanics, because the collapse is not a unitary transformation, and therefore it involves a kind of dynamics which does not appear in the theory in any other situation. The theory also doesn't provide any clear explanation of when and how this collapse happens, beyond the basic stipulation that it occurs during a 'measurement.' It seems we must simply presuppose an observer external to the theory who performs the measurement and precipitates the collapse, which is a somewhat unsatisfactory state of affairs – as Chandrasekaran et al. (2023, p. 17) put it, '*In a satisfactory theory, we are not entitled to introduce an observer from outside. The observer should be described by the theory.*' Ultimately we would like to be able to formulate a fundamental physical theory without making essential reference to measurements or else to understand the meaning of the term 'measurement' in physical terms: do measurements have to involve conscious observers? Do they require certain sorts of measuring devices? Does the wavefunction collapse at some particular scale or under some particular circumstances? Without answering these kinds of questions, the theory seems incomplete; we will never know how to actually apply it to any real situation unless we start off with some prior specification about what parts of that situation count as a 'measurement.'

One of the first authors to articulate difficulties relating to quantum measurement was Max Born (1926), who noted that the quantum-mechanical description of a measurement or 'collision' suggests that after the measurement the instrument and system are in a superposition of all of the possible outcomes, whereas experimentally we, of course, find only one outcome. Subsequently a suite of views known as the 'Copenhagen interpretation,' associated in particular with Bohr (1928) and Heisenberg (1955), came to be regarded as the 'orthodox' approach to the problem – although it should be emphasised that Bohr and Heisenberg had substantial disagreements, so arguably the Copenhagen approach is not really a single unified view (Howard, 2004). Broadly speaking, Copenhagen-style approaches place emphasis on the fact that we must describe our measurement apparatus in classical terms and insist that we should not try to directly model the relation between the classical apparatus and the quantum world. For example, Bohr argued that '*The unavoidable interaction between the objects and the measuring instruments sets an absolute limit to the possibility of speaking of a behavior of atomic objects which is independent of the means of observation*' (Bohr, 1939, p. 269). Similarly, Heisenberg famously advocated instituting a separation between systems described by the quantum formalism and systems which can be described classically – this putative boundary at which quantum reality gives way to classical reality has become known as the 'Heisenberg cut' (Compton and Heisenberg, 1984; French, 2023).

A pioneering analysis of measurement in quantum mechanics was given by von Neumann (1932), who separated quantum dynamics into Process I and Process II – the former referring to a non-unitary, probabilistic evolution occurring during measurements, i.e. a 'collapse,' and the latter referring to unitary evolution in between measurements. Importantly, von Neumann showed that due to the linearity of quantum mechanics, the transition between these two processes can be placed at many different points without affecting the predictions of the theory – as von Neumann puts it, '*This boundary can be pushed arbitrarily into the interior of the body of the actual observer*' (von Neumann, 1932, p. 421). This result provides a rigorous proof of Heisenberg's suggestion that we can choose more or less arbitrarily where to put the 'Heisenberg cut.' Von Neumann, therefore, did not necessarily agree with Bohr that any measurement apparatus always has to be described classically since he allowed that the Heisenberg cut might be pushed past the apparatus and into the body of the observer. But nonetheless, he still subscribed to a Copenhagen-style view in which the interaction between quantum and classical realities and the bringing-about of a single outcome is considered unanalysable. Indeed, it has been argued that von Neumann himself did not see the transition between Process I and Process II as involving some kind of physical collapse but rather as indicating a change in the observer's relation to the measured system (Becker, 2004; French, 2023).

This Copenhagen-style orthodoxy was challenged by Schrödinger (1935) with a famous thought experiment in which he notes that the linearity of quantum mechanics appears to predict that cats can in principle be put in superpositions of being 'alive' and 'dead' at the same time. It seems that Schrödinger intended to '*(draw) attention to the problems that arise from the Copenhagen "duality postulate" that macroscopic systems (such as measurement devices and cats) admit both a classical and quantum-mechanical description*' (Landsman, 2017, p. 439).

Later, Wigner (1961) posed another challenge to the Copenhagen-style approach, in the form of another thought experiment. Wigner envisions an observer, Alice, who performs a measurement on a quantum system inside a closed laboratory, whilst another observer, Bob, watches from outside of the laboratory. Alice will presumably see a definite outcome to her measurement. And yet if Bob describes the entire laboratory using unitary quantum mechanics, he will predict that Alice and the system end up in a superposition of all the possible measurement outcomes – for example, if the system is originally in the state $|\psi\rangle = c_0|0\rangle + c_1|1\rangle$, then unitary quantum mechanics predicts the following state for Alice, her detector, and the measured system after their interaction:

$$|\psi\rangle_{SDA} = c_0|0\rangle_S|0\rangle_D|0\rangle_A + c_1|1\rangle_S|1\rangle_D|1\rangle_A$$

Here, $|0\rangle_A$ and $|1\rangle_A$ are states of Alice corresponding to 'observed measurement outcome 0' and 'observed measurement outcome 1,' respectively, and similarly $|0\rangle_D$ and

$|1\rangle_D$ are states of her detector corresponding to 'displaying measurement outcome 0' and 'displaying measurement outcome 1.' So a Copenhagen-style interpretation of these events seems to lead to a contradiction – Alice has surely seen a definite outcome of her measurement, and yet it seems to Bob that there has not been any definite outcome.

Such considerations have motivated a search for alternatives to the Copenhagen approach, and ongoing research over the intervening years has yielded a plethora of putative solutions to the measurement problem, many of which we will explore in this book.

2.2.1 Decoherence

An important recent development in the debate over the measurement problem is an improved understanding of the phenomenon of 'decoherence' (Bacciagaluppi, 2007; Wallace, 2012; Bacciagaluppi, 2020). This refers to the fact that the natural evolution of large quantum systems typically results in them becoming coupled to their environment in such a way that the 'off-diagonal' elements of their density matrix rapidly become very small. The result is that distinct terms in the quantum state of the system become orthogonal – that is, they cease to interact with one another, and therefore the state of such a system ends up looking like a probabilistic mixture of states which are essentially classical in the sense that they no longer undergo interference. The process of decoherence is usually most evident in a description using some particular basis which happens to be favoured by the dynamics of the relevant systems, and often this is the coarse-grained position basis, so we typically end up with what looks like a probabilistic mixture of classical states in which the system has different classical positions.

For example, in the case of the measurement described in section 2.1.2 which results in a joint state $\psi_{SD} = c_0|0\rangle_S|0\rangle_D + c_1|1\rangle_S|1\rangle_D$ of the system and the measuring instrument, the measuring instrument is macroscopic and thus will become coupled to its environment, and then as a result of decoherence it can be expected to end up in a mixed state of the following form:

$$\rho = |c_0|^2|0\rangle\langle 0|_D + |c_1|^2|1\rangle\langle 1|_D$$

This mixed state looks like a classical probabilistic mixture of the states $|0\rangle_D$ and $|1\rangle_D$, each corresponding to a quasi-classical reality in which either outcome 0 or 1 is obtained; thus the process of decoherence results in a state of the measuring device containing distinct representations of the possible classical outcomes of the measurement, which are stable in the sense that they do not undergo interference. Note that the states $|0\rangle_D$ and $|1\rangle_D$ correspond to different coarse-grained positions for the detector – for example, if the detector includes a dial with a needle pointing to the outcome, then the needle will have different positions in states $|0\rangle_D$ and $|1\rangle_D$ – and thus we can see that decoherence is indeed taking place in the coarse-grained position basis.

It has sometimes been suggested that the existence of decoherence means we can do without the collapse postulate because decoherence on its own explains how we get stable classical outcomes out of the unitary quantum description. However, this is not quite right, because although decoherence ensures that we now have distinct and stable representations of the outcomes 0 and 1 as part of the state ρ, they are both still present: as Maudlin (1995, p. 9) puts it, '*It should be obvious that the problem has not been touched ... (the mixed state* ρ*) does not represent exactly one of the outcomes as occurring. The argument goes through unscathed.*' Somehow we still need to move from this mixed state ρ to a single definite outcome, so we still apparently need a collapse postulate or some other way of solving the measurement problem – decoherence alone will not finish the job.

2.2.2 Extended Wigner's Friend Experiments

Another important recent development is the realisation that it is possible to come up with experiments in which putting the Heisenberg cut in different places *does* make a difference to the empirical predictions we obtain. A key example is given by the class of Extended Wigner's Friend scenarios (Frauchiger and Renner, 2018; Bong et al., 2020; Schmid et al., 2023). Such scenarios are extensions of Wigner's original thought experiment: again we imagine observers inside closed laboratories performing measurements on entangled particles and other observers external to these laboratories, but now in addition we imagine that the external observers can keep the whole laboratory-observer-particle system under very sensitive quantum control, such that they can subsequently perform quantum measurements on the entire composite system.

In such an Extended Wigner's Friend scenario, if we imagine that the wavefunction collapses inside the laboratory when the internal observer makes a measurement, we get one prediction for the results of the measurements made by the external observers; but if instead we imagine that the wavefunction only collapses when one of the external observers makes a measurement or does not collapse at all, we get a *different* prediction for the external measurements. So now we appear to obtain not only a theoretical tension, as in the original Wigner's Friend scenario, but a concrete difference in empirical predictions depending on how the measurement problem is resolved.

Now, these Extended Wigner's Friend experiments have not yet been performed, and indeed they may never be within our technological capabilities. But nonetheless they seem possible in principle, and thus most scientific realists – indeed, even most empiricists – would presumably agree that there must be some fact of the matter about what would happen if we were to perform such experiments, or else there must be some clear physical reason why such experiments are not possible. And this indicates that the question of where we should put the Heisenberg cut, or whether there is, in fact, a Heisenberg cut, is not just a matter of idle intellectual curiosity. It is a concrete scientific question which has real empirical consequences and might one day be empirically testable, thus demonstrating that the measurement problem has very real scientific significance. I will discuss the significance of Extended Wigner's Friend experiments in more detail in section 9.7.

2.3 Measurement

Evidently there is some kind of problem in the vicinity of measurement in quantum mechanics, but it turns out to be quite non-trivial to say exactly what the problem is and what it would take to solve it. The literature contains many different descriptions of the measurement problem – indeed, nearly everyone who works in this area seems to have a slightly different conception of what the problem is! – and naturally these different conceptions of the problem lead to different ideas about what a viable solution looks like.

In this book, I will suggest that one useful way to understand the measurement problem is to see it as a problem of *epistemology*. This presentation of the problem is a little different from some popular accounts, but I hope to make a case that this epistemic construal of the problem does indeed highlight a serious gap in our understanding of quantum mechanics, and moreover that conceptualising the problem in this way ultimately improves the prospects that it can actually be solved.

Before presenting this epistemic account of the measurement problem, let us start with a simpler question: what is a *measurement*? In this chapter I will offer a somewhat schematic answer to this question; later, in chapter 4, I will take a more detailed look at what recent work in epistemology has to say about the matter, fleshing out the details of the account I sketch below.

First, note that my concern in this discussion is not to demarcate 'measurements' from more general kinds of observations. The term 'measurement' is most commonly associated with scientific experiments performed in specially designed laboratories, but in a sense every time we open our eyes and look at the world around us or perceive it in some other way, we can be regarded as carrying out a rudimentary 'measurement' – indeed, science presumably has its origins in attempts to systematise the results of these simple kinds of measurements. So in this book I will use the term 'measurement' quite generally to refer to any observation or interaction which can be regarded as yielding some kind of empirical knowledge – so, for example, measurements may include things like looking at the sky, reading a book, conversing with another person, or even merely consulting one's memories. Part of the reason for using the term in this way is that, as we will see in chapter 3, certain interpretations of quantum mechanics do actually suggest that all of these kinds of interactions involve the same kind of process as a more conventional 'quantum measurement,' so it is sensible to treat measurement in the most general possible sense here.

Thus construed, 'measurement' is a quite a special kind of category. On the one hand, it is an ontological category: a measurement is a physical process and so, at least in the context of physicalism or any other flavour of scientific realism, it is natural to think we should be able to describe what goes on during a measurement in purely physical terms. But measurement is also an epistemic category: it is nothing more or less than the ultimate contact point between our experience and physical reality. As van Fraassen (2008, p. 143) puts it, '*Measurement is an operation by whose means we gather information: but this is of course done by way of a physical interaction between apparatus and object, to play the information-providing role*,' and thus measurement has both '*physical and intentional aspects*.'

As a result of its dual nature, measurement occupies a complex position in the epistemology of science. We perform measurements, assuming that they yield meaningful information about reality, and then we use the results of those measurements to come up with a description of the physical world. But how do we know that our measurements are *actually* telling us something meaningful? We started off without really knowing anything about what a measurement is or what kind of information it is providing, so how do we ultimately arrive at reasonable beliefs about reality by means of such measurements?

Versions of this question have been posed in the philosophy of science ever since the 'problem of coordination' first appeared in the work of Mach (1893) and was subsequently studied by Poincaré (1907), Reichenbach (1965), Schlick (1978), and others. As Tal (2013, pp. 2–3) explains it, '*The problem is that the empirical adequacy of the theory or model and the reliability of measuring procedures appear to presuppose each other in a circular way. To establish a theory of mass, for example, it is necessary to test its predictions, a task which requires a reliable method of measuring mass. Testing the reliability of measurements of mass, however, presupposes background theoretical knowledge about mass and its relations with other physical properties such as force and motion.*' Similarly, Chang (2004, p. 221) writes that '*Empirical science requires observations based on theories, but empiricist philosophy demands that those theories should be justified by observations.*' The worry is that neither theory nor experiment is a suitable starting point for scientific practice, so how can empirical science ever begin or make progress?

In chapter 3 I will look in more detail at various possible responses to the problem of coordination, but for now let us just note that the answer seems to be connected to the fact that a scientific theory will typically tell us something about what is going on in the events we describe as 'measurements.' As van Fraassen (2008, p. 122) puts it, '*We have no problem identifying what is measured by a particular procedure if we look at it 'from above,' that is,* ***in retrospect****, from within a theory that is already stable and established, and deals with that physical magnitude.*' That is, we can simply allow the theory itself to tell us what our measurements are measuring and how it is that these measurements provide evidence for the theory. The result, as described by Ryckman (2005, p. 64), is an '*inevitable epistemological holism*' associated with a theory '*in principle capable of explaining its own measurement appliances, and so its ties to observation.*'

Now, one might worry that there is something dodgy about this whole procedure, rather as if you read some paper by Smith which supports a controversial claim by citing a paper by Jones, and when you look up the paper by Jones you find that that it supports the same claim by citing the original paper by Smith. Similarly for measurement: we are first assuming that measurements are accurate and then using the conclusions we reach on the basis of those measurement results to justify our original assumptions, so how can we be confident that we are really learning something about the world in this process, rather than getting caught in a self-justifying circle that never makes contact with reality?

As a first response, it should be emphasised that we should not ask too much of the epistemology of science – it would be wholly unreasonable to expect that our scientific theories should somehow *prove* that our measurements are making contact with reality in the way that the theory suggests. No measurement and no theory of measurement can ever rule out various kinds of sceptical scenarios, such as the possibility that you are just a brain in a vat and all the results of your 'measurements' are generated by a computer program aiming to deceive you about the true nature of reality. So in order to do science we do have to allow ourselves to disregard sceptical possibilities: to some degree it will always be an article of faith that our measurement outcomes are not deceiving us too significantly.

But what we *should* demand is coherence. For example, Shimony (1987, p. 62) advocates an approach to the coordination problem that he refers to as 'closing the circle': '*The metaphysics should be capable of understanding the existence and the status of subjects like ourselves who are capable of deploying the normative procedures of epistemology. And the epistemology should suffice to account for the capability of human beings to achieve something like a good approximation to knowledge of metaphysical principles, in spite of the spatial and temporal limitations of our experience and the flaws and distortions of our sensory and cognitive apparatus.*' Shimony focuses here on metaphysics but the same can be said of physics and science more generally: as van Fraassen (2008, p. 145) puts it, '*The theoretical characterization of the measurement situation is required to be coherent with the claims about the existence of measurement outcomes, their relation to what is measured, and their function as sources of information ... for this coherence it is required that the theoretical characterization of the measurement interaction allows for a coherent story about how its outcomes provide information about what is being measured.*' That is, we start out the process of scientific enquiry with rudimentary assumptions about the nature and reliability of our means of enquiry, and as we cumulatively build our account of reality based on the empirical results of those enquiries, we hope that account will both affirm and refine our original assumptions by telling us how our measurements are able to provide meaningful information about the objects or structures appearing in that account. So we don't have to start from measurement results and then move to theory or vice versa; rather '*the joint evolution of measurement and theory can happen and come to a stable resting point*' (van Fraassen, 2008, p. 123), i.e. we should ideally reach a situation in which the theory itself is able to tell us how the physical process of measurement as characterised in this theory could possibly provide us with knowledge about a reality of the kind described by the theory.

To some extent this coherentist structure can be seen within all domains of scientific enquiry. If I am studying sea-turtles, I need to assume that the turtles behave in roughly the same way when I am observing them and when I am not, and the results of my enquiry will hopefully affirm this assumption, in that it will lead me to a theory of turtle behaviour which suggests that the turtles are not intelligent enough to mount an organised campaign of deception. But there are certain domains in which this circular structure is particularly relevant because the functioning of the means of enquiry is itself part of the domain of enquiry. Shimony (1987) notes that this is clearly the case in empirical psychology and the study of human perception; but as Duhem (1954, p. 259) points out, it is also very much the case in fundamental physics because in physics there is no more fundamental science to rely upon, and thus '*it is impossible to leave outside the laboratory door the theory we wish to test.*' In my study of sea-turtles, it is reasonable for me to simply trust the physicists who tell me that my binoculars are yielding accurate images of the turtles because the functioning of the binoculars is not the subject of my enquiry; but in physics, the detailed workings of our instruments are usually strongly dependent on physics in the very domain that we are trying to study. So coherence becomes a particularly pressing concern in this context, and arriving at an overall understanding the nature of measurement becomes a much more central part of the process of scientific enquiry: a key measure of success involves arriving at a 'stable resting point' in which our theory of fundamental physics is able to explain what the measurements we used to arrive at it are actually doing and why they are able to yield reliable information about the objects or structures described by the theory.

2.4 The Measurement Problem

We can now give a more precise account of 'the measurement problem' in the context of scientific epistemology. In fact, it will be helpful to break the problem into two parts. The first problem is that standard textbook quantum mechanics, interpreted in the most literal way, seems to say almost nothing about the physical nature of measurement. If we consider only the unitary part, then there appear to be no measurements at all since measurement-like events don't have unique outcomes; and if we consider quantum mechanics with a collapse upon measurement, then 'measurement' is treated as a special dynamical process completely outside of the framework of the standard theory and incapable of being modelled within it, and therefore the theory tells us nothing at all about the nature of this process. So in neither case can we say much about what is going on in a measurement or how the physical nature of measurement leads to measurement outcomes which can be expected to accurately reflect some feature of the reality we are trying to learn about.

Thus it appears that standard textbook quantum mechanics, construed in a literal way, fails to answer the crucial epistemological question '*What does this theory, or an accepted background theory, imply about the measuring process whence the theory reaps its credentials?*' (van Fraassen, 2008, p. 145). Therefore the theory cannot reassure us that our measurements are yielding reliable information about the structures that they are supposed to be latching onto. Shimony (1993b) emphasises the threat posed by the measurement problem to scientific epistemology, noting that '*If the world view of quantum mechanics precludes the definiteness of events involving particle detectors, then the evidential underpinnings of the theory are removed*,' and thus a solution to the measurement problem '*would contribute greatly to "the closing of the circle," which is one of the chief requirements for a coherent worldview, for it would explain in principle how there are definite outcomes of experiments.*' So from the epistemological point of view, what is needed to shore up the theory's credentials is an account of the contents of reality which both reproduces the quantitative results of quantum mechanics but also tells us, in clear physical terms, what occurs in the process of a measurement and how it is that this kind of process can connect us with the theoretical structures postulated by unitary quantum mechanics.

But here is the second part of the problem: it transpires that quantum mechanics has special features which make it very difficult to give an account of quantum measurement according to which measurements still have all the features that we might have expected prior to the advent of quantum mechanics. Of course, it is normal for some parts of the interpretation of measurement processes to be revised in the course of theory change – for example, moving from the caloric theory of heat to the energy theory involved a revision in the theoretical account of the way in which thermometers operated since the caloric theory attributed the expansion of the liquid or gas in the thermometer upon heating to the fact that caloric fluid took up space, whereas the energy theory understood this as a consequence of more energetic molecules being on average further apart (Brown, 1949). But, nonetheless, up until the advent of quantum mechanics there were certain basic assumptions about measurement which had never been seriously challenged by any scientific discovery, and which, presumably, no one really imagined would ever be challenged by any scientific discovery. Van Fraassen (2008, p. 149) describes some of these basic assumptions: '*Value definiteness: Each physical parameter always has some value, namely one of the values which may be found by measurement*' and '*Veracity: measurement of a parameter faithfully reveals the value it really has.*' We can add other similar assumptions to this list, such as the assumption that measurements reveal the values of quantities that already existed *prior* to the measurement; that measurements always have only one outcome; that in principle any observer could have access to any measurement outcome; that records and memories typically provide accurate information about past measurement outcomes; and so on. These assumptions together provide a rudimentary understanding of the relation between measurements and physical reality which allows us to get started on the progressive, iterative process of coordinating theory and practice.

But in the context of quantum mechanics, something seems to go wrong in the course of this joint evolution of experiment and theory because it turns out that it is most likely impossible to give an account of physical reality which is consistent with the empirical predictions of quantum mechanics and which affirms the correctness of *all* aspects of this intuitive physical picture of measurement. Below I discuss one key example of this in the form of the phenomenon of contextuality, and more generally, in chapter 3 we will see that a number of putative solutions to the measurement problem are incompatible with other aspects of our intuitive picture of measurement, such as the assumption that measurements have only one outcome or the assumption that observers can share information about their measurement outcomes. Thus in these cases it seems significantly more difficult to tell a sensible story about the epistemology of science; at best these approaches make scientific knowledge somewhat fragile, and at worst they make scientific knowledge impossible. Moreover, unfortunately these epistemically problematic interpretations also seem to be the most empirically successful solutions to the measurement problem, so it turns out to be not at all straightforward to tell a sensible story about quantum mechanics which reproduces all the empirical evidence while preserving the robustness of scientific knowledge!

Thus we probably cannot get away with demanding that any viable solution to the measurement problem should maintain *all* features of the intuitive metaphysical account of measurement. We are going to have to do something more complicated: we will have to consider each individual interpretation in detail in order to understand the story it has to tell about the physical nature of measurements and then decide whether this story is '*coherent with the claims about the existence of measurement outcomes, their relation to what is measured, and their function as sources of information*' (van Fraassen, 2008, p. 145). So this is the lens through which we will consider the measurement problem in this book – the central challenge is to find an empirically adequate solution which is also capable of accounting in a coherent, reasonable way for the epistemology of science.

2.4.1 Contextuality

The phenomenon of quantum contextuality is one of the main reasons to think that no possible account of quantum measurement can maintain all aspects of the intuitive physical picture of measurement. The existence of contextuality in quantum mechanics is demonstrated by the Kochen-Specker theorem (Kochen and Specker, 1975), which tells us that for quantum systems of dimension greater than two, we can identify sets of possible quantum measurements on the system such that, if we think of the possible outcomes of these measurements as corresponding to pre-existing properties of the system, we will arrive at a contradiction.

To see this, note that if we think the results of quantum measurements simply reveal pre-existing properties of the systems we are measuring, then we ought to believe that for any quantum system S and any set of measurements that we can perform on that system, it will be possible to assign the numbers 1 and 0 to all of the possible outcomes for all of the measurements in such a way that every measurement has exactly one outcome which is assigned 1. That is, we assign '1' to the measurement outcome A if S has the property corresponding to A, meaning that any measurement on S to which A is a possible outcome will definitely produce outcome A. And we assign '0' to the measurement outcome B if S does not have the property corresponding to B, meaning that any measurement on S to which B is a possible outcome will definitely *not* produce outcome B. Then since every measurement must produce exactly one outcome, each measurement should include exactly one outcome which is assigned 1.

But the problem is that for quantum systems of more than one dimension, the same measurement outcome can appear in two or more different measurements – for example, one possible measurement might be defined by the operators $\{|A\rangle\langle A|, |B\rangle\langle B|, |C\rangle\langle C|\}$ and another measurement could be defined by the operators $\{|A\rangle\langle A|, |D\rangle\langle D|, |E\rangle\langle E|\}$, so $|A\rangle\langle A|$ is a possible outcome for both of these measurements. And it turns out that for certain sets of measurements, outcomes are shared between the measurements in such a way that it simply isn't possible to ensure that every measurement has exactly one outcome which is assigned 1, unless we allow that an outcome may be assigned 1 in the context of one measurement and yet also assigned 0 in the context of another measurement.

Thus it seems that quantum mechanical measurement outcomes are 'contextual' – a given outcome may be certain to occur in the context of one measurement and yet certain *not* to occur in the context of some other measurement, and therefore that outcome can't correspond to a context-independent property of the system. This indicates that we can't in general think of quantum measurements as simply revealing pre-existing values of quantities associated with measurement outcomes, so we will need some other way of understanding what such measurements are giving us information about. In chapter 11 I will discuss in more detail the consequences of contextuality for the epistemology of quantum mechanics.

2.5 Scientific Knowledge

Throughout this book I will be concerned with the status of various kinds of scientific knowledge, so let us pause to specify what exactly the term 'scientific knowledge' refers to. In this book I will adopt a very minimal kind of realism: I take it that there exists something that can be referred to as physical reality and that scientific knowledge is knowledge about physical reality. This is a necessary but perhaps not sufficient condition; I don't claim that all knowledge about physical reality is scientific, but I do take it that all scientific knowledge pertains to physical reality. This does not exclude empiricists or operationalists because 'reality' could just mean 'empirically accessible reality' or 'the outcomes of performing operations' or similar. Nor does it exclude those who believe that physical reality is relational or perspectival: knowledge of relational or perspectival facts about physical reality is still knowledge about physical reality. It does, however, exclude anyone who believes that there is no physical reality or that everything we typically consider to be 'scientific knowledge' is just subjective or normative and has nothing whatsoever to do with physical reality.

In this book, I will use the term 'empirical scientific knowledge' to refer to the knowledge that science delivers to us about the empirical world, e.g. about observable regularities pertaining to past and/or future observations. Meanwhile I will use the term 'theoretical scientific knowledge' to refer to the knowledge that science may deliver to us about unobservable parts of reality, e.g. quarks, quantum fields, pilot-waves, and so on. This terminology makes it possible to state more precisely the epistemic concerns addressed in this book: I will be concerned primarily with the threat that quantum mechanics poses to *empirical* scientific knowledge rather than theoretical scientific knowledge. This is important because a sizeable proportion of physicists and philosophers are relatively unconcerned about theoretical scientific knowledge, taking the view that empirical knowledge is the most important product of science; and thus although it is fairly obvious that failing to solve the measurement problem means there are some gaps in our theoretical scientific knowledge, many people may not mourn this loss too deeply. But as I will argue throughout this book, failing to solve the measurement problem is also a threat to *empirical* scientific knowledge because it limits our ability to affirm the reliability of the methods of enquiry we use to arrive at our empirical scientific knowledge. And furthermore, we will see that failing to solve the measurement problem not only threatens the empirical knowledge associated specifically with quantum mechanics but also potentially undermines many other kinds of empirical scientific knowledge that one might have thought to be completely secure against innovations in theoretical physics, like the knowledge we have about Newton's laws or the cetacean species of the Pacific!

Thus part of the point I want to make in this book is that even if you do not care at all about theoretical scientific knowledge, you still ought to be concerned about the measurement problem. And therefore I will largely avoid engaging with ongoing discussions about whether and to what extent science can, in fact, give us knowledge about the unobservable parts of reality. For after all, if science can't even give us knowledge about the things we *do* directly observe, there seems little hope that it can yield knowledge about things we cannot directly observe! In particular, I will not be concerned with the problem of how to choose between various solutions to the measurement problem which are empirically equivalent with respect to all of the current evidence: for given the issues raised in this book I think it would be quite a coup to come up with even one fully satisfactory approach which reproduces quantum mechanics in all the regimes in which it has been tested. There are worse problems to have than underdetermination!

Of course, as has often been noted (Churchland, 1992), the line between empirical scientific knowledge and theoretical knowledge is not perfectly well-defined, but this only serves to emphasise my point – if there is in any case no sensible way of dividing our scientific knowledge cleanly into empirical and theoretical parts, then it is all the more clear that epistemic problems arising from our inability to fully specify the content of theoretical scientific knowledge will inevitably bleed into empirical scientific knowledge as well.

2.6 Theories, Accounts of Reality, and Package Deals

I have been referring to quantum mechanics as a 'theory.' To be more specific, in this book I will largely understand a theory in the semantic sense, so a scientific theory is simply a collection of models. I will take it that a 'theory' may be a purely instrumental device, so the only assertions that a theory makes about reality are in its empirical content.

However, given a scientific theory which successfully predicts certain empirical results, we are often interested in coming up with a description of some underlying reality which could give rise to and explain the empirical predictions of the theory. In this book, I will refer to such a thing as an 'account of reality.' An account of reality typically includes assertions about things which are not directly observable, such as unobservable entities or unobservable structures. Note that what I have called an 'account of reality' is also sometimes referred to as an 'interpretation' of a theory, but I will not use that term in this book since it invites confusion with a different set of questions about the interpretation of the quantum formalism, as I will explain in chapter 5.

Finally, I will use the term 'package deal' to refer to the combination of a theory and an account of reality which could explain the empirical predictions of that theory. It should be noted that in non-technical contexts the term 'theory' is often really referring to what I call a package deal. For example, when we talk about 'the atomic theory,' there is often an implicit assumption that atoms are not just convenient tools for making predictions but that they are in some sense part of the furniture of reality, so in such cases we are really referring to a package deal rather than just a theory.

Evidently in order to solve the epistemic version of the measurement problem we need not just a theory but also an associated account of reality. This is because a theory on its own has only empirical content, and therefore a theory can't fully account for the physical nature of measurement, given that the process of performing an observation is not itself directly observable. For example, much of this book will focus on a class of solutions to the measurement problem that I will refer to as the 'unitary-only' solutions because they make use of only the mathematical formalism of pure unitary quantum mechanics, i.e. they do not allow the addition of wavefunction collapses, hidden variables, postulates about the relations between observers, or any other kind of extra structure that is not already there in unitary quantum mechanics[1]. The unitary-only approaches thus represent putative accounts of reality which could be associated with the theory of unitary quantum mechanics. These approaches have the considerable advantage that they automatically reproduce all the existing mathematical structure of unitary quantum mechanics and quantum field theory since they retain all the same models. But as we will see in chapter 3, this comes at the cost of significant epistemic problems, and thus it is unclear that any unitary-only approach can solve the epistemic version of the measurement problem.

Meanwhile, there also exists another class of approaches to the measurement problem, which I will refer to as the 'primitive ontology' approaches, involving models that are mathematically distinct from what one would see in a standard quantum mechanics textbook. For example, although these models predict all the same things as standard quantum mechanics for all the non-relativistic experiments that have so far been performed, they typically make different predictions for Extended Wigner's Friend experiments. So these approaches are not merely proposing a account of reality which could be attached to unitary quantum mechanics but rather putting forward what is essentially a whole new theory. The primitive ontology approaches are thus package deals since they offer both a new theory and an account of reality associated with that theory. As we will see in chapter 11, the primitive ontology approaches don't typically encounter the severe epistemic problems that appear in unitary-only approaches. But on the other hand, because the primitive ontology approaches are mathematically distinct from standard quantum mechanics, it's more challenging for them to reproduce the existing predictions, and in particular, no such approach has yet been completely successful in reproducing all the predictions of quantum field

[1] Note that there is some ongoing discussion about whether or not evolution is typically unitary for quantum field theory in curved spacetime – see, for example, Colosi (2011). And one might naturally think that if the evolution is not always unitary, this would be a problem for the unitary-only interpretations. However, the central point of the unitary-only interpretations is not unitarity per se but rather the fact that they employ only the formalism of standard quantum mechanics without wavefunction collapse or any addition like hidden variables. One may conjecture, therefore, that when applied to quantum field theory in curved spacetime they will continue to employ the standard formalism and hence will simply cease to be unitary in just the same way as the formalism itself ceases to be unitary. The term 'unitary-only' is thus potentially something of a misnomer since unitarity is not really the key issue. However, there has not been much discussion about what the unitary-only interpretations would look like when applied to QFT in curved spacetime, so I can't provide a definitive account of what happens in that case; therefore, in this book I will focus on the better-understood cases of non-relativistic quantum mechanics and QFT in flat spacetime, in which case these approaches are indeed always unitary and hence it is reasonable to describe them as unitary-only.

theory. So at present, it's unclear that we have *any* approach to quantum mechanics which both reproduces all successful quantum predictions and also solves the epistemic version of the measurement problem.

2.7 Looking Forward

So, what would it take to solve the epistemic version of the measurement problem? There are a few possible options. Most ambitiously, we could seek a successor theory to quantum mechanics – perhaps a version of quantum gravity or a grand unified theory – in the hope that it will be easier to make sense of measurement within the successor theory. That would certainly be a pleasing resolution, but I think we should not lean too much on this hope. After all, it seems likely that a successor theory to quantum mechanics will be even less classical than quantum mechanics itself, so it may be even *more* difficult to make sense of measurement within such a theory! Furthermore, I will argue in chapter 13 that finding the correct solution to the measurement problem might help guide the formulation of a successor theory, so it may be a mistake to defer consideration of the measurement problem until after the successor theory has been constructed.

Alternatively, we could seek to convince ourselves that despite the epistemically curious features of the existing unitary-only approaches, nonetheless there is after all some way to arrive at a satisfactory epistemology in the context of one of these approaches. In chapters 6, 9, and 13 I will consider a number of attempts to achieve this in the context of several different unitary-only approaches. In very broad summary, my conclusion is that this problem is not completely hopeless, but certainly all such approaches seem to come at the cost of making scientific knowledge more fragile than we would ideally like it to be.

Finally, we could look for an interpretation of quantum mechanics which does not have these epistemic problems – most likely an approach outside of the unitary-only category. There are two possibilities here. First, we could take one of the existing primitive ontology approaches and show how to make this approach empirically adequate by extending it successfully to all of QFT. In this book I will not spend too much time on this possibility because at the moment I don't see any obvious path to achieve it, but this would certainly be a valuable area for further research. Or alternatively we could explore a completely different kind of interpretation – for example, an interpretation which is not unitary-only but is also not a primitive ontology picture in the standard sense. In chapter 13 I will discuss a number of such approaches and assess the prospects for their further development.

3
Some Solutions

In this chapter, I will describe several attempted unitary-only solutions to the measurement problem and then sketch some reasons why they may cause problems for scientific epistemology. It is not my intention in this chapter to argue that there is no way of solving these problems – indeed I will later set out to evaluate various proposed solutions to these problems. But here the aim is simply to get a sense of the landscape of the epistemic difficulties associated with these solutions, in order to better motivate the epistemic version of the measurement problem that I am presenting.

Because all of these approaches are unitary-only, they can all be understood as accounts of reality intended to accompany the standard theory of unitary quantum mechanics (QM). And in particular, they all postulate that unitary quantum mechanics is universal and complete. Here, the term 'universal' is intended to mean that unitary quantum mechanics is universally applicable: '*The evolution of an isolated system's quantum state is always unitary, with no physical 'collapses' or 'jumps', where this requirement holds in any special relativistic inertial frame when quantum theory is applied in Minkowski spacetime*' (Healey, 2024, p. 3). Note that this formulation leaves room for the possibility that quantum theory may not be applicable in extreme physical regimes where spacetime is not Minkowski – that is, we allow that unitary quantum mechanics may not apply in the standard way when quantum-gravitational effects become significant. But that possibility is not relevant to any experiment we have so far been able to perform, nor to any scenario that I will consider in this book, so for our purposes it will be sufficient to think of the term 'universal' as requiring that the predictions of unitary quantum mechanics are always correct. In particular, unitary-only approaches necessarily refrain from postulating an additional dynamical mechanism involving something like a wavefunction collapse, so the only dynamics in such an approach are the unitary evolution dynamics.

Meanwhile, Maudlin (1995, p. 8) explains the meaning of 'complete' as follows: '*When we say that the wave-function is complete, we mean simply that all physical properties of a system are reflected, somehow, in the wave-function. It follows that two systems described by identical wave-functions would be physically identical in all respects. We do not assert in any deep sense that the wave-function is real.*' So completeness does not necessarily mean that the quantum state of a system at a time must always be taken as a literal description of the occurrent physical properties of the system at that time – for example, some presentations of unitary-only relational quantum mechanics appear to suggest that systems do not have occurrent physical properties at all in between measurements, even though we can still assign them quantum states during those times (Di Biagio and Rovelli, 2022). However, completeness does mean that the unitary-only approaches cannot postulate any *additional* physical properties or structures which are not included in the unitary quantum description. In particular, unitary-only approaches to quantum mechanics maintain that quantum systems

Saving Science from Quantum Mechanics. Emily Adlam, Oxford University Press. © Oxford University Press (2025).
DOI: 10.1093/9780197808887.003.0003

do not have 'hidden variables' characterising their physical properties beyond what is included in their quantum state. Thus, for example, if a system is in a quantum state which takes the form of a superposition of two different values 0 and 1 for some variable V, i.e. $\psi = c_0|0\rangle + c_1|1\rangle$ with c_0 and c_1 both non-zero, then we can say nothing more about the value of V for that system – we cannot say that the system actually has some 'true' value for the property V, since that would amount to postulating a hidden variable beyond what is included in the state ψ.

3.1 The Everett Interpretation

The Everett interpretation, originally proposed by Hugh Everett in 1957 (Everett, 1957), is in some ways the simplest possible solution to the measurement problem, since it amounts to leaving the unitary quantum formalism completely unchanged.

As noted in Chapter 2, the unitary formalism tells us that during a measurement, the system and measuring instrument will generally evolve into a superposition of all of the possible measurement outcomes. For a measurement with two possible outcomes, 0 and 1, the resulting state will be something like this:

$$|\psi\rangle_{SD} = c_0|0\rangle_S|0\rangle_D + c_1|1\rangle_S|1\rangle_D$$

Here, $|0\rangle_S$ and $|1\rangle_S$ are states of the measured system corresponding to the two possible outcomes of the measurement, and $|0\rangle_D$ and $|1\rangle_D$ are states of the measuring device corresponding to 'displaying measurement outcome 0' and 'displaying measurement outcome 1,' respectively.

Now we can, if we like, also include an observer in this description, and what the unitary formalism tells us is that the observer will also end up as part of the superposition, leading to a state which looks something like this:

$$|\psi\rangle_{SDO} = c_0|0\rangle_S|0\rangle_D|0\rangle_O + c_1|1\rangle_S|1\rangle_D|1\rangle_O$$

Here, $|0\rangle_O$ and $|1\rangle_O$ are states of the observer corresponding to 'observed measurement outcome 0' and 'observed measurement outcome 1,' respectively. So as discussed in Chapter 2, we appear to have a problem – the theory provides us with a representation of a measurement being performed, but since all outcomes are present in the resulting state, there appears to be nothing in this representation which corresponds to some particular outcome actually being obtained.

The Everettian response to this situation is straightforward: we simply say that all of the outcomes occur. That is, during a measurement the system, measuring instrument and observer together undergo a branching process such that every possible outcome of the measurement is observed by one post-measurement copy of the observer. The Everett interpretation is therefore also known as the 'many-worlds' interpretation, since it suggests that in some sense we can think of the universe as branching into multiple 'worlds' during quantum measurements – though it should be reinforced that in most versions of the Everett approach these worlds are still part of the same universe, and in principle they can still be in causal contact.

A particularly important feature of the modern Everett interpretation is the use it makes of decoherence. For as noted in chapter 2, macroscopic quantum systems typically undergo very rapid decoherence, meaning that the branches of the wavefunction describing such systems quickly start to look like stable quasi-classical worlds which no longer interfere or interact at all with each other. This allows Everettians to argue that after a measurement event, the branches of the wavefunction containing the different possible outcomes of the measurement will quickly separate into distinct, non-interacting, stable classical realities, so the 'multiple worlds' posited by the approach emerge in a natural way from the existing quantum formalism, without any need for additional structure to define the branches or worlds. The Everettian branches and worlds are therefore usually regarded as being approximate and emergent, rather than well-defined fundamental structures in their own right (Wallace, 2003).

The Everett interpretation has been criticised in a number of different ways (Saunders et al., 2010), but there is a sense in which some of the criticisms seem to be largely a matter of taste. For example, it is often argued that the Everett approach is ontologically excessive, since it must postulate a continuous infinity of other worlds just to explain the events occurring inside a single world. But proponents of the Everett approach have replied that, in fact, the Everett approach is ontologically very parsimonious (Carroll and Singh, 2019): it only postulates one kind of thing, i.e. the wavefunction, and therefore it is an improvement on alternatives which add some further ontology in addition to the wavefunction. It's quite unclear how to adjudicate this debate, as ultimately the question of what kind of ontological excess is more worthy of condemnation seems somewhat subjective.

However, the Everett approach also appears to face severe *epistemic* problems which are not so easily dismissed. The basic point has been made many times: in an Everettian world we know that every outcome of every measurement actually does occur and is witnessed by someone, so what could it possibly mean to assign probabilities to the outcomes? Indeed, as noted by Wallace (2006), there are two related problems in this vicinity – first the 'incoherence problem,' which is about making sense of what it could mean to assign *any* probabilities other than 1 to the outcomes, and second the 'quantitative problem,' which is about showing why we should assign specifically the mod-squared amplitude probabilities predicted by quantum mechanics to the outcomes.

Now, there are a variety of reasons why one might think it is important for the Everett approach to make sense of probabilities. Research on 'the Everettian probability problem' often focuses on the need to rationalise the role that probabilities play in our ordinary reasoning about the future. For example, one common challenge involves the game of 'quantum roulette,' in which you are offered the following bet: I will perform a quantum measurement which has two possible outcomes, 'up' and 'down,' each assigned equal probabilities by the Born rule, and if the result is 'up' then you get a million dollars, but if the result is 'down' you will be immediately killed (Price, 1996). Would you take that bet? At least naïvely, it seems that if you believe the Everett interpretation then you should, since the Everett approach predicts that both outcomes definitely occur in different branches of the wavefunction. Thus in an Everettian version of the quantum roulette, you know with certainty that one version of you will survive and get a million dollars, and no other version of you will survive, so in effect you are guaranteed to win. Somewhat less exotically, in an Everettian world it is plausible that for almost any logically possible course of events there will be a branch of the multiverse in which that course of events occurs, so naïvely it seems that if you believe the Everett interpretation, then you should never put any thought at all into deciding between mutually exclusive alternatives: every decision is just a branching event, and thus for any possible option there will always be a branch in which a version of you chooses that option, so you know that you always end up choosing *all* the options.

Many Everettians find these kinds of consequences distasteful, and thus various strategies have been proposed to explain why Everettian agents should still reason based on probabilities in just the way that we currently do, even though they know that for almost any uncertain event all of the possible outcomes will, in fact, occur. For example, the decision-theoretic approach pioneered by Deutsch (1999) and Wallace (2010, 2012a), which we will return to in chapter 6, is an influential approach to this question. However, ultimately these issues relating to how agents use probabilities in ordinary decision-making are also in some sense a matter of subjective preference. For even if the decision-theoretic program were to be judged untenable, it would be open to the Everettian to simply bite the bullet of the quantum roulette and say that physics has now revealed to us that our ordinary decision-making behaviour has been wrong all along. We have been making decisions on the basis of a false picture of reality, and now that we know we are really in a branching Everettian multiverse, we should give up reasoning this way – we should all be willing to play the quantum roulette! This would certainly be a strange consequence of the Everett approach, and it may well be psychologically impossible to actually live one's life as if no choice has any meaning, but the idea that our ordinary decision-making is somehow severely misguided isn't a paradox and doesn't in and of itself undermine the theory.

But the *epistemic* aspects of the probability problem do plausibly lead to a kind of self-undermining because these difficulties with probability pose a serious threat to the very possibility of performing any kind of empirical confirmation in the context of the Everett interpretation. For in order to empirically confirm a theory, we must start from the assumption that the empirical evidence available to us generally provides some meaningful information about the structure of the theory. So in particular, if the theory is deterministic, we must assume that the outcomes we have seen generally match the deterministic predictions of the theory, and if the theory is probabilistic we must assume that in the long run we will see a sequence of outcomes which is assigned a reasonably high probability by the theory. Indeed, when we confirm probabilistic theories we usually appeal to the law of large numbers, which states that in a long series of trials the average value of some variable will tend towards its expected value as the number of trials tends towards infinity (Greaves and Myrvold, 2010), thus allowing us to take for granted that after a large number of measurements the relative frequencies we observe should be very close to the theoretical probabilities predicted by the theory. And in fact, the empirical predictions of quantum mechanics are mostly probabilistic – as we saw in chapter 2, most of the central predictions of the theory appear in the form of mod-squared amplitudes, interpreted as probabilities for measurement outcomes as stipulated by the Born rule. So as Greaves and Myrvold put it, '*It is incumbent upon the Everettian to provide an interpretation in which the statistical analysis of the outcomes of repeated experiments provides empirical support for the theory*' (Greaves and Myrvold, 2008, p. 1). That is, in the context of the Everett interpretation, the empirical confirmation attached to quantum mechanics depends crucially on an assumption that I will refer to as the 'Everettian Statistical Assumption,' or ESA, in which an Everettian observer assumes that the relative frequencies she herself observes in experiments are, after a sufficiently large enough number of experiments, very close to the mod-squared-amplitudes for the corresponding outcomes.

So we can now state the central epistemic problem for the Everett interpretation: it is not at all clear that an Everettian observer has any good reason to assume the ESA (Myrvold, 2002; Adlam, 2014). After all, we know that in an Everettian world every possible sequence of outcomes does actually occur and is experienced by someone, so it's unclear what grounds the Everettian observer could possibly have for assuming that she just happens to be in one of the branches where the relative frequencies do approximately match the mod-squared amplitudes associated with the outcomes. Yet if she *can't* justify an assumption like the ESA, then she will have no idea whether or not she is in a branch of the wavefunction in which the relative frequencies reflect the underlying mod-squared amplitudes, and thus she will not be able to maintain the quantum mechanics has been empirically confirmed by her observations. And, of course, if quantum mechanics is not empirically confirmed, then there is no good reason to believe in the Everett interpretation. As Myrvold puts it, '*Any interpretation of quantum mechanics, or any theory that takes quantum mechanics as its starting point, is deserving of our serious consideration to an extent that cannot exceed the extent to which quantum mechanics itself is so deserving*' (2002, p. 537) So, using a term introduced by Barrett, Everettian quantum mechanics is in danger of being *empirically incoherent* (Barrett, 1999; Myrvold, 2002) – that is, it may undermine its own empirical evidence.

In chapter 6 we will explore a range of possible approaches the Everettian might consider in order to solve the probability problem in a way that provides justification for the ESA. However, for now let us just note that if no suitable solution to the probability problem is forthcoming, then it looks like the Everett interpretation cannot possibly be accepted – empirical incoherence so severe would be the death-knell for any putative solution to the measurement problem. So unlike more general questions about behaviour and decision-making in an Everettian universe, the epistemic aspects of the probability problem cannot simply be disregarded: if we are ever to have grounds to take the Everett approach seriously, the Everettian must be able to offer some story about how the empirical evidence could still confirm a theory like quantum mechanics in an Everettian context.

3.2 Observer-Relative Interpretations

In this book I will use the term 'observer-relative interpretations' to refer to putative solutions to the measurement problem which satisfy the following three postulates:

1. Unitary quantum mechanics is universal and complete
2. Every measurement has a unique outcome, and the probabilities for these unique outcomes are given by the Born rule
3. All quantum-mechanical descriptions are relative to individual observers

There are a large class of putative solutions to the measurement problem which appear to uphold these three postulates, including perspectival interpretations (Dieks, 2018; Glick, 2019), some neo-Copenhagen interpretations (Zeilinger, 1999; de Muynck, 2004; Bub, 2012; Demopoulos, 2012; Brukner, 2017; Janas et al., 2021), some versions of relational quantum mechanics (Rovelli, 1996; Di Biagio and Rovelli, 2022), and possibly QBism (Fuchs et al., 2014). Many observer-relative interpretations are strongly inspired by the Copenhagen interpretation pioneered by Bohr and Heisenberg, but the Copenhagen interpretation itself arguably is not an observer-relative interpretation in the above sense – in section 9.8 I will return to the question of how the Copenhagen approach fits into this picture.

Let us examine these three postulates in a little more detail. Postulate one tells us that these approaches are unitary-only approaches, which means that they don't postulate a collapse mechanism or any hidden variables. Meanwhile, postulate two tells us that these approaches are not many-worlds theories – they maintain that measurements have unique outcomes, at least for any given observer. Thus it is easy to see the appeal of such interpretations: like the Everett interpretation, they allow us to maintain the entire theoretical structure of unitary QM, but they also avoid the postulation of multiple worlds. Most importantly for our current purposes, this means they don't have a 'probability problem,' since they tell us that measurements have unique outcomes and therefore probabilities can be understood in the standard way.

Now, this may at first seem too good to be true, for the unitary evolution of the quantum wavefunction clearly contains representations of what appear to be multiple outcomes for measurements, so how can an interpretation be unitary-only if it does not postulate multiple outcomes? To sharpen this problem, consider what the observer-relative interpretations would have to say about a case where some observer Alice performs a measurement on a system S with two possible outcomes. Postulate two says that the measurement has a definite, unique outcome, so Alice now knows a definite value for the variable she measured. But now suppose that as in the Wigner's Friend thought experiment discussed in section 2.2 (Wigner, 1961) there is a second observer, Bob, witnessing all of this. From postulate one, unitary quantum mechanics is universal and complete, so Bob must describe this whole interaction unitarily, meaning that he sees Alice and S as ending up in a superposition state ψ_{SA} which includes both of the possible outcomes of the measurement, 0 and 1:

$$|\psi\rangle_{SDA} = c_0|0\rangle_S|0\rangle_D|0\rangle_A + c_1|1\rangle_S|1\rangle_D|1\rangle_A$$

Here $|0\rangle_S$ is the state of S corresponding to outcome 0, $|0\rangle_D$ is the state of the detector corresponding to it displaying outcome 0, $|0\rangle_A$ is the state of Alice corresponding to her seeing outcome 0, and likewise mutatis mutandis for $|1\rangle_S$, $|1\rangle_D$, and $|1\rangle_A$.

Clearly the state ψ_{SDA} does not represent S as having some definite unique value for the variable that Alice measured. And postulate one entails that it cannot be the case that there exists some additional hidden variable which specifies the actual state of Alice, the detector, and/or S, over and above what is included in the state ψ_{SDA}. So from Bob's point of view it must be the case that system S does not, in fact, have a definite, unique value for the relevant variable, even though from Alice's point of view it does. This looks like a contradiction!

The observer-relative interpretations resolve this dilemma by means of postulate three, which stipulates that quantum mechanics does not describe an observer-independent reality. That is, quantum states and quantum measurement outcomes don't have values in and of themselves; they only have values relative to 'observers,' and different observers can disagree about what those values are. So it can be true both that S has a definite value of V relative to Alice and that S does *not* have a definite value of V relative to Bob. It should be reinforced that this is not to be interpreted as merely lack of knowledge; the observer-relative interpretations are not simply saying that Bob doesn't know the definite value of V, which would entail accepting that quantum mechanics is not truly complete. Rather they are necessarily committed to saying that V *really* doesn't have any value for Bob, and moreover, this can in a sense be verified operationally. For Bob could in principle perform quantum operations on the joint system of Alice, the detector, and S, and if quantum mechanics is really universal, as postulate one requires, it follows that if he did so he would obtain results confirming that the two of them really are in the superposition state ψ and not actually in either the state $|0\rangle_S|0\rangle_D|0\rangle_A$ or the state $|1\rangle_S|1\rangle_D|1\rangle_A$.

Now, different observer-relative interpretations have different ways of specifying what they mean by 'observers.' In particular, relational quantum mechanics (RQM) stipulates that any physical system can play the role of an 'observer' in a generalised sense, and thus every physical system has its own relational description of reality (Rovelli, 1996b); whereas some other orthodox interpretations suggest that only conscious beings like humans can play the role of observers (Brukner, 2017). There are also more general views in this vicinity which take measurement outcomes to be relativised not to an individual observer but to sets of systems (Ormrod and Barrett, 2024) or to a decoherence environment (Healey, 2022), but I will not include these in the class of observer-relative interpretations, since they do not necessarily face the same epistemic issues – I will discuss such approaches further in Section 13.1.2.

Different observer-relative interpretations also have different ways of describing the relation between observers and the values of variables. For example: '*A QBist takes quantum mechanics to be a personal mode of thought – a very powerful tool that any agent can use to organize her own experience ... quantum mechanics itself does not deal directly with the objective world; it deals with the experiences of that objective world that belong to whatever particular agent is making use of the quantum theory*' (Fuchs et al., 2014, p. 3). Meanwhile, Brukner (2017, p. 2 (online version)) writes of his neo-Copenhagen interpretation that '*measurement records ... can have meaning only relative to the observers; there are no "facts of the world per se,"*' Zeilinger (2002, p. 252) emphasises that '*it is information about possible measurement results that is represented in the quantum states,*' and Timpson (2003, p. 463) summarises the view of Zeilinger as the statement that '*a physical system literally is nothing more than an agglomeration of actual and possible sense impressions arising from observations.*' But in most of this book I will treat the observer-relative interpretations all together and simply refer to them as describing a set of perspectives or relative facts which are relativised to observers; this terminology is supposed to capture all the different ways in which these approaches may be relational, perspectival, observer-relative, and so on, as well as the different stipulations they make about what counts as an observer.

Let us now explore one important consequence of the observer-relative approach. Returning to the example of Alice, Bob, and S described above, suppose that after Alice has measured S, Bob now performs a measurement on Alice in the basis $\{|0\rangle_A, |1\rangle_A\}$, which can be thought of as asking the question 'is Alice in state $|0\rangle_A$ or $|1\rangle_A$'? For example, one practical way of implementing such a measurement would be for Bob to simply ask Alice what outcome she saw, so the outcome of the measurement would be her response. Of course, we would naturally expect that under these circumstances the result of Bob's measurement will match the result of Alice's measurement – that is, Bob will correctly hear what Alice tells him about her measurement outcome. But, in fact, in the context of an observer-relative interpretation, this cannot reliably be the case! For if quantum mechanics is really complete and universal, then relative to Bob, a complete description of Alice, the detector, and S is given by the state $|\psi\rangle_{SDA}$ as defined above. And thus when Bob measures Alice to see what outcome she obtained, the probability for him to discover that she obtained 0, as predicted by the Born rule applied to the state ψ, is necessarily $|c_0|^2$, and likewise the probability for him to discover that she obtained 1 is necessarily $|c_1|^2$, and all this is true *regardless of what outcome Alice actually obtained.* That is, applying unitary quantum mechanics from Bob's point of view entails that Alice's own outcome is completely irrelevant to Bob's outcome probabilities, and therefore he simply has no meaningful way of getting any information about what outcome Alice actually obtained.

One might hope to solve this problem by adding a third observer, Chidi, who consults both Alice and Bob and thus confirms that there is, in fact, some specific relation between their measurement outcomes. But what Chidi learns about Alice and Bob's outcomes is itself a measurement outcome which is relativised to Chidi, and thus it is not an absolute, observer-independent fact either – it tells us about the relation between Alice and Bob's outcomes *relative to Chidi*, but it cannot tell us anything about the relation between the measurement outcomes for Alice and Bob as they themselves have perceived them, so it cannot offer any reassurance to Bob himself that the outcome that he saw bears any systematic relation to what Alice actually saw in her original measurement, or vice versa. Thus relations between observers which are relativised to other observers appear inadequate to ground any meaningful exchange of information between observers.

This point can be generalised. In an observer-relative approach, all quantum-mechanical descriptions are relative to an observer; but also quantum mechanics is complete, so there are no physical facts outside of quantum-mechanical descriptions, and therefore in such an approach there are simply no observer-independent facts at all. Thus such approaches do not allow that there exists any third-person, observer-independent reality which could contain a value of a variable that both Alice and Bob have access to. So in an observer-relative approach, the only way for Alice and Bob to share any information would be for us to explicitly specify some theoretical structure which connects their perspectives in some way. Moreover, postulate one says we cannot add anything to unitary quantum mechanics, so any such structure would have to come directly from unitary quantum mechanics. However, no such structure can possibly be provided by unitary quantum mechanics, since the theory without wavefunction collapses does not even provide any mechanism to select and actualise a single measurement outcome relative to any one observer, so it certainly cannot provide a mechanism to select and actualise a single measurement result in a global way that holds for all observers who try to access it. Thus it is inevitable that in an observer-relative approach, observers will not be able to share information. This is also the conclusion reached by van Fraassen (2010, p. 43), who notes that '*we have no basis or law (within unitary quantum mechanics) on which to connect the outcomes of measurements by different observers, no matter how intimately they may be related.*'

Moreover, it should be reinforced that Bob's inability to gain information about Alice's measurement outcomes applies even in the case in which his 'measurement' simply amounts to asking Alice what her outcome was. And thus in an observer-relative approach, when Bob asks Alice about her outcome the words that he hears her say will not reliably match what Alice herself thought she said – a phenomenon that Adlam (2024) refers to as 'Type-III disaccord'[1]. Furthermore, the argument for the inevitability of Type-III disaccord in observer-relative approaches applies not only to interactions in which Alice and Bob discuss quantum measurement results but to *any* interaction in which they might attempt to exchange information on any subject. So in the context of an observer-relative interpretation, there appears to be no way for Alice and Bob to share any information at all – each of them is trapped in their own '*island universe*' (Pienaar, 2021) unable to make meaningful contact with any other observer.

This is, in the first instance, a deeply horrible vision of reality: every one of us completely alone in our own little world, constantly attempting to communicate with other observers and constantly failing to reach them. But putting the human element aside, we should consider whether it is possible within this kind of picture to give a coherent account of epistemology of science – and, in fact, there are good reasons to think that it is not.

The central issue is that the practice of science, and empirical confirmation in particular, requires us to make use of reports from other observers – or at the very least from past versions of ourselves, who, as we will see in chapter 9, seem likely to count as 'other observers' in the context of an observer-relative interpretation. I will discuss the role of intersubjectivity in science in more detail in chapters 9 and 10, but it is not difficult to see why it matters. Scientific theories are not usually understood as instantaneous descriptions of the experiences of one observer at a single time; they postulate regularities which are understood to be at least somewhat universal. Thus in order to empirically confirm them we must make use of reports from other observers as well as memories and records inherited from past versions of ourselves, in order to verify that the regularities we have observed are at least somewhat generalisable rather than mere flukes observed for no particular reason in our own little pocket of the universe.

Yet in the context of an observer-relative interpretation we cannot expect to obtain any veridical information from reports obtained from other observers or past versions of ourselves. Therefore these approaches entail that we cannot possibly find out anything about whether our observations fit into wider regularities, making it impossible to perform any meaningful empirical confirmation. Thus in particular,

[1] Adlam (2024) offers a classification of three different types of disaccord, but only Type-III will be relevant for us in this book, so I will not give details on the others.

observer-relative interpretations appear to tell us that quantum mechanics cannot be empirically confirmed, and so just as with the Everett interpretation, we appear to be facing empirical incoherence: it cannot be reasonable to believe some interpretation of quantum mechanics which has the consequence that we are not even justified in believing in quantum mechanics itself. As Stein (1963, p. 24) reportedly noted in personal communication to Shimony, '*If quantum mechanics implies solipsism then it must be self-contradictory, since manifestly no person could have discovered it alone!*'

Note that the proponents of observer-relative interpretations are sometimes inclined to promote as a virtue of their approach that it does away with the notion of an intersubjectively accessible objective reality, arguing that those who wish to keep such a thing are simply holding sentimentally onto naïve classical ideas. For example, Cuffaro and Hartmann (2024, p. 23) suggest that '*the ideal of an observer-independent reality is not methodologically necessary for science and ... modern physics (especially, but not only, quantum theory) has taught us ... that there is a limit to the usefulness of pursuing this ideal.*' This dispute therefore appears to be in danger of becoming merely a comparison of subjective preferences regarding the aims and ideals of science. However, when we consider the issue in light of the epistemic construal of the measurement problem, it becomes clear that the absence of intersubjectivity in observer-relative approaches does actually pose a serious threat to empirical scientific knowledge, and thus, against Cuffaro and Hartmann (2024), I believe an observer-independent reality may well be methodologically necessary for science – not in order to hold onto a naïve classical picture, but because it is not epistemically rational to accept an interpretation of a scientific theory which undermines the entire practice of science. If we are ever to have grounds to the take the observer-relative approaches seriously, their proponents must be able to offer some story about how the empirical evidence shared by the scientific community could still confirm a theory like quantum mechanics even in the absence of the usual mechanisms for intersubjective sharing of information.

3.3 Consistent Histories

As noted in chapter 2, the standard way to extract empirical content out of quantum mechanics is to use it to predict probabilities for the outcomes of a measurement performed by an external observer. Clearly, then, we are going to have difficulties if we would like to use quantum mechanics to describe the universe as a whole – for example, for the purpose of doing cosmology – because there is presumably nobody outside of the universe who could measure the whole thing at once.

The consistent histories formalism (Griffiths, 1984; Dowker and Kent, 1996; Craig and Singh, 2010) was originally proposed as a way of addressing this problem. In this formalism we define a 'history' as a chain of time-indexed projection operators $E^i_{t_1}, E^j_{t_2} ... sE^k_{tn}$. Recall that quantum measurements are defined by sets of operators, so each of these operators $E^j_{t_i}$ can be thought of as representing a possible outcome of a measurement taking place at time t_i. However, the measurement is not necessarily performed by a person – the operators are simply thought of as corresponding to certain kinds of events taking place in definite ways at definite times, so we can think of a history as encoding a sequence of events, as one might expect from the term 'history.'

To assign probabilities to these histories, we specify an initial state ρ and the probability for the history $E^i_{t_1}, E^j_{t_2} \dots E^k_{tn}$ is then given by $Tr(E^k_{t_n} \dots E^j_{t_2} E^i_{t1} \rho E^i_{t1} E^j_{t_2} \dots E^k_{t_n})$. Notice that this is essentially a generalisation of the Born rule for mixed states as set out in section 2.1.2, so the probabilities obtained from this procedure will match the usual predictions we would make using the Born rule[2]. We say that a set of histories is 'consistent' when this algorithm results in probabilities such that the probabilities for mutually exclusive events can be added up in the same way as ordinary classical probabilities. This will be the case provided the set satisfies a condition known as the weak decoherence condition, which requires that $Tr(E^k_{t_n} \dots E^j_{t_2} E^i_{t1} \rho E^m_{t1} E^n_{t_2} \dots E^o_{t_n}) = 0$ for any two distinct histories $E^i_{t_1}, E^j_{t_2} \dots E^k_{tn}$ and $E^m_{t_1}, E^n_{t_2} \dots E^o_{tn}$ in the set (Bacciagaluppi, 2007).

In general, consistent sets of histories contain histories largely made up of 'macroscopic' happenings, since the states of macroscopic objects are generically well-decohered, so events involving them are much more likely to obey the weak decoherence condition. Using this formalism, we can extract many different consistent sets of histories from the same initial state ρ, each coming along with a classical probability distribution over its histories. So the formalism allows us to extract empirical content from quantum mechanics in the form of macroscopic histories rather than individual measurement outcomes, and thus it could plausibly be used to describe the universe as a whole.

Now, a rewritten formalism for quantum mechanics in and of itself isn't a solution to the measurement problem; the consistent histories formalism is simply a reformulation of the theory, or arguably a new theory, whereas in order to solve the epistemic version of the measurement problem we need an account of reality. Thus if we hope to extract a solution to the measurement problem from this approach, some account of reality must be added. One natural way to do this would be to say simply that all of the consistent sets and all of the histories in each set are physically real, which would lead us to something that looks somewhat like the many-worlds interpretation (Saunders, 1992). This version of the consistent histories formalism would therefore give rise to the same kinds of epistemic issues as the Everett interpretation. Alternatively one could always imagine probabilistically choosing one history from each set, as in the 'many-histories' approach described by Dowker and Kent (1996) – but since we can extract many different consistent sets from the same initial state ρ, in this approach we will still end up with a selection of incommensurable but equally real histories, so the epistemic difficulties encountered in the Everettian context are likely to persist. Thus in order for the consistent histories approach to avoid the difficulties of the Everett picture, we need an account of reality in which just *one* history is selected, which would presumably involve first choosing a single consistent set and then probabilistically selecting exactly one history from that set. Dowker and Kent

[2] There also exists a time-symmetric version in which we use both an initial and final state to determine the probabilities (Dowker and Kent, 1996), but here I will focus on the more common initial condition version.

(1996) refer to this as 'The Unknown Set Interpretation,' and Adlam (2023) refers to it as the 'Single-World-Realist Consistent Histories' (SWRCH) approach.

Assuming for the moment that this kind of approach can be made to work at a technical level, is it an improvement on the Everettian picture? Since it postulates just one course of history, it does not have a probability problem – but it may encounter other severe epistemic problems.

In particular, Dowker and Kent (1996) argue that the consistent histories formalism has problems with the reliability of records of the past and the persistence of quasi-classicality. For example, note that when we define a set of consistent histories, that set will include every possible combination of outcomes to all of the projections in the set. That is to say, a consistent set can include histories in which some measurement has outcome O_1 and that outcome gets recorded in a lab book, but then at a later moment of time the configuration of the lab book records some outcome O_2 rather than O_1, and at a later moment of time the configuration of the lab book records some other outcome O_3, and so on. Obviously one would hope that the probability distribution predicted by quantum mechanics will assign probability close to zero to all histories in which the states of macroscopic records fluctuate in this way, but although this will usually be the case if a record is observed twice in quick succession, it is not generically true. The difficulty is that the projections defining 'events' in the history do not necessarily have to be chosen to correspond to events that look like the ordinary kinds of classical events we are familiar with: there is nothing in the formalism to prevent us from choosing extremely complex non-local projections which can't be interpreted in classical terms. So we can come up with consistent sets of histories which describe the world using projections corresponding to ordinary classical events up until a certain point in history, and then suddenly switch to using extremely complex non-local projections, and then switch back to the quasi-classical projections; and the results of the classical projections before and after the non-classical projection will not generally be related in any reliable way, so macroscopic records cannot be expected to say the same thing as they did before the non-classical interruption. Even worse, we will not generally be able to tell that something of this kind has occurred, because if we have multiple records of the same event, they are necessarily all entangled and thus each time we perform a projection, the results obtained for all of them will be correlated – and yet this joint result might not be the same result as obtained in previous projections! Similarly, it's also possible to construct consistent sets in which the histories use only complex non-classical projections up to some point, then at some time switch to quasi-classical projections, and in this case the state of the world produced by one of these later quasi-classical projections will contain sets of correlated records which appear to tell a mutually consistent story about a well-defined classical history which never actually took place.

So in a SWRCH picture, it seems we would not in general be able to rely on records to report the past accurately to us, and indeed the existence of records of a quasi-classical history doesn't even guarantee that *any* quasi-classical history actually occurred. Moreover, within the standard consistent histories formalism we can't even hope to make the argument that histories with these undesirable properties are unlikely; for the consistent histories formalism assigns probabilities *within* consistent sets, but it doesn't assign any probabilities to the sets themselves, so it gives us no reason to expect that the actual history will be selected from a 'well-behaved' consistent set rather than a maverick one. Therefore it seems that agents who believe they are in a world as described by SWRCH would have very little justification for thinking that the records they're looking at give them any reliable information about what happened in the past.

And this looks like bad news for the practice of science, since we can't perform any meaningful empirical confirmation without relying on records of past measurement results. On pain of incoherence, we need the account of reality that we ultimately arrive at on the basis of our empirical enquiries to affirm that records of the past typically reflect events that really happened. If records of the past *don't* reflect what has really happened, then the theory we confirm won't be a theory of measurement results as they actually occur: it will simply be a description of the contents of records at some particular point in time, which will clearly be a drastic reduction in our empirical scientific knowledge. Moreover, if we can only confirm quantum mechanics as a description of some instantaneous set of records, then we have no good reason to adopt the consistent histories formalism in the first place – for the whole point of the formalism was to enable us to apply quantum mechanics to large-scale cosmological scenarios which are outside the scope of the observations of an individual observer, and thus its formulation was motivated by the belief that quantum mechanics has, in fact, been empirically confirmed as a somewhat universal theory, rather than as a description of the experiences of one individual. So again, we encounter what looks like empirical incoherence: the SWRCH formalism appears to tell us that we should not trust the very evidence that we have used to arrive at it. And thus again, if we are to ever have grounds to take the SWRCH approach seriously, its proponents must be able to offer some story about how the empirical evidence could still confirm a theory like quantum mechanics in a consistent histories context.

3.4 Beyond Quantum Mechanics

We have now seen that a number of popular approaches to solving the measurement problem appear to have fairly drastic epistemic consequences, potentially leading to a picture in which it is not possible for quantum mechanics to be empirically confirmed. Thus all of these approaches to quantum mechanics are potentially empirically incoherent. But I want to close by emphasising that it is not only quantum mechanics which is endangered here – putative solutions to the measurement problem which undermine the empirical confirmation attached to quantum mechanics are also liable to undermine the empirical confirmation for many other parts of science as well.

To see this, note that an important feature of the unitary-only interpretations is that they don't allow us to add anything to the standard formalism of unitary quantum mechanics, and thus they are committed to the view that ultimately everything in the physical world must be in some sense derived from unitary quantum mechanics. That is, although the unitary-only approaches do affirm that we get an effective quasi-classical world emerging at large scales due to the effects of decoherence, they still must deny that there is some scale at which we pass absolutely from 'the quantum world' to 'the classical world': in their picture of reality all physical systems are ultimately composed of quantum systems, and all physical interactions emerge out of an underlying quantum structure, meaning that it is possible for them to retain some of the features that we would traditionally describe as 'quantum.'

Now, in many ways the continuity of the quantum and classical world is a desirable feature, since it avoids the dilemma discussed in chapter 2: we don't have to provide a Heisenberg cut or a demarcation criterion singling out the set of interactions which count as quantum measurements if we say that nothing physically distinctive happens during quantum measurements. But on the other hand, this also makes the measurement problem into a much bigger problem. For now the measurement problem no longer applies only to experiments testing quantum mechanics; in the unitary-only picture every interaction could in principle have the same character as a quantum measurement, and therefore if we can't give an account which assures us that quantum measurements provide us with meaningful information about physical reality, then we risk also being unable to give such an account for many other kinds of measurements across all of science. So in a unitary-only context, we need to be worried not only about the empirical confirmation attached to quantum mechanics itself but also about the empirical confirmation attached to any scientific theory whatsoever; not only quantum mechanics but all of our scientific knowledge is at risk.

To be clear, it is not my claim that any unitary-only interpretation is inevitably forced to treat all scientific theories as quantum theories – one would naturally hope that in such approaches it will be possible to appeal to decoherence or take some appropriate limit in order to arrive at macroscopic physics which is relatively autonomous from the underlying quantum description. But the point is that this is not *guaranteed* to occur – if we are not postulating a classical realm which is distinct from the quantum realm, then we need to be able to demonstrate that classical reality emerges robustly from the underlying quantum substratum in such a way that it does not inherit the epistemic difficulties appearing in the quantum realm. And certainly in the context of the unitary-only approaches discussed in this chapter, it is not obvious that this is the case.

For example, consider what the Everett approach has to say about an ordinary classical observation such as observing the position of a billiard ball at the end of an experiment. Observations of this general form constitute most of our evidence for Newtonian mechanics and other important parts of classical physics. If we describe this scenario using Everettian quantum mechanics, we will find that there is a branch of the wavefunction in which all of the particles in the billiard ball simultaneously undergo a phenomenon known as 'quantum tunneling', meaning that the ball ends up in a position different to the one which classical physics predicts. Now, the mod-squared amplitude for branches in which something like this occurs is minuscule, and thus if we interpret the mod-squared amplitudes as probabilities, this possibility can be ignored for all practical purposes. However, we saw in section 3.1 that in the Everettian context it is unclear that mod-squared amplitudes *can* be interpreted as probabilities. So if we can't come up with a suitable solution to the Everettian probability problem, then we will have no reason to disregard the tunneling branch. And if we can't disregard that branch, we will have no reason to expect that the billiard ball will be found in the position predicted by classical mechanics, and therefore we won't be able to use our observation of its position to empirically confirm classical mechanics. That is, if Everettians cannot solve the probability problem, not only will they be unable to use quantum-mechanical observations to empirically confirm quantum mechanics, they will also be severely limited in their ability to use observations to confirm any other scientific theory.

Similarly, in observer-relative approaches the epistemic difficulties also extend beyond quantum mechanics. In particular, we have already noted that in such approaches the act of having a conversation with another observer must be modelled as a 'quantum measurement' in the sense that it produces an outcome which is relativised to an observer, meaning that all ordinary acts of communication are potentially subject to Type-III disaccord. And there is nothing here to suggest that Type-III disaccord obtains only when the conversation is about a quantum measurement outcome, so if it is impossible in this picture for observers to reliably share information about quantum measurement outcomes, it is presumably also impossible for them to reliably share information about any other kind of observation. Thus not only will we be unable to get any evidence about whether or not other observers see regularities consistent with quantum mechanics, we also won't be able to get any evidence about whether or not they see regularities consistent with *Newtonian* mechanics, or even with the theories of genetics, evolution, and so on. Of course, it may be that decoherence or some other such limiting procedure can overcome such problems, but the point is that this is not automatic – it should be a central task for proponents of such approaches to explain how they can recover a sensible epistemology for the rest of science, in addition to quantum mechanics.

Indeed, even if we don't assume some particular solution to the measurement problem, we still must face up to these difficulties. For as we saw in chapter 2, as things currently stand we have a great deal of freedom about where to put a Heisenberg cut or indeed about whether we need such a cut at all – quantum mechanics itself does not appear to suggest any natural cutoff point. There do exist experiments which can probe the location of the Heisenberg cut, because for a given system we can in principle determine whether or not it belongs to the quantum world or the classical world by attempting to put it in various quantum states and then performing measurements on it to establish whether or not the attempt has succeeded. But current experimental techniques only enable us to perform these kinds of experiments for fairly small systems – at the time of writing, the record for largest quantum superposition was held by the Hybrid Quantum Systems Group at ETH Zurich, who put a sapphire crystal weighing 16 micrograms into a superposition (Bild et al., 2023). So on the basis of current empirical evidence the Heisenberg cut could occur almost anywhere above the scale of a sapphire crystal, or it could be the case that there is no Heisenberg cut at all. Consequently, it is quite difficult to stop quantum mechanics from 'leaking' into our descriptions of the rest of reality. That is, until we actually have a solution to the measurement problem in hand – or until we manage to show experimentally that systems above a certain scale cannot enter into superpositions – we have no choice but to take seriously the possibility that our ordinary interactions may inherit some of the epistemic difficulties associated with ordinary quantum interactions, just as they appear to do in the Everettian case and the observer-relative case.

Now, one might object that decoherence is relevant here – even if we are not sure where the Heisenberg cut occurs or whether there exists any such cut, surely in the limit as decoherence effects take hold everything becomes quasi-classical, and then we can just treat all interactions as ordinary classical observations? That is, one might hope that decoherence assures us that our knowledge of classical phenomena is reliable, even if we can't make the case that measurements of individual quantum particles are reliable.

However, this strategy has a number of problems. First, as we will see in more detail in subsequent chapters, at least in the case of the Everett and observer-relative approaches, it is not clear that decoherence gets rid of the epistemic difficulties. Second, it must be emphasised that many phenomena which seem entirely classical to us have been shown to be partly explained by quantum mechanics. For example, the fact that the sky is blue probably seems like a macroscopic fact which has nothing to do with quantum mechanics, but it turns out that, in fact, we must appeal to quantum electrodynamics to explain it, so in a sense any time we inspect the sky we are actually making a quantum measurement (Loudon, 2000; Wallace, 2022). Indeed, even the very stability of matter is ultimately explained by a quantum-mechanical mechanism (Lieb, 2002). Therefore in order to decide which particular macroscopic events are insulated from quantum mechanics by decoherence, we have to perform detailed empirical and theoretical enquiries, so if we don't have good reason to believe that the formalism of quantum mechanics is right, we'll never have robust grounds for thinking that some particular kind of classical process is not affected by the quantum measurement problem. And more generally, the theory of decoherence comes from quantum mechanics in the first place, so if we can't use quantum measurements to empirically confirm quantum mechanics at microscopic scales, we will have no grounds for believing that the decoherence mechanism really functions in the way required to ensure that macroscopic measurements are safe from the quantum measurement problem, and therefore the threat of epistemic instability cannot be averted merely by appeal to decoherence.

This leaves us with something of a dilemma. We have good empirical reasons to believe that the unitary-only approaches may be correct, given that they currently appear to have better prospects than any other kind of approach of reproducing all the empirical evidence for both non-relativistic and relativistic quantum mechanics. And indeed, many well-informed people currently believe that some unitary-only approach *is*, in fact, correct. Yet the unitary-only approaches appear to undermine not only the empirical confirmation of quantum mechanics itself, but also the empirical confirmation attached to scientific knowledge in general. So if we take seriously what quantum mechanics is telling us about reality, right now there is a sense in which our empirical evidence seems to be telling us that empirical evidence cannot in general be relied upon!

Of course, we should not allow rhetoric to get in the way of clarity, so let me be clear about the scope of the problem. Although I do think there is a real sense in which the quantum measurement problem represents an epistemological emergency, it is not a *practical* emergency – although we don't have a fully satisfactory solution to the problem right now, presumably no one seriously doubts that there is a solution to the measurement problem, so no one seriously imagines that these considerations threaten the usefulness and applicability of science at a practical level. Certainly, the absence of a solution to the measurement problem has not stopped scientists from doing measurements and building theories on the basis of them, and nor will it ever. So I am not suggesting that anybody should lose sleep over the possibility that all our scientific knowledge might be wrong; I simply want to suggest that, as (aspiring) epistemically rational agents, we ought to aim for an account of reality which is consistent with the beliefs that we all hold about the epistemic status of our empirical scientific knowledge. In this book I will argue that facing this epistemic challenge head-on can ultimately make our scientific knowledge more secure, and that it is potentially a valuable way of narrowing down the options for future research in order to facilitate further scientific progress on issues like quantum gravity, cosmology and so on.

4
The Epistemology of Measurement

Having seen in broad outline some of the epistemic problems associated with various putative solutions to the measurement problem, we are now in a position to consider in more detail how these issues relate to the epistemology of measurement. So let us return to the observation that 'measurement' has a dual nature: '*A measurement outcome is something physical: an event, the end-state of the apparatus, or an object (photo, graph, list of numbers) produced by the measurement process. On the other hand, measurement is information-gathering, so a measurement outcome has a meaning*' (van Fraassen, 2008, p. 157). That is, measurement is in some sense both a physical process and also an epistemic category – indeed, it is the central element of the epistemology of science.

And yet discussions of the measurement problem do not typically treat it primarily as a problem of epistemology. In particular, epistemic difficulties of the kind described in chapter 3 are not often discussed explicitly by proponents of these putative solutions – in general it seems to be taken for granted that the empirical evidence available to us confirms the basic mathematical formalism of unitary quantum mechanics *regardless* of the account of reality that goes along with it. That is, it seems that 'empirical confirmation' is often implicitly understood in this literature as a direct relation between a mathematical hypothesis and a set of empirical experiences which confirm the hypothesis, meaning that all unitary-only solutions to the measurement problem necessarily have an equal degree of confirmation, since they are all associated with the same theory and the same set of mathematical models.

This view of confirmation looks similar to the approach adopted by the logical positivists or logical empiricists, a group of philosophers most active in Vienna and Berlin in the 1920s (Ayer, 1961; Feigl, 1981). A key feature of the logical positivist programme was the idea that sentences are 'significant' only if it is possible to translate them unambiguously into a set of possible experiences which would verify them or at least empirically confirm them; so, for example, Carnap envisions sentences about physical objects being translated into a set of experiences involving colours and shapes at various parts of one's field of vision (Carnap and George, 2003). So from the logical positivist point of view, empirical confirmation is regarded a two-place relation between a hypothesis and a set of possible empirical observations which would provide confirmation for that hypothesis, as encoded in the hypothetico-deductive approach to confirmation advocated by Hempel and Oppenheim (1966).

Alternatively, the view of confirmation as a direct relation between a mathematical hypothesis and a set of empirical experiences could also be compared to the

Saving Science from Quantum Mechanics. Emily Adlam, Oxford University Press. © Oxford University Press (2025).
DOI: 10.1093/9780197808887.003.0004

approach adopted by operationalists such as Bridgman (1980)[1]. A key feature of the operational program was the idea that quantity-concepts should be defined directly by the procedures that measure them, so again, from the operationalist point of view we can understand confirmation as a two-place relation between a hypothesis and the procedures which are defined to be the means of checking it empirically.

However, this view of confirmation as a two-place relation between a hypothesis and a set of empirical evidence has turned out to be unsustainable. For example, operationalism faces a number of serious problems (Chang, 2013) including '*the automatic reliability operationalism conferred on measurement operations, the ambiguities surrounding the notion of operation, the overly restrictive operational criterion of meaningfulness, and the fact that many useful theoretical concepts lack clear operational definitions*' (Tal, 2013 p. 3). Similarly, the positivists' efforts to lay out in detail their criteria of empirical significance were plagued with serious problems, and subsequently in the post-war years a movement led by Hempel (1950, 1951), Quine (1976), and Duhem (1954) offered powerful reasons to think that no such thing could ever possibly work because statements on their own do not have empirical consequences – we obtain empirical consequences from a hypothesis only in conjunction with a large number of background assumptions fixing the conditions in which the hypothesis is understood to impinge on our experience. As Quine (1976, p. 39) put it, '*Total science is like a field of force whose boundary conditions are experience. A conflict with experience at the periphery occasions readjustments in the interior of the field. Truth values have to be redistributed over some of our statements ... But the total field is so undetermined by its boundary conditions, experience, that there is much latitude of choice as to what statements to re-evaluate in the light of any single contrary experience. No particular experiences are linked with any particular statements in the interior of the field, except indirectly through considerations of equilibrium affecting the field as a whole.*'

The refutation of operationalism and logical positivism makes it clear that confirmation is not and cannot be a straightforward map from hypothesis to evidence: it is really a *three-place* relation, in which a hypothesis is confirmed by some evidence relative to our other background beliefs (Douven and Meijs, 2006). As Longino (1990, p. 44) puts it, '*A state of affairs will only be taken to be evidence that something else is the case in light of some background belief or assumption asserting a connection between the two.*' This explains what is wrong with imagining that the mathematical formalism of quantum mechanics is confirmed by our empirical evidence regardless of interpretation. For the confirmation attached to the theory relative to one set of background beliefs will not necessarily transfer across to a different set of background beliefs, so when we adopt a solution to the measurement problem which involves significant changes to foundational beliefs about reality, the confirmation attached to the theory will need to be reassessed in light of the new context – a step that proponents of revisionary responses to the measurement problem do not always seem to take very seriously.

[1] The philosophical view known as 'operationalism' should be distinguished from the approach to the practice of physics that goes by the same name – for example, the practice of defining and studying 'operational theories' (Abramsky and Heunen, 2013; Hardy, 2016; Di Biagio et al., 2021) is not dependent on the philosophy of operationalism, and need not suffer from the same problems.

4.0.1 Non-Locality

The holistic nature of empirical confirmation, as described by Quine, is well illustrated by the familiar debate around locality in quantum mechanics. As noted in chapter 2, the three postulates of unitary quantum mechanics give rise to a phenomenon known as 'entanglement,' which appears to violate locality. In this context, 'locality' refers to the idea that events at one place can only influence events at another place if some physical signal passes between them to mediate that influence. For example, the macroscopic world generally exhibits a high degree of locality: if I want to influence events in Auckland from my current location in California, I cannot instantaneously cause events to take place there, so I have to communicate my intent via some physical mechanism which travels at or below the speed of light, such as an email, a telephone call, or a carrier pigeon. Thus when we say that quantum mechanics exhibits 'non-locality,' we mean that there exist processes in quantum mechanics which appear not to require this kind of mediation – they seem to involve influences passing instantaneously from one point to another, sometimes faster than the speed of light, without any intervening physical process.

The argument that quantum entanglement results in non-locality of this kind is based on an important theorem due to John Bell (1990). In this theorem, Bell considers an experiment in which two quantum-mechanical systems are prepared in a joint entangled state, and then the systems are separated and one is sent to Alice and another is sent to Bob, with Alice and Bob very far apart in space. Alice and Bob then each choose a measurement and perform it on their half of the entangled system, and it turns out that the results of these measurements are correlated.

Now, the existence of correlations at a distance is not in and of itself particularly surprising because we can easily arrange for such correlations to be produced even in an entirely classical scenario. For example, Bell (1981) notes that we could produce a similar effect by simply separating a pair of socks and sending one to Alice and one to Bob – clearly in this case it would not be surprising to find that the colours of the socks are correlated. But in classical scenarios, correlations observed at a distance can always be explained in a local way by attributing them to some common cause in the past, such as the fact that the socks came from the same pair. Bell therefore set out to prove that any correlations which can be explained locally by a common cause in the past must obey a certain inequality, which is known as Bell's inequality.

And the punchline is that quantum mechanics predicts violations of this inequality – that is, in quantum-mechanical experiments of the kind envisioned by Bell, the entangled quantum systems can exhibit correlations too strong to be explained by a common cause in the past, so it appears that Alice's choice of measurement can have some kind of direct, non-local influence on the result of Bob's measurement. Thus quantum mechanics exhibits something like 'spooky action at a distance' in Einstein's words (Einstein et al., 1971). This non-locality is limited – in particular, it turns out that quantum non-locality can never be used to send messages faster than light – but nonetheless it is a very surprising result.

Now, the experiment envisioned by Bell has subsequently been performed and checked many times (Aspect et al., 1981; Hensen et al., 2015), so it is clear that Bell's inequality can indeed be experimentally violated. This, one might think, is a conclusive falsification of the hypothesis of locality. But some physicists are still unwilling to give up that hypothesis. Instead they have argued that there exist 'loopholes' in Bell's theorem which can be exploited to make the experimental results compatible with locality after all, for example, by denying the assumption that experimenters can freely choose their measurements (Hossenfelder and Palmer, 2020), denying that causation always goes forward in time (Costa de Beauregard, 1953), or denying that experiments have only one outcome (Bacciagaluppi, 2002); I will discuss these options in more detail in chapter 12. For now, let me simply point out that these 'loopholes' correspond to ways in which the set of background beliefs relative to which the results of the Bell experiments disconfirm the hypothesis of locality can be adjusted in order to accommodate the experimental results without abandoning the hypothesis of locality, in exactly the way that Quine envisioned. And note that the background beliefs that these authors are proposing to adjust are very foundational ones, pertaining to important and almost universal beliefs about the nature of our reality, so this episode demonstrates that physicists are in principle willing to revise even the most foundational of background beliefs in order to save a particularly cherished hypothesis. Thus when we are addressing this kind of proposal, it will be important to be cognizant of the holistic features of scientific knowledge, in order to fully understand the consequences of revising such foundational beliefs.

4.1 Confirmation

Once we appreciate that empirical confirmation is necessarily relativized to a set of background beliefs, it immediately follows that confirmation can no longer be attached in a definite way to specific individual beliefs: as Quine puts it, '*Our statements about the external world face the tribunal of sense experience not individually but only as a corporate body*' (Quine, 1976, p. 38). And thus, for a time after the revolution instituted by Quine, Hempel, and Duhem, the field of epistemology swung towards a kind of radical holism, in which nothing could be confirmed or falsified other than our entire set of beliefs all at once. But as pointed out by many authors (Glymour, 1981; Urbach and Howson, 1993; Sober, 2004), this is clearly not a reasonable description of real scientific practice: for example, it would have the consequence that the results of the double slit experiment are just as relevant to our theories about the life cycles of Amazonian tree-frogs as they are to the formalism of quantum mechanics. In real scientific contexts, experiments are almost always understood to be relevant to some particular hypothesis, rather than to our set of beliefs as a whole: one performs an experiment with the intention that the hypothesis will be confirmed if some particular observation is obtained, and rejected if some other observation is obtained.

In light of the obvious unsuitability of radical holism to account for real scientific practice, it is perhaps not surprising that many people working on the measurement problem are inclined to think about confirmation along similar lines to the logical positivists, for it is certainly the case that in many scientific scenarios we do have in mind a straightforward mapping from a hypothesis to relevant empirical evidence. So in such contexts, a two-place confirmation relation as encoded in the hypothetico-deductive method or a naïve Bayesian approach seems entirely reasonable. However, this feature of scientific practice does not contradict the observation that scientific knowledge is ultimately holistic in character. It simply indicates that in most ordinary experimental contexts we start out with some partition between hypothesis and background beliefs in mind, so we have already committed to leaving the background beliefs fixed – but confirmation is still implicitly relativised to the set of background beliefs, even if we don't normally need to take that into account.

There are many ways to account for the choice of partition – for example, Glymour offers a systematic account of the relevance of evidence to hypotheses in which theories are split into individual hypotheses and then each hypothesis is supposed to be confirmed relative to the other hypotheses in the theory, thus ultimately confirming the theory as a whole (Glymour, 1981; Douven and Meijs, 2006); Chang offers an account grounded in the cumulative nature of scientific progress (Chang, 2004); and Longino argues that in any given scientific setting a system of background assumptions are simply taken for granted, and therefore the relation of confirmation must be relativised to a cultural, historical, and psychological context (Longino, 1990). But regardless of one's preferred account of the way in which the choice of partition comes about, it's clear that in much of what we might call 'ordinary science' our scientific knowledge progresses by the simple accumulation of new hypotheses, which are confirmed by the evidence relative to a fixed, static set of background beliefs. And a number of well-known accounts of empirical confirmation can be understood as describing the ways in which we add hypotheses to our belief set in the course of this ordinary science. For example, as discussed by Glymour (1981), despite the weaknesses of the hypothetico-deductive model espoused by the logical positivists, it is clear that many ordinary scientific inferences *do* employ hypothetico-deductive reasoning because it works well in cases where background knowledge is fixed and static. Similarly, as we will see in more detail in chapter 8, Bayesian updating can also be understood in this way – Bayes' law tells us how to add credences to our belief set in a coherent way using priors and conditional probabilities determined by a fixed set of background beliefs, and thus it is a good description of the dynamics of belief in periods when the set of background beliefs is largely unchanging.

So what kinds of circumstances force us to depart from 'ordinary science' and take into account the holistic nature of our scientific knowledge? Well, one obvious problem case for ordinary science arises when the experiments that we perform in the course of doing ordinary science appear to confirm, relative to our set of background beliefs, a hypothesis which is incompatible with some of those background beliefs. In that case we will not be able to simply go on updating our beliefs in the usual way; instead we will have to go back and reconsider our whole belief set, assessing the consequences that may follow if we change some of our background beliefs to accommodate the new hypothesis. As Chang (2004, p. 226) puts it, '*The concern is that the whole process might become a morass of self-contradiction. What we need to ensure is that the changes in the initially affirmed system do not invalidate the very outcomes that prompted the changes.*'

And this appears to be exactly the scenario that we face with regard to the quantum measurement problem. For example, prior to the advent of quantum mechanics it was generally accepted that we are located in a single world, performing experiments that produce a single outcome; and this was clearly a central element of the background beliefs relative to which experimental results were interpreted. So if we accept for the moment that the Everettians are right to say that the Everett interpretation is the natural result of taking the quantum formalism literally, then it seems that our experiments have empirically confirmed a hypothesis which blatantly contradicts the background beliefs we started out with. We can, of course, choose to change our background beliefs in order to allow the possibility of a multiverse picture, but this will significantly change the context in which we are interpreting the empirical evidence for quantum mechanics, such that the evidence may no longer confirm the theory to the same extent.

Moreover, making alterations to foundational background beliefs may cause changes to propagate out towards beliefs and theories that initially seem completely unrelated to the empirical evidence under consideration. By analogy, imagine that you have almost completed a crossword puzzle and then you realise that the answer to one of the last clues is incompatible with one of the words you have already filled in. Clearly under those circumstances you cannot just change one letter in that word and leave it at that; at the very least you will have to re-evaluate your solution for the word which is incompatible with your newest answer, and that may lead you to re-evaluate your solutions for other words which intersect this one, and so on, and therefore the effects of updating even just a single letter can potentially spread a long way across the crossword. Quine made this point in his famous paper refuting logical positivism: '*Re-evaluation of some statements entails re-evaluation of others, because of their logical interconnections – the logical laws being in turn simply certain further statements of the system, certain further elements of the field. Having re-evaluated one statement we must re-evaluate some others, whether they be statements logically connected with the first or whether they be the statements of logical connections themselves*' (Quine, 1976, p. 39).

So it must be emphasised that we cannot have one set of background beliefs relative to which quantum-mechanical experiments are interpreted and a completely different set of background beliefs relative to which all other experimental results are interpreted – if we choose to accept the hypothesis of a multiverse as a response to the empirical evidence for quantum mechanics, this will potentially change not only the empirical confirmation attached to the quantum mechanics itself, but also the empirical confirmation attached to many other scientific theories. Thus, as foreseen in section 3.4, the changes to our background beliefs required by revisionary approaches of this kind may have significant consequences not only for the empirical confirmation attached to quantum mechanics but also for the confirmation attached to many other scientific theories.

Of course, there are no doubt many possible changes we could make to our background beliefs which would not lead to any particularly profound changes in the rest of our belief system, and therefore one might still optimistically hope that the changes required by putative solutions to the measurement problem might be of this kind. However, there are reasons to think that the changes needed to solve the measurement problem will indeed have significant effects on the rest of our belief system. This is because the measurement problem is, obviously, about *measurement*, and therefore the changes proposed in solutions to this problem are typically changes to our conception of the nature of measurement and of observation itself. Evidently such changes can be expected to have a significant effect on the empirical confirmation attached to a wide variety of scientific theories, for no scientific theory can be empirically confirmed without some kind of measurement or observation. Of course, one might hope to impose some kind of cutoff such that the effects of the changes in our conception of observation are confined to measurements specifically relating to quantum mechanics, but the prospects for making this work are not very good – such a cutoff would effectively be a Heisenberg cut, and as we saw in chapter 2, the difficulty of making such a cut in a principled way is a major part of the reason we have a measurement problem in the first place.

4.2 Measurement

In light of the discussion of the previous section, one may be tempted to suggest that we are better off not solving the measurement problem at all – if all the plausible solutions to it end up wreaking havoc on the totality of our scientific knowledge, perhaps it is better to simply leave unspecified what exactly is going on during the process of a quantum measurement! However, declining to model measurement at all will also lead to problems for scientific epistemology. To see this, we may turn to recent scholarship on the epistemology of measurement, including work by van Fraassen (2008), Chang (2004), Tal (2019), Curiel (2020), and Stein (1994), amongst others, which emphasizes the importance of having a detailed understanding of real measurement processes, rather than simply treating measurements as some kind of vague idealisation. This line of work may perhaps be understood as a branch of naturalised epistemology (Quine, 1968; Laudan, 1990), a meta-epistemic thesis which holds that epistemic claims are to be assessed by scientifically studying our processes of enquiry, although, of course, believing that we need to understand our processes of enquiry scientifically does not necessarily commit us to believing that *all* epistemic claims can be understood in purely naturalistic terms.

For example, Tal argues that in order to know what a measurement is measuring or indeed to know whether some procedure is a measurement at all, we must have in mind some model of what is going on during the measurement process and how the 'indications' produced by the measuring device are related to the quantities of the theory that we are attempting to measure. In particular, Tal (2019) contends that in order to do 'quantity individuation,' i.e. to determine whether or not two distinct measurement procedures are measuring the same quantity, we must appeal to some model of the measurement process in order to resolve underdetermination relating to the way systematic error is distributed. He shows it's possible that '*an effect that requires systematic correction under one representation of the apparatus is deemed part of the correct operation of the apparatus under another*,' illustrating this with the example of the Michelson-Morley experiment (Tal, 2012, p. 54). Similar arguments follow from his detailed consideration of the process of calibrating measurement instruments, which again is shown to require a model of the actual measurement.

A similar line of thought is pursued by Stein (1994) and Curiel (2020), who argue that any viable scientific theory must involve a 'schematisation of the observer' which '*identifies the junctions where meaningful connections can be made between the [theory and experience] and embodies the possibility of the epistemic warrant we think we construct for our theories from such contact and connection*' (Curiel, 2020, p. 6). Curiel argues that an understanding of the physical process of measurement by which we gain access to a theory cannot be separated out from the theory itself: '*We should not assume without argument that the epistemic content of a theory can be cleanly and exhaustively divided into two parts, one consisting of what is encoded in the pure formalism alone and the other a catch-all for the rest, including the mess of real application. Much less should we assume without argument that two such parts, no matter how characterized in detail, can be cleanly and exhaustively segregated from each other in an adequate analysis of that epistemic content*' (Curiel, 2020, p. 8). And thus, much like Tal, Curiel emphasises the importance of modelling measurements in order to calibrate instruments, correct errors, and even to simply understand what it is that we are measuring.

This is a theme repeated across much modern work on the epistemology of measurement – we cannot treat measurements as a 'black boxes' which spit out operational probability distributions, for the practice of science requires a concrete physical model of what is actually going on in a measurement, up to and including the observer herself. Measurement results do not have meaning and cannot confirm anything in and of themselves; they become meaningful and useful for empirical confirmation only in the context of such a 'model' of the measurement process, i.e. a set of background beliefs specifying what it is that the measurement is measuring and how the correlation between theoretically defined quantities and device indications come about. As Tal puts it, '*All measurement outcomes are relative to an abstract and idealized representation of the procedure by which they were obtained*' (Tal, 2012, p. 81).

Indeed, Carrier argues that Einstein himself advocated such a view – as Carrier (1990, p. 370–371) puts it, Einstein seemed to feel that '*a theory should be able to cope with the measuring devices needed to connect its theoretical terms with experimental data ... A complete theory manages to explain why certain empirical processes are reliable indicators for certain theoretical states specified within its framework.*' Now, this is perhaps too strong a requirement – as emphasised by Curiel (2020), a theory can fully explain the reliability of the empirical processes only in combination with other theories describing aspects of the operation of the devices which fall outside the domain of applicability of the theory in question, so this completeness criterion should perhaps be thought of as applying to science as a totality, rather than to individual theories. But certainly it seems reasonable to demand that a theory should at least explain the parts of the measurement process falling within its own domain of applicability, and should allow us to see how this description of the measurement process could in principle connect up with parts of the process modelled in other theories, in order to yield a total measurement process across various different domains which does indeed reassure us that it is a reliable indicator for the structures modelled in the theory.

These ideas in the epistemology of measurement make it clear that a fully fleshed-out epistemology of science demands a solution to the measurement problem, because the set of background beliefs to which the empirical confirmation of a hypothesis by a measurement outcome is relativised must include beliefs about the nature of the measurement process itself. Ultimately, a measurement model and schematisation of the observer are necessary to reassure us that the theoretical structures which are supposed to be confirmed by a given outcome of a measurement are in some meaningful sense responsible for this observation rather than some other observation being made, otherwise the measurement cannot be regarded as providing information about such structures.

Now, at this point one might object, as does Dickson (2007), that the absence of a solution to the measurement process does not mean we are incapable of modelling measurements in quantum mechanics. And indeed, it should be emphasised that in many ways our modern understanding of quantum measurement is highly quantitative and sophisticated – Jordan and Siddiqi (2024) survey many applications in which we can both model theoretically and test empirically certain key elements of the measurement process. In particular, much progress can be made on modelling measurements by appealing to the process of decoherence (Wallace, 2012; Bacciagaluppi, 2020). For as noted in chapter 2, since measuring instruments are macroscopic objects, it can be shown that as a result of decoherence, measuring instruments reliably end up after an experiment in a state which looks roughly like a distribution of classical probabilities over macroscopically distinct states associated with the various possible outcomes of the measurement, with the probabilities attached to the states proportional to the mod-squared amplitudes for the corresponding measurement outcomes. Moreover it can be shown that, if we interpret mod-squared amplitudes as probabilities, the probability that the different branches of this state will ever meaningfully interact again is very low, so we get what look like stable, quasi-classical measurement outcomes.

So in a sense decoherence *does* provide us with a model of the physical process of measurement, and indeed decoherence-based models can be used to answer many of the questions one would naturally wish to address using a model of the measurement process. For example, a decoherence-based model of measurement could play a role in answering some of the quantity individuation questions that concern Tal, since we may be able to show that the differences between two different measurement procedures get 'washed out' by decoherence effects such that they end up producing the same final distribution over macroscopically distinct outcomes, and therefore can be regarded as measuring the same quantity. Therefore it is not unreasonable that a number of physicists believe that the measurement problem is solved by decoherence, since decoherence does indeed provide us with a detailed account of what happens during a quantum measurement.

However, despite their practical utility, these decoherence models leave something to be desired from an epistemic point of view. For although such models can offer a detailed account of the process in which the states of the microscopic quantum system get magnified up to macroscopically distinct states of a measuring instrument and even an observer, the result of this process is that the observer ends up in a decohered state which looks roughly like a distribution of classical probabilities over macroscopically distinct states associated with witnessing the various possible outcomes of the measurement. And while such a model is a useful starting point, it is clearly not a complete schematisation of the observer, for it says nothing at all about how it is that the observer actually comes to see one specific outcome. This is hardly a trivial point, since most of the information we obtain from a measurement is linked to *which* measurement result we obtain. We don't access the decoherence process or the resulting probability distribution over outcomes directly – the only thing we have access to is the specific measurement result that we actually obtain, so if we are trying to understand how our experience connects up with the reality we are studying, the one crucial thing a measurement model needs to do is show us how the result that is actually obtained is related to the rest of the model's structure.

Now, of course, one option here is to adopt the Everettian view and suppose that all of the outcomes represented in this decohered state are witnessed in different branches; but then we still have the problem of justifying the probabilistic interpretation of the amplitudes, as discussed in section 3.1. Alternatively, we can address this issue in a hand-wavy manner by simply stipulating that we probabilistically obtain one of the results appearing in the decohered branches, with the specific result selected at random according to the probability distribution defined by the mod-squared amplitudes. At a practical level this approach works perfectly well and allows us to use the decoherence approach to do a variety of important calculations; and yet it is virtually silent about the physical nature of the putative probabilistic process that is taking place when we obtain our result, so it can offer us only very weak reassurance about the reliability of our means of enquiry.

Note that the problem here is not indeterminism – it is perfectly reasonable for a measurement result to be related in a probabilistic way to the rest of the model's structure, since then it can provide us with meaningful information about the probabilities arising from the underlying structure. But in the hand-wavy 'probabilistic selection' approach we really have no description at all, either deterministic or probabilistic, of the way in which some specific outcome from the set of decoherence-stabilised outcomes actually comes to be observed; the selection is simply imposed ad hoc, without any attempt to describe how it is realised by the underlying theoretical structures. Similarly, the problem is not that the epistemology is only approximate – after all, almost all approaches to quantum mechanics agree that the appearance of macroscopic observers involves some kind of emergence, and therefore any complete account of scientific epistemology will likely involve some categories which cannot be precisely defined in fundamental terms. (In chapter 11 I will return to this issue.) But accepting that central elements of our epistemology emerge in approximate ways from the underlying theoretical structures is very different from accepting an epistemology whose central process cannot be given an account in terms of the underlying theoretical structures at all. For example, in an Everettian picture observers are emergent and hence not defined precisely at a fundamental level, but we are nonetheless offered a quantitative account of the decoherence processes by which observers and measuring devices emerge from the underlying wavefunction description, thus giving us a concrete understanding of the relation between observations and underlying theoretical structures. Whereas the hand-wavy 'probabilistic selection' approach offers no such account: the selection process is simply imposed directly at the macroscopic level, so this approach offers no concrete understanding of the relation between our macroscopic observations and underlying theoretical structures.

Indeed, the absence of a model of the process by which the outcome of a quantum measurement outcome actually comes about leaves us unable to address various kinds of important questions about the epistemology of the theory. For example, in the spirit of Tal's arguments about the need for measurement models to understand the distribution of systematic error (Tal, 2019), one might worry that without a model of the probabilistic selection of one of the decohered branches, we will have no way to draw conclusions about possible systematic error occurring in this process. That is, in principle one could imagine a scenario in which decoherence produces the expected range of branches as in the usual models, but then there is some kind of systematic error in the probabilistic process by which the outcome we actually see is selected, such that some of the outcomes are systematically more likely than they ought to be given the underlying mod-squared amplitudes. Such systematic error would result in us being systematically misled about the theoretical structures we are attempting to find out about. And without a model of how it is that we actually end up seeing one measurement outcome rather than another, it would be impossible for us to identify the presence of such systematic error, since the error inheres in exactly the part of the process of which we have no theoretical understanding.

Now, in fact, whether or not this kind of systematic error is conceptually possible depends on one's understanding of the meaning of the quantum wavefunction. For those who think the wavefunction is an observer-independent element of physical reality, it seems natural to think systematic errors of this kind are possible, but for those who think the wavefunction is nothing other than a theoretical device recording probabilities for observed outcomes, it would not make sense to say that the observed outcomes could systematically mislead us about the actual underlying wavefunction. However, this simply reinforces the point: before we can even say whether or not such systematic error is a coherent possibility we must first have some account in hand of the relation between measurement outcomes and the theoretical object known as the 'wavefunction,' so a hand-wavy model involving some unspecified probabilistic selection process appears to be inadequate to fully resolve the detailed questions for which we typically require measurement models.

Moreover, measurement models are *particularly* crucial in the context of quantum mechanics precisely because quantum measurements don't appear to have all the properties that we expect from classical measurements. For example, Tal's measurement models usually involve a procedure in which a measuring device interacts with a system and in the course of the interaction the 'indications' produced by the device end up being correlated with the value of some pre-existing quantity of the system, which according to Tal is what justifies us in thinking of the measurement as being a measurement *of* that quantity (Tal, 2012). Naïvely, this would seem to be in tension with evidence coming out of quantum mechanics, such as the contextuality theorems, which as noted in section 2.4.1 are often interpreted as telling us that quantum measurement outcomes cannot always be understood as simply revealing the value of some pre-existing quantity. One might then worry that if Tal's account is right, it seems that quantum measurements are not really 'measuring' anything at all, so how can we meaningfully use them to do empirical confirmation?

One possible response here would be to simply insist that proper consideration of the epistemology of measurement indicates that we have no choice but to postulate some hidden variable model underlying quantum mechanics such that measurements *can* be understood as revealing the pre-existing value of some quantity. However, I think that even if measurement in quantum mechanics cannot ultimately be understood in that way, nonetheless the spirit of Tal's approach could be preserved. Perhaps in a quantum context a model of measurement might no longer involve showing how indications become correlated with some pre-existing quantity of the system, but the model could still confer empirical authority on the measurement process in virtue giving some *other* kind of description of what is happening during a measurement, thus elucidating what it is that the results of that procedure are giving us information about. For example, one might naturally hope that in the context of quantum mechanics, a measurement model will demonstrate that in the course of a measurement the 'indications' produced by the device become correlated with the wavefunction of the system in such a way that the probabilities of seeing certain indications become roughly proportional to the mod-squared amplitudes assigned by the wavefunction – and indeed, in chapters 11 and 13 we will see examples of models which seem to achieve this. So it is vital to have a concrete model in this context – if our previous understanding of measurement is really inapplicable, it must be replaced with an alternative understanding, or else we have not succeeded in arriving at a stable resting place where theory and experiment are fully compatible.

4.3 Bootstrapping

At this point, there is a natural worry one might have about the idea that measurements provide meaningful information only relative to some model of the measurement process: where does the model initially come from? After all, our models of reality are typically inspired by and justified by observations and measurement results, so if we can't even regard a process as a measurement without first having a model for it, how can any of this ever get started?

One might imagine that perhaps the regression halts with a set of 'observations' which do not have the formal epistemic authority of measurements and which therefore need not be modelled but which can still serve as the grounds for a rudimentary model which we can then use to designate certain processes as measurements. However, if it is permissible to use 'observations' which are not measurements to arrive at and empirically confirm a rudimentary model, one might wonder why we need 'measurements' at all, rather than just relying on the raw information provided by unmodelled observations. And, in fact, there doesn't seem to be a well-defined line between 'observations' and measurements, since even the simplest kinds of observation require at least some rudimentary theoretical beliefs about the connection between observation and reality before we can interpret them as providing evidence for some hypothesis. So one may worry that neither measurement nor theory provides a suitable starting point for scientific enquiry – and yet clearly we must start somewhere!

In fact, this dilemma is really just a modern version of the problem of coordination discussed in chapter 2, which is itself a variant on problems that have long been discussed in the context of foundationalist accounts of justification. Traditionally, epistemologists have considered that beliefs are justified in virtue of being inferred from some other beliefs, and thus foundationalists have argued that the regress of justification must terminate at some set of 'basic' beliefs which have some kind of non-inferential justification – for example, often basic beliefs are said to be related in some particularly direct way to observation: '*The underlying idea here is that of confrontation: in intuition, mind or consciousness is directly confronted with its object, without the intervention of any sort of intermediary*' (Bonjour, 1978, p. 11). The positivist search for a 'criterion of empirical significance' can be thought of as a version of foundationalism because the empirical experiences associated with a given sentence are supposed to act as a foundation which directly justifies our belief in the sentence in question.

The obvious difficulty for the foundationalist approach is that we know observations can give us unreliable information – either in localised ways, as when we see temporary illusions or mirages, or in global ways associated with sceptical possibilities like the idea that we could all be brains in vats. Thus a belief's being related in a direct way to observation does not guarantee that it is true, nor even well-justified. As Bonjour (1978, p. 12) puts it, '*Even if empirical knowledge at some point involves some sort of confrontation or seeming confrontation, this by itself provides no clear reason for attributing epistemic justification or reliability, let alone certainty, to the cognitive states, whatever they may be called, which result.*' So surely it cannot be epistemically rational to accept beliefs on the basis of observation without having some reason for believing that the observations in question provide reliable information.

And in a sense the quantum measurement problem appears to make this problem worse. For one influential approach to the foundationalist dilemma is the 'relevant alternatives' account of knowledge, which suggests that in order to know p we need only be able to rule out all the relevant alternatives where p does not obtain: as Dretske puts it, '*To know that x is A is to know that x is A within a framework of relevant alternatives, B, C, and D. This set of contrasts, together with the fact that x is A, serve to define what it is that is known when one knows that x is A*' (Dretske, 2000, p. 1022). This approach allows us to say that certain kinds of beliefs may be able to act as 'basic beliefs' even though we can't rule out various kinds of sceptical possibilities which would entail the falsity of those beliefs simply because the sceptical hypotheses are not 'relevant alternatives.'

From this point of view the quantum measurement problem presents us with a novel and unusually serious form of scepticism because it has the effect of making certain kinds of dramatic sceptical alternatives relevant to a wide variety of scientific contexts, even though we would never normally have taken them seriously. For example, when you report to me that you observed a certain measurement result, then if I consider you trustworthy I would typically suppose that I now know what you observed – I will not usually consider as a relevant alternative the possibility that what I hear you reporting to me may not be what you thought you said. But as we saw in chapter 3, the observer-relative interpretations of quantum mechanics appear to have the consequence that Type-III disaccord is generic, meaning that when you report a measurement result to me, I don't usually hear the same thing that you thought you said. And this is not a far-fetched scenario dreamed up by stubborn philosophers bent on making trouble for foundationalism: it follows in a straightforward way from one very natural class of interpretations of humanity's most successful scientific theory, and many well-informed people really believe that some interpretation in this class may be correct! Thus it seems that in many scientific scenarios we will not be able to treat this possibility as irrelevant, so it will potentially undermine any knowledge claims we might have been inclined to make on the basis of observations reported to us.

But fortunately, we are not obliged to accept the foundationalist idea that that scientific enquiry must always start from some set of self-justifying basic beliefs, and indeed this idea does not reflect real scientific practice. For, in fact, as Tal explains, the problem of coming up with a model of measurement and the problem of empirically confirming the theory within which the model is posed can be '*addressed together in a process of mutual refinement. It is only when one adopts a foundationalist view and attempts to find a starting point for coordination free of presupposition that this historical process erroneously appears to lack epistemic justification*' (Tal, 2013, p. 6, online version). Similarly, Chang rejects both foundationalism and traditional coherentism in favour of an approach that he calls progressive coherentism (Chang, 2007, 2004), writing '*the real potential of coherentism can be seen only when we take it as a philosophy of* ***progress****, rather than* ***justification***' (Chang, 2004, p. 224). Chang's approach is based on '*the possibility of making progress by first accepting a certain system of knowledge without ultimate justification, and then using that system to launch lines of inquiry which can in the end refine and correct the initially affirmed system*' (Chang, 2007). Shimony (1987, p. 5) advocates a similar approach to epistemology, which he refers to as the dialectical method: '*Tentative suppositions are accepted at the beginning of inquiry, but they are subject to criticism and may be revised and refined on examination. Fundamental principles are, then, the end rather than the beginning of inquiry.*'

The basic idea suggested by these various approaches is that in the course of doing science, we begin with some loosely sketched ideas about physical reality which allow us to use empirical experiences to confirm hypotheses, and then these empirically confirmed hypotheses are subsequently used to refine and correct the initial assumptions, thus allowing a cumulative growth in scientific knowledge. In this way we are able to both correct and refine but also retrospectively justify our initial use of our empirical evidence to confirm hypotheses, since the account of reality we ultimately arrive at will offer some account of the physical mechanism giving rise to the experiences on the basis of which it is confirmed; and if things work out well this account will affirm that the kind of empirical evidence we started out with does indeed reliably provide information about exactly the kinds of things which our more sophisticated theories go on to describe. Shimony (1987, pp. 21–22) notes that this approach to epistemology has rich philosophical roots: '*The paradigm is provided by Aristotle ... The program of closing the circle is explicit or implicit in the philosophies of Plato, Descartes, Spinoza, Leibniz, Hegel, Peirce, and Whitehead, and in variant form in that of Berkeley.*'

For example, Ohnesorge (2023) examines the problem of arriving at a model of measurement via a detailed account of attempts by seventeenth century scientists to measure the earth's ellipticity using various different theoretical frameworks, demonstrating that even in the absence of a shared model of measurement, scientists can still engage in common enquiries and enter into debates over sources of error. Ohnesorge thus argues that the epistemic status of measurement need not always be based on an explicit measurement model shared across the scientific community, but instead can be based on epistemic commitments to the *stability* and *nomicity* of measurement, where the former refers to the expectation that measurements of the target property have the same outcome under the same conditions, and the latter refers to the expectation that variations between outcomes measured under different conditions should be determined by laws of some kind or attributable to external distortions.

But as Ohnesorge himself notes, this account need not be in competition with model-based accounts like Tal's. Rather we can think of Ohnesorge's broad commitments and Tal's measurement models as occurring at different points in the development of a physical theory: we start out with loosely defined commitments to the idea that measurements provide meaningful information in virtue of exhibiting features like stability and nomicity, and then we perform empirical enquiries in the hope and expectation that our enquiries will vindicate our original assumptions by providing us with a physical theory which affirms that measurements of this kind do indeed have properties like stability and nomicity. This is the implicit promise of the scientific method: the account of reality that we ultimately arrive at will, if all goes well, reassure us as to the reasonableness of the methods we used to arrive at it.

While it need not be the case that this kind of progressive, coherentist approach is necessarily the right way of addressing the foundationalist dilemma in all possible domains of knowledge, it does appear to be a sensible way of thinking about scientific knowledge, and particularly scientific knowledge, in the domain of physics. For as noted in chapter 2, in the practice of physics our understanding of the operation of our measuring devices often depends crucially on the very physics we are trying to find out about, and so a distinctive feature of fundamental science is that it contains within itself the means to learn something about the physical nature of its associated means of enquiry, giving it a natural circular structure which is clearly a good fit for this progressive coherentist conception of justification. Thus in the development of a physical theory we are engaging in a kind of bootstrapping, where our assumptions about the reliability of measurement are to be retrospectively refined and justified by our success in arriving at a coherent set of beliefs based on the results of our empirical enquiries.

4.4 No-Lose Investigations

With this in mind, one way to understand the quantum 'measurement problem' is to say that this progressive bootstrapping procedure is not working as we would usually hope in the context of quantum mechanics. For if it is really true that the only empirically adequate interpretations of quantum mechanics are the unitary-only ones, and if it is true that these interpretations undermine our usual beliefs about the reliability of the measurement process, then it appears that our initial assumptions have not, in fact, led us to a picture of reality which affirms the assumptions we started with; quite the reverse! And if we are not able to achieve a coherent account of the physical nature of measurement in this context, that should surely shake our confidence in our initial assumptions about the reliability of the measurement process. So until we can find a way of resolving this problem by arriving at a coherent picture of the epistemology of quantum measurement, the epistemic status of our scientific beliefs as a whole will remain somewhat unresolved. Indeed, as Chang (2004, p. 227) puts it, '*Such repeated self-destruction is as close as we can get to an empirical falsification of the initially affirmed system.*'

On the other hand, the fact that it is *possible* for this bootstrapping process to go wrong is in some sense good news for the epistemology of science. For as emphasised by epistemologists such as Glymour (1981) and Titelbaum (2010), bootstrapping is a sensible way of arriving at beliefs only if the bootstrapping process could conceivably have gone wrong because otherwise we learn nothing when we see that it does not go wrong. Titelbaum thus argues that a sensible epistemology cannot permit 'no-lose investigations': to test a hypothesis we must do something that could possibly result in evidence against the hypothesis. And one might initially worry that assuming the reliability of measurements and then seeking to arrive at a coherent account of reality which vindicates that view was quite unlikely to result in evidence against the reliability of the measurement process.

However, as Chang (2004, p. 95) emphasises, '*There is no guarantee that observations enabled by a particular theory will always validate that theory*' – for example, Grünbaum (1960, p. 82) gives the example of Euclidean geometry: '*The initial stipulational affirmation of the Euclidean geometry G_0 in the physical laws P_0 used to compute the corrections [for measurement rod distortions] in no way assures that the geometry obtained by the corrected rods will be Euclidean.*' So it is, in fact, possible for empirical enquiries to undermine the assumptions we initially used in the analysis of our empirical results; and the long saga of the measurement problem illustrates a particularly subtle way in which this can happen, for the fact that it has been so hard to find a version of quantum mechanics which accommodates all of the empirical evidence but also provides a coherent epistemology reveals that, in fact, successfully arriving at an account of reality which tells a coherent story about its own epistemology is highly non-trivial. There are many ways our reality could be which would have the consequence that measurements would not yield reliable information, and it turns out that it *is* entirely possible to start out assuming that measurements are providing us with reliable information and still end up with a picture of reality which entails that measurements are not reliable – for the empirical results of quantum mechanics have led many people to arrive at various pictures of reality which, at least naïvely, appear to entail exactly that! So the existence of the measurement problem serves to reassure us that the bootstrapping process involved in empirical confirmation is not, in fact, a no-lose investigation.

Moreover, we can also see clearly why it's inadequate to solve the problem by just postulating a hand-wavy, ad hoc 'collapse of the wavefunction' that somehow selects out one of the branches produced by decoherence without being properly explained or modelled. Here, I will follow Schindler (2018, p. 59) in arguing that to be 'ad hoc' is precisely to cohere poorly with other beliefs: '*A hypothesis H, when introduced to save a theory T from empirical refutation by data E, is ad hoc, iff (i) E is evidence for H and (ii) H appears arbitrary in that H coheres neither with theory T nor with background theories B, i.e., neither T nor B provide good reason to believe that H (possibly specifying a particular value of a variable), rather than non-H (or some value other than the one specified by H).*' From this point of view, clearly an 'ad hoc' solution to the measurement problem will not do well at solving the epistemic version of the problem – our success in accommodating the empirical evidence in a coherent way is what gives us confidence in the overall belief system, but as the structure becomes less coherent, our reasons for confidence become accordingly weaker.

We will have occasion to return to this point in future chapters because several putative solutions to the epistemic problems faced by the Everett interpretation, the observer-relative interpretations, and the consistent histories approach involve making a highly ad hoc stipulation which has no purpose or justification other than the need to rescue the possibility of scientific knowledge. It should now be clear why I think such stipulations should be avoided. The whole point of the bootstrapping procedure is that we should be able to conclude that something meaningful has been achieved when we arrive at coherence, and this suggests that in seeking to solve the measurement problem we should set high standards of rigour and as far as possible avoid making use of ad hoc stipulations designed purely to secure the results we want about the reliability of measurement. The bootstrapping procedure will not be worth much if we just keep adding ad hoc postulates until the theory does what we want it to do; we must be willing to risk something, and thus we must be willing to let our solutions to the measurement problem fail.

4.5 A Priori Assumptions versus Coherent Sets of Beliefs

The broadly coherentist approach to empirical confirmation developed in this chapter also gives us the resources to respond to an obvious response that one might make to the epistemic concerns raised in chapter 3. Namely, one might worry that there is a double standard being applied. For example, consider the Everettian case: in section 3.1 I argued that the Everettian Statistical Assumption (ESA) needed by the Everettian to make sense of empirical confirmation seems unfounded, and I suggested that if no justification can be given for it, then we must reject the Everettian approach – that is, we must assume that our reality is not a branching universe. But is this not also an unfounded assumption? Why is 'our reality is not a branching universe' a more acceptable assumption than the ESA? If we're going to have to assume something unfounded either way, why not just assume the ESA, so we can accept the Everett interpretation and thus at least be done with the measurement problem?

Now, as a matter of fact I think 'our reality is not a branching universe' is preferable to the ESA in a number of ways – for example, the former is a fairly simple claim about the nature of our physical reality, whereas we will see in chapter 6 that the ESA is imprecisely defined and relies on contentious assumptions about consciousness and indexicality. However, the more important point is that this objection misconstrues the nature of the choice we are making between branching and non-branching accounts of reality. The idea is not that we should simply take as a foundational belief that the universe is not an Everettian branching universe. Rather, the fact that the Everettian picture (arguably) leads us to an account of reality which cannot reassure us as to the reliability of the measurement process is motivation for us to reject this system of beliefs in toto and to seek a more coherent system of beliefs. If such a system can be found, our commitment to that system will not be predicated on an unfounded assumption that the world is not an Everettian branching universe – rather the new belief system will offer some positive account of what the world is like instead, including a clear picture of how the physical process of measurement can be expected to give us reliable information about reality.

Of course this positive account will presumably entail that the world is not an Everettian branching universe and so, for those who hold that one should generally believe all of the implications of one's beliefs, it will be reasonable to say that once we adopt that belief system we should thereby end up believing that the world is not an Everettian branching universe. But the reason for choosing the non-Everettian system of beliefs rather than the Everettian system is not because we just decided from the outset that the world is not branching; rather, we are following a well-motivated strategy of rejecting one complete system of beliefs in favour of another complete system of beliefs on the grounds that the latter is more coherent. And much the same point can be made about the other epistemically problematic approaches to the measurement problem that I will discuss in this book. We are not simply swapping one unfounded belief for another; rather we are adopting a principled approach which ultimately aims at holding a system of beliefs which is as coherent as possible, particularly with regard to its own epistemology.

At this point one might object that there is not yet on the table any fully-formulated alternative to the interpretations discussed in this book which can reproduce all of the empirical predictions of quantum mechanics, so it is premature to advise the abandonment of these belief sets in favour of a more coherent belief set which is at present purely hypothetical. As a first response, I would contend that if the coherence of the belief systems associated with the approaches discussed in this book falls below some minimum reasonable level, then we are entitled to reject these belief systems even if no alternative is yet available. For we need some minimal degree of coherence in order to have any reason to think that measurements and observations are providing useful information about reality – if the coherence of the belief system is too low to offer any meaningful reassurance on this point, it would not be epistemically rational to accept the associated belief system, and that remains true even if there is no alternative. Furthermore, as I will argue in chapter 13, even if we do not yet have any fully-developed alternative routes to solving the measurement problem, there are a number of promising possibilities, so there are good reasons to be optimistic that a more coherent set of beliefs can be found.

5
Strategy

In this chapter I will explore what this epistemic conception of the measurement problem suggests about the kind of strategy we ought to adopt as we try to solve the problem, or judge proposed solutions to it.

Let me begin by noting that it has become common, and indeed somewhat banal, to associate quantum mechanics with 'weirdness.' And in fairness, the reputation is to some degree deserved – nearly every putative solution to the measurement problem has something about it which seems 'weird,' or at least strongly at odds with ordinary classical intuitions about reality. For example, there are solutions to the measurement problem which postulate 'spooky action at a distance,' retrocausality, the existence of multiple branching universes, contextuality, and so on.

As a result, discussions about the measurement problem sometimes have a tendency to devolve into arguments about which kind of weirdness is more acceptable. But this 'pick your poison' approach to solving the measurement problem often leads to stalemate, for it is very unclear what standards we could possibly use to judge which kinds of weirdness are more likely to be realised in the actual world. Frequently the standard actually invoked is intuition, but this is not a very helpful standard, because people have different intuitions about which kind of weirdness is worse, and we cannot easily argue one another into changing intuitions. And even if we *could* determine which kind of weirdness is least objectionable to the intuitions of the majority, there is no obvious reason to think that the intuition of the majority should tell us anything about the fundamental structure of reality, except insofar as certain features of reality might explain why we have developed the intuitions that we have.

For example, consider the debate over whether or not quantum mechanics exhibits non-locality. As noted in section 4.0.1, the macroscopic world of our experience is almost entirely local and thus many of us have very strongly-held intuitions about locality, so it is not surprising that many generations of physicists have hoped to exorcise the spectre of non-locality from quantum mechanics. And this would be entirely reasonable if microscopic locality were a good explanation for the apparent locality of the macroscopic world – but it's unclear that this is true. For quantum mechanics undeniably produces effects which look like subtle instances of non-locality, in the form of violations of Bell inequalities, and therefore any empirically adequate interpretation of quantum mechanics, even if it is, in fact, microscopically local, must nonetheless provide us with some means by which it is possible to *simulate* violations of locality. And then, as argued by Adlam (2022b), there is no reason in principle why we could not use that same mechanism to simulate much more widespread apparent violations of macroscopic locality.

In particular, some physicists have proposed using retrocausality, i.e. causation oriented backwards in time, to produce violations of the Bell inequalities in a theory which is really local at the microscopic level (de Beauregard, 1953; Sutherland, 1983;

Saving Science from Quantum Mechanics. Emily Adlam, Oxford University Press. © Oxford University Press (2025).
DOI: 10.1093/9780197808887.003.0005

Wharton, 2018). But if we allow retrocausality in this context, then there is nothing in principle to stop us from using retrocausality to simulate non-locality in more dramatic ways as well, which would result in a macroscopic world which appears wildly *non*-local despite the underlying microscopic locality. In particular, if it were possible to send controllable signals backwards in time, then we could straightforwardly produce effects which would look like superluminal signalling: if I wanted to send a 'faster than light' signal from California to Auckland, I could just send the message backwards in time, and then my collaborator three days ago could take that message and send it forwards in time such that it would arrive in Auckland at the very same time as I sent it, relative to my own reference frame. So if we are willing to allow retrocausal influences, it is possible to produce very dramatic instances of *apparent* macroscopic non-locality, even though, in fact, these instances can be fully explained in terms of microscopically local processes.

More generally, as soon as we introduce some means of simulating macroscopic non-locality within a microscopically local theory, then microscopic locality does not by itself explain the apparent locality of the macroscopic world – in addition we must postulate some restrictions on our means of simulating non-locality in order to prevent more dramatic and widespread *apparent* macroscopic non-locality. So in this kind of picture, the apparent locality of the macroscopic world ends up being in large part the result of the restrictions imposed on the method of simulating non-locality, rather than being the result of microscopic locality in and of itself. This suggests that, whether or not our world actually exhibits microscopic locality, nonetheless microscopic locality by itself probably cannot be the primary explanation for the apparent locality of the macroscopic world, given that such microscopic locality must be compatible with apparent failures of non-locality. And therefore intuitions based on our experience of macroscopic locality may not offer any good reason to think that the quantum world must really be local.

Another common criterion used to judge solutions to the measurement problem involves considering how compatible they are with Einstein's theories of special and general relativity. Clearly a viable solution to the measurement problem must be compatible with these theories at least at the empirical level, because empirically quantum mechanics and relativity never contradict each other. But many people feel we should also be seeking a deeper kind of compatibility, particularly if we hope that our solution will ultimately also work for a theory of quantum gravity which unifies quantum mechanics and relativity. And in particular, one important founding principle of special relativity is the idea there are no preferred reference frames. That is, there is no fact of the matter about whether or not two distant events take place at the same time because the definition of 'the same time' will be different for different observers, who each use their own reference frames to decide on a standard of distant simultaneity (Einstein, 1905). However, a number of putative solutions to the measurement problem, such as the original non-relativistic Ghiradi-Rimini-Weber (GRW) approach (Ghirardi et al., 1986), implement non-locality in the form of a collapse of the wavefunction occurring everywhere at the same time, which means that these approaches require us to choose a preferred reference frame on which this collapse takes place. Approaches incorporating this kind of non-locality are still compatible with relativity at the empirical level, since they typically tell us that it is impossible for observers to know which frame is the preferred one, but nonetheless one may feel that the very existence of a preferred reference frame, even if it is unobservable, violates the spirit if not the letter of relativity. It is therefore often suggested that we should prefer approaches which either are local, or implement non-locality in a global, 'all-at-once' way that does not require a preferred reference frame, as in the relativistic version of the GRW flash approach (Tumulka, 2006, 2021).

However, this criterion is also not as conclusive as one might hope because there are different things that one might mean by 'compatibility with relativity' – no preferred reference frames? Strict locality in which all influences are mediated by signals travelling along continuous paths? All structures defined in a relativistically covariant way? Respecting relativistic causal structure? Respecting relativistic conservation principles? One physicist may feel that preferred reference frames are a clear violation of the guiding principles of special relativity, while another may consider that it's acceptable to have a preferred reference frame in a relativistic context if that reference frame is defined in a relativistically acceptable way – for example, rather than just stipulating a preferred reference frame in an ad hoc manner, we could instead derive a preferred reference frame in a more relativistic way by allowing the global structure of the wavefunction to determine the chosen frame (Dürr et al., 2014). So although 'compatibility with relativity' seems like a reasonable thing to demand, there is still room for disagreement about what exactly that ought to look like, and thus to some extent this approach still leads to stalemate.

In light of the widespread disagreement on what a satisfactory solution to the measurement problem would look like, it is perhaps unsurprising that some physicists have come to regard the measurement problem as unscientific or unanswerable, and that even amongst those physicists who are willing to admit it as a genuine scientific problem there are persistent worries that we may have reached a permanent impasse. So it is helpful to seek alternative ways to characterise the problem and the criteria for a successful solution. And understanding the measurement problem as a problem of *epistemology* offers an alternative to the 'pick your poison' approach, and thus potentially a new way of moving past the long-standing impasse.

In particular, appealing to epistemic considerations allows us to divide the selection of 'weirdnesses' which may arise in solutions to the measurement problem into two categories. First, we have what I will call 'benign weirdness,' referring to features which may seem somewhat in tension with our classical intuitions, but which do not lead to any empirical incoherence: if reality had these features, the usual methods of science would still be a reliable way of learning about reality. Second, we have what I will call 'epistemic weirdness,' referring to features which *do* lead to empirical incoherence: if reality had those features, standard scientific methods would cease to be a reliable way of learning about reality, and the very possibility of scientific enquiry would be endangered. This categorisation offers us a useful way of evaluating solutions to the measurement problem: benign weirdness may sometimes be unappealing but nonetheless we can rationally accept a solution to the measurement problem which involves benign weirdness, whereas it would simply not be epistemically rational to accept a solution to the measurement problem which involves severe epistemic weirdness. Therefore by framing the measurement problem as a problem of *epistemology* we can also set out a clearer program for solving it.

Of course, this is still not an exact science. As noted in chapters 2 and 4, our goal in seeking a physical account of measurement must be coherence rather than conclusive refutation of all sceptical possibilities, so there is a certain amount of judgement to be exercised in deciding whether or not some approach is coherent *enough* to meet reasonable standards of epistemic rationality. Therefore simply invoking epistemology is not a decisive algorithm which will spit out a definite answer as to whether some putative solution to the measurement problem is acceptable. But nonetheless, appealing to epistemic rationality offers a whole new range of criteria that we can use to assess solutions to the measurement problem, thus giving us a new lens through which to consider the problem.

For example, although it isn't possible to give a completely definitive classification of a kind of weirdness as benign or epistemic, here is a first pass at classifying some features relevant to the measurement problem. The category of benign weirdness may include features like spatial and temporal non-locality; contextuality (Kochen and Specker, 1975; Spekkens, 2005); retrocausality; wave-particle duality; non-dynamical approaches (Wharton, 2015; Adlam, 2018). The category of epistemic weirdness may include features like measurements failing to produce unique outcomes (see chapter 6); dynamics which forbid intersubjective sharing of information (see chapter 9); dynamics which make macroscopic records of the past unstable (see chapter 13); and 'fine-tuned' correlations in the initial state of the universe (see chapter 12). Of course there is a discussion to be had about whether or not these features really undermine standard scientific methods, and indeed much of the next few chapters of this book will be devoted to discussing this very question, but for now it is enough to note that all of the features in this second set are at least superficially problematic from an epistemic point of view, and thus the epistemic construal of the measurement problem gives us new questions to ask about putative solutions which exemplify these features.

With that said, I reinforce that in many cases whether or not a kind of weirdness is benign or epistemic may come down to the details of the way it is implemented in a specific model. For example, Einstein famously objected to non-locality in quantum mechanics on the grounds that '*an essential aspect of this arrangement of things in physics is that they lay claim, at a certain time, to an existence independent of one another, provided these objects "are situated in different parts of space." Unless one makes this kind of assumption about the independence of the existence (the "being-thus") of objects which are far apart from one another in space ... physical thinking in the familiar sense would not be possible. It is also hard to see any way of formulating and testing the laws of physics unless one makes a clear distinction of this kind*' (Einstein, 1948, p. 2). Einstein's worry here is precisely that 'non-locality' may belong to the category of what I have called 'epistemic weirdness,' since it may threaten our ability to formulate and empirically confirm laws of physics; and he argues, just as I am arguing in this book, that a sufficiently serious form of epistemic weirdness is a valid reason to reject a hypothesis.

But as it turns out, the specific kind of non-locality exhibited by quantum mechanics does not, in fact, prevent us from formulating and testing physical laws. Einstein is undoubtedly right that if we lived in a world which exhibited generic, unrestricted, and random non-locality, it would be next to impossible to empirically confirm anything; however, the non-locality of quantum mechanics is not unrestricted and random, but rather precisely regulated by the laws of quantum mechanics, so, in fact, we can learn about quantum mechanics in just the same way as we learn about any other kind of regularity. This episode demonstrates that it is not enough to simply gesture at some feature of an interpretation which seems likely to be epistemically problematic: we must consider the specific details of how a given interpretation proposes to implement that feature in the context of quantum mechanics, and decide whether or not that particular implementation is compatible with the possibility of empirical scientific knowledge.

I also want to emphasise that this epistemic approach to the measurement problem is not intended to be used as a catch-all excuse to dismiss any and all solutions to the problem which lead to a non-intuitive conception of measurement or the role of observers. For we have already noted that it is likely impossible to give an account of reality compatible with quantum mechanics which preserves *all* of our usual expectations about measurement, so if we are too conservative we could easily end up with no solutions left at all! And the goal here is not to find a solution to the measurement problem which accords with our classical intuitions about measurement. Rather the goal is to find a solution which can reassure us of the stability of empirical scientific knowledge – and it is possible that this could be done in very unexpected ways, so we should keep an open mind about what exactly it means to arrive at a coherent scientific epistemology in the context of quantum mechanics.

5.1 Other Formulations of the Measurement Problem

It will now be useful to put the epistemic formulation of the measurement problem in context by comparing it to several other versions of the measurement problem. Now, of course, everyone is free to decide for themselves what exactly constitutes a 'problem,' so I don't mean to suggest that there is one correct way to think about the measurement problem. However, I do want to argue the epistemic version of the measurement problem is *rationally compelling* in a way that some other versions are not. Thus I would argue that the epistemic construal of the measurement problem has a special urgency, such that even those who profess to be unmoved by other versions of the measurement problem have reason to be concerned by this one.

So let us take a brief tour through a few other formulations of the measurement problem. This tour is, necessarily, far from complete, but I will try to cover some of the main themes in the discussion, in order to demonstrate how the epistemic standpoint can meaningfully contribute to these themes.

First, it is often suggested that the measurement problem pertains to the fact that it is unsatisfactory to have imprecise anthropocentric terms like 'measurement' playing a primitive, irreducible role in what is apparently supposed to be a fundamental physical theory. As Bell (1990, p. 34) puts it in his oft-quoted remark: '*What exactly qualifies some physical systems to play the role of 'measurer'? Was the wavefunction of the world waiting to jump for thousands of millions of years until a single-celled living creature appeared? Or did it have to wait a little longer, for some better qualified system ... with a PhD?*' Similarly, Albert (1994, p. 80–81) notes that '*what these laws actually* ***amount*** *to ... will depend on the precise meaning of the word* ***measurement*** *... and* *it happens that the word **measurement** doesn't have any absolutely precise meaning in ordinary language*. Now, this presentation of the problem is already quite compelling, but nonetheless, adopting a specifically epistemic standpoint is useful to emphasize the significance of the point being made here about the word 'measurement.' For one main reason why it is inadequate to treat 'measurement' as a fundamental primitive is precisely because properly understanding the epistemology of our theories requires us to have a model of what is actually going on during the events that we refer to as measurements, as the modern scholarship on the epistemology of measurement reminds us. So the epistemic standpoint helps to further explain why it seems inadequate to many of us to accept a theory which works 'for all practical purposes,' as Bell puts it, without demanding of that theory a physical account of the measurement process.

Second, it is sometimes suggested that the measurement problem is about how to provide an ontology for quantum mechanics – for example, Esfeld (2019, p. 2) characterises the measurement problem as pertaining to '*the question of ontology, that is, the question of what the wave function refers to – in other words, the question of what the objects in nature are to which the wave function dynamics relates.*' And indeed, it would undoubtedly be very useful to have a viable ontology for the full theory, not least because specifying a full ontology and explaining how measurements are instantiated within it would be one way of solving the epistemic version of the measurement problem. But it is not the *only* possible way to address those epistemic issues. For example, one could potentially give a satisfying account of what is going on in the process of measurement by appealing to a purely structural or functional account, which demonstrates how observations and measurements are able to latch onto real structures in a way which ensures they will exhibit meaningful universal regularities, but which does not say anything specific about the nature of the entities instantiating those structures. Indeed, as structural realists have urged for some time (Maxwell, 1970), there is a sense in which scientific evidence can never really give us direct access to theoretical entities as they are in themselves; we get to know about theoretical entities only through the roles they play in our theories, and thus as we move further and further from empirically accessible regimes, it is to be expected that we will increasingly have to rely on structures rather than intuitively comprehensible ontologies. So I think it is a mistake to put too much emphasis on ontology in discussions of the measurement problem – like everyone else I would be happy to be provided with a satisfying ontology for the theory, but ontology is not indispensable and thus should not be the primary focus.

Third, it is sometimes suggested that the measurement problem is about how to make quantum mechanics look more classical, or to force it to conform to our classical intuitions. This summary of the problem is perhaps most common amongst those who think it is not a real problem – for example, Fuchs (2011, p. 12, online version) contends that '*the "measurement problem" is purely an artefact of a wrong-headed view of what quantum states and/or quantum probabilities ought to be.*' And indeed, it seems entirely reasonable to be dismissive towards this version of the problem, for there is no guarantee that a version of quantum mechanics which conforms to our classical intuitions will ever be found, nor any particular reason to think it would be of much help with open scientific problems. Thus although a number of the putative solutions to the measurement problem that I discuss in this book are frequently criticised on the grounds that they are too counterintuitive or too far removed from classical intuitions, in this book I will not consider this as a valid criticism.

A fourth formulation of the measurement problem, due to Maudlin (1995), is known as the 'problem of outcomes.' Maudlin notes that there is an inconsistency between the following three propositions: a) the wavefunction of a system is complete (completeness) b) the wavefunction always evolves unitarily (universality), and c) measurements always or usually have unique determinate outcomes. Maudlin's formulation here is similar to the account of section 2.1.2, where we saw that in pure unitary quantum mechanics there appears to be no mechanism to bring about an actual outcome to measurements. So in a sense the difference between Maudlin's presentation and the epistemic approach is mainly one of emphasis. Maudlin observes that there appears to be a contradiction in this vicinity, but has little to say about why we should consider the resolution of the contradiction a pressing matter, and thus his presentation may appear to leave it open to us to simply remain agnostic, or to adopt a compromise which at least superficially retains all three propositions. In particular, Maudlin's formulation of the problem makes it tempting to move to an observer-relative view of quantum mechanics, which as we have seen in section 3.2 allows us to maintain completeness and universality whilst also having unique determinate outcomes at least relative to individual observers. Whereas from the epistemic standpoint, we can see that the existence of this apparent contradiction ought to worry us precisely because it points to an inadequacy in our understanding of the epistemology of the theory. And we can see why the compromise proposed by the observer-relative positions is unlikely to be adequate. For one key reason why we ought to *care* about recovering determinate outcomes from the theory is because we need to make sense of the role played by determinate outcomes in the empirical confirmation of the theory – and as we saw in section 3.2, determinate outcomes relativised to individual observers are arguably not capable of doing justice to this role, which suggests that the observer-relative solutions do not really get at the heart of the problem posed by Maudlin. Thus taking a specifically epistemic approach here serves to strengthen the case made by Maudlin, since it gives further support to the claim that we will have to reject at least one of Maudlin's three propositions.

Finally, two other well-known formulations of the measurement problem are due to Pitwosky (Pitowsky, 2005; Bub and Pitowsky, 2008). First, his 'big measurement problem' refers to the fact that quantum mechanics does not usually produce deterministic predictions for the results of measurements. However, as various people have argued – including Pitowsky himself – it is not obvious that this is really a problem. Prima facie there is no obvious reason why the world could not simply be indeterministic – although we might well prefer to have a deterministic theory, reality is not obliged to respect our preferences on this point! Meanwhile, Pitwosky's 'small measurement problem' refers to the fact that in many quantum-mechanical scenarios, '*the classical probability distributions that can be associated with the system's observables (conditional on various measurements that one could perform) ... cannot be embedded into a global classical probability distribution over all of the system's observables*'(Cuffaro, 2023a, p. 8). That is, as discussed in section 2.4.1, the phenomenon of quantum contextuality means that it is not possible to think of quantum mechanical observables as standing in a simple one-to-one map with underlying properties which are simply revealed by measurements – whatever is going on in quantum measurements, it is more complicated than simply revealing pre-existing values of properties. Now, this may be a surprising revelation, but again it is not entirely evident why it should be regarded as a *problem*. Why should we not just simply say that we have discovered something new about measurements – i.e. that they do *not* just reveal the pre-existing values of properties?

Again, we can better understand the significance of these two problems if we approach them from an epistemic standpoint. For example, as I will discuss in chapter 7, the structure of empirical confirmation for probabilistic theories is more complex than for deterministic theories, so one might try to make the case that indeterministic theories cannot be empirically confirmed, or at least that they cannot be confirmed as effectively as a deterministic theory. This would be a controversial claim, but in any case, if one buys this kind of approach, then the 'big measurement problem' could be construed as an aspect of the epistemic version of the measurement problem.

Similarly with regard to the small measurement problem: if measurements don't reveal pre-existing values of properties, what do they reveal? Can we still understand them to be giving us information about some real structural feature of the world? If we can't provide some sensible account of what measurements *are*, in fact, doing and how precisely they relate to the features of the world that we are trying to find out about, one might worry that measurements are just delivering some kind of random value which actually reveals no meaningful information at all. Thus expressing the 'small measurement problem' in this way strengthens the case that it is a genuine problem that must be solved.

The big picture here is that some discussions of 'the measurement problem' seem to simply point to the fact that the quantum world exhibits some kind of weirdness, but if the weirdness is merely benign, then it is not clear that this is a real problem – after all, science frequently reveals that the world has features which are not intuitively obvious from our flat-footed coarse-grained experience of it! But while we can always choose to respond to scientific advances by renouncing naïve classical intuitions about the nature of reality, we *cannot* rationally respond to scientific advances by giving up the idea that measurements are revealing meaningful information about reality that can be used to construct and confirm a scientific theory, and therefore we must find a way of accounting for the process of measurement in quantum mechanics without epistemic weirdness, otherwise we will be left with an internally incoherent system of beliefs. Thus whatever one may think of the other formulations of the measurement problem described here, there is at least one problem in this vicinity which ought to be taken seriously by anyone who has a stake in maintaining the status of scientific knowledge, and therefore I think epistemic considerations of the kind that we will be surveying in this book deserve to be given more prominence in this discussion than they have been afforded thus far.

5.2 Interpreting Quantum Mechanics

At this point I want to distinguish what I have called 'the measurement problem' from the question of how quantum mechanics should be interpreted. As I understand it, the question of how quantum mechanics should be interpreted involves asking questions like 'What is the meaning of this mathematical formalism? What is it *for*? Which parts of it play a representational role and which do not?' Answers to these interpretational questions will, by definition, not involve adding anything to the quantum formalism. Instead they will likely involve making reference to the nature of the agents who came up with quantum theory, their interests and purposes, and their limitations. For example, one possible way of answering questions about the interpretation of the quantum formalism is to say that quantum mechanics is an agent-centric tool designed by human beings to advise us on how to form credences about the results of future measurements (Healey, 2012; Fuchs and Schack, 2014) – and indeed, there is something that is clearly right about this. Quantum mechanics *was* designed, first and foremost, to predict the results of experiments: there is no doubt that it is an agent-centric tool. Thus these kinds of perspectival and pragmatic ways of thinking about quantum mechanics invite us to reflect on the nature of human science and scientific representation, leading to many interesting insights.

But the question answered by these kinds of approaches is not necessarily the same as 'the measurement problem' as I have construed it here. For the epistemic version of the measurement problem is not about understanding the nature of a particular formalism developed by human scientists, but about how to give a coherent account of the physical process of measurement which we have used to arrive at that formalism. To see that the problems are distinct, note that solving the measurement problem may well require adding something to the quantum formalism, for it is entirely possible that in order to fully understand the physical nature of quantum measurements we must appeal to something *outside* of standard quantum mechanics (Dickson, 2007).

Of course, the measurement problem and the question of how to interpret quantum mechanics are not completely unrelated. For example, if your approach to these questions about interpretation is to say that the entire quantum formalism plays a literal representational role, as in some versions of the Everett interpretation, then your solution to the measurement problem may well be the same as your answer to the question of how quantum mechanics should be interpreted. But if your answer to the problem of interpretation is to say that the formalism should be understood largely as a pragmatically useful tool, then you will have to come up with some other solution to the measurement problem, because saying that the formalism is largely a pragmatically useful tool does nothing to address the epistemic version of the measurement problem – it simply indicates that in order to solve the measurement problem we ought to be trying to understand what it is about physical reality and our access to it that makes this tool pragmatically useful.

Of course, this is not a criticism of pragmatic and perspectival interpretations of quantum mechanics, since there is no particular reason to think these two problems must have the same solution. However, proponents of these kinds of views are sometimes prone to argue that they *have* solved the measurement problem, or indeed that there is no measurement problem at all (Fuchs, 2011; Mermin, 2022). Or alternatively, they may extrapolate from their account of the nature of the mathematical formalism of quantum mechanics to argue that *reality itself* is intrinsically perspectival, or that there is just no such thing as objective reality (Brukner, 2017; Di Biagio and Rovelli, 2022; Fedrizzi and Proietti, 2023). It's likely that this is partly the result of conflating the question of how to interpret the quantum formalism with the measurement problem, leading to the idea that if your preferred way of interpreting the formalism is perspectival, so too must be your solution to the measurement problem. But hopefully it is now clear that this is not correct: the question of interpretation and the measurement problem are distinct issues which could have entirely different solutions.

Of course, this is not to say that pragmatic and perspectival interpretations of the formalism cannot *inform* possible solutions to the measurement problem. They certainly call our attention to interesting features of quantum mechanics and, in particular, to the fact that there is something going on in quantum regimes which somehow makes the fact that the theory is formulated from a specific human perspective more relevant than in other areas of physics. After all, classical physics was also formulated from the specific perspective of human observers, but in classical physics it seemed quite easy to eliminate the perspective and arrive at what, at the time, was understood to be an observer-independent description of the world; whereas in quantum mechanics the human perspective seems much harder to remove. This surely tells us something important about the nature of quantum measurement. However, it still does not necessarily entail that reality itself is intrinsically perspectival; and indeed, as I will discuss in chapter 10, I do not think it even provides any particularly strong reason to think that reality *might* be intrinsically perspectival. So although pragmatic and perspectival approaches are surely in many ways correct in what they say about the *interpretation* of the quantum formalism, it's unclear that they offer any particular insight into the epistemic version of the measurement problem.

5.3 Epistemology and Reality

An important caveat to the discussions in this book is that, of course, the world does not have to be such that it is possible for us to do science and perform empirical confirmation. So the empirical considerations discussed in this book do not constrain ways the world could possibly be; rather they constrain the things that we can reasonably *believe* about the way the world is.

For example, I have argued that if the epistemic part of the Everettian probability problem is unsolvable, the Everett interpretation cannot be accepted as a viable solution to the measurement problem. But this does not entail that the world *could not be* the way that the Everett interpretation suggests. Rather the point is that if the probability problem cannot be solved, it could never be epistemically rational for us to understand quantum mechanics in the way that the Everett interpretation suggests, even though in principle the Everett interpretation could be right.

Now, one might at first find this conclusion disturbing: isn't there something inconsistent about saying that we could never rationally believe the world has some feature, even though we admit it nonetheless *could* have that feature? Surely there must be something wrong about this reasoning? But, in fact, the claim is not quite so strong. I am not arguing that we could never rationally believe that we live in an Everettian branching world, but rather that it is not rational to believe this *on the basis of quantum mechanics*. For the central thrust of my argument is that, if there is no solution to the probability problem, then Everettian quantum mechanics is empirically incoherent – that is, Everettian quantum mechanics entails that quantum mechanics cannot possibly be empirically confirmed, even if we regard it as nothing more than a description of observable regularities. So Everettian quantum mechanics entails that we should not use quantum mechanics as a reason to believe any account of reality, Everettian or otherwise. This does not preclude the possibility that there may be some other kinds of evidence for a branching-world picture that are immune to this kind of undermining: for example, if a god came down from heaven and personally informed you that you live in an Everettian branching universe, you might reasonably start believing in an Everettian branching universe, regardless of your views on quantum mechanics. So the claim is merely that we cannot reasonably arrive at a belief in Everettian quantum mechanics *on the basis of our belief that quantum mechanics has been empirically confirmed*, not that there could never possibly be any reason to believe it.

Now, one might object at this juncture that ultimately we want to know the *true* solution to the measurement problem, which has nothing to do with the question of what we can reasonably believe. Indeed, one important motivation for working on the measurement problem is the idea that a solution to the problem might help us make progress towards a successor theory, for example, by helping understand how to implement quantum gravity. But surely it would be ridiculous to suppose that we can get some insight into how to do quantum gravity by means of philosophical reflections on epistemology!

However, I actually do not think this idea is so ridiculous. For the point of discarding solutions to the measurement problem which give rise to epistemic weirdness is that in a universe with these features there would be no possibility of obtaining any substantive scientific knowledge at all – not even purely empirical scientific knowledge! So in a universe of this kind there would really be no hope of successfully doing science; our impression that we have, in fact, successfully done some science would be just a momentary illusion arising from the fact that we briefly happen to be at some unusual place in the Everettian branching multiverse, or inside some unusual perspective, or looking at some unusual records, or so on. And therefore, if we are interested in actually doing science, we lose nothing in assuming that epistemic weirdness does not obtain – for if, in fact, some kind of sufficiently severe epistemic weirdness *does* obtain, our attempts at doing science are doomed to failure in any case. Shimony (1987, p. 6) refers to this kind of argument as 'vindicatory,' referring to cases in which our reason for adopting a certain belief is that '*nothing is lost and possibly something is gained*' by adopting this belief.

By comparison, note that physicists working on a quantum theory of gravity do not habitually lose much sleep over the possibility that we could all be brains in vats, even though the brain-in-vat hypothesis would probably undermine a lot of the reasoning behind their theories of quantum gravity, since the higher entities programming the vats would not be obliged to create our illusory experiences in accordance with a properly worked out and consistent theory of quantum gravity. There's just no point in worrying about this possibility too much because it almost certainly won't lead anywhere scientifically fruitful. And in the same way, physicists working on a quantum theory of gravity should feel free to assume that our universe does not instantiate severe epistemic weirdness because if it did, then most likely no attempt at formulating a theory of quantum gravity would be successful. After all, if we *are* in an epistemically weird universe, we have no good reason to think that quantum mechanics is approximately correct even in its usual domain of application, and if quantum mechanics is not approximately correct in its usual domain of application, then it is highly unlikely that a theory of quantum gravity will be an approximately correct description of any domain of application at all.

So although the epistemic issues I discuss in this book may appear to be somewhat orthogonal to questions about real physics research, as a matter of fact, it is perfectly reasonable to use these kinds of issues as a criterion for theory selection. Thus there is the potential for these issues to shape research on quantum gravity and other new physics in interesting ways. Indeed, epistemic considerations may turn out to be quite a powerful selection criterion; as we will see, they arguably rule out a number of popular interpretations of quantum mechanics, and even if they do not *completely* rule out these possibilities, they certainly place strict constraints on the way in which we should implement these approaches.

6
The Everett Interpretation

We saw in chapter 3 that of all the concerns one might have about Everettian probability, the one which seems most directly relevant to the epistemic conception of the measurement problem concerns the role of probabilities in the empirical confirmation of a probabilistic theory like quantum mechanics. That is, according to the bootstrapping approach to scientific epistemology set out in chapter 4, we cannot rationally believe Everettian quantum mechanics unless we are eventually able to give an account of the physical process of measurement in an Everettian universe which affirms that, if we believe the world is really Everettian, it still makes sense for us to use probabilistic empirical data to confirm a theory in the way that we have actually done to empirically confirm quantum mechanics. In this chapter, I will discuss in more detail what it might take to achieve this, and consider whether various proposals for solutions to the Everettian probability problem have the resources to address the epistemic part of the problem.

The task at hand has two parts, analogous to the incoherence problem and the quantitative problem as described by Wallace (2006). First, our account of the physical process of measurement in an Everettian universe must have the consequence that in an Everettian universe it is possible to learn *something* relevant to confirming a probabilistic theory from observing a measurement outcome. Second, once we have achieved that, our account must also have the consequence that in an Everettian universe it still makes sense for observers to update their beliefs in the *specific quantitative ways* that we have actually employed in the process of arriving at quantum mechanics – that is, we must explain why it makes sense for Everettian observers to believe that the relative frequencies they have observed in experiments are, after a large enough number of experiments, reliably close to the mod-squared amplitudes assigned by the theory. I will address each of these tasks in turn.

6.1 Task One: Learning From Measurement Outcomes

The discussion around probabilities an Everettian context is often framed in terms of self-locating information, referring to information which tells you about where or when you are located, but does not tell you anything new about reality as a whole. For example, if you know the locations of all the people in the world but you don't know which of those people *you* are, and then you learn which person you are, you have gained only self-locating information: you now know where you are, but you have not learned anything new about the state of the world as a whole.

Self-locating information is relevant to the Everettian context because an Everettian agent already knows prior to a measurement that all of the outcomes will definitely occur, and therefore it seems natural to say that what the agent learns when she

Saving Science from Quantum Mechanics. Emily Adlam, Oxford University Press. © Oxford University Press (2025).
DOI: 10.1093/9780197808887.003.0006

actually sees an outcome is purely a piece of self-locating information – she learns which branch of the wavefunction she is in, but she already knew that such a branch would exist, so she has not gained any substantive new information about her world from an external point of view. Appealing to the concept of self-locating information helps make sense of some of our common-sense intuitions about measurement in an Everettian universe. For example, it explains why an observer might feel there is something uncertain about the outcome of an Everettian measurement even though she knows that all the outcomes will occur, because in the period after performing the measurement but before looking at the outcome, she experiences self-locating uncertainty about which one of the branches she is currently located in, and thus when she sees the outcomes she feels like she has gained information, because she has indeed gained *self-locating* information about which branch she is in.

However, invoking self-locating information is potentially problematic if the aim is to make sense of empirical confirmation in an Everettian context. A number of authors have expressed scepticism about the use of self-locating probabilities in empirical confirmation: for example, Norton (2010, p. 518) argues that we should not base any substantive conclusions on such probabilities because there is a fact of the matter about the right probabilities to use only if we can '*find a grounding in the facts of the particular case for probabilities of the logic*' and in a multiverse-style scenario there cannot be any such facts unless there is some '*factual analog of a randomizer over the multiverses.*' Similarly, Friederich (2021, p. 103–104) expresses the concern that using self-locating credences in empirical confirmation involves treating '*phenomena that are in fact fundamentally uncertain as risky*' which is problematic because '*there is a widespread view according to which the ascription of probabilities is warranted only in "risky" epistemic situations.*'

One specific reason to be cautious about the use of self-locating probabilities in empirical confirmation is encoded in the 'Relevance-Limiting Thesis,' which claims that since self-locating information by definition does not tell you anything new about the state of the world as a whole, it cannot cause you to change any of your non-self-locating credences[1]. This is important because the process of performing empirical confirmation is concerned with updating *non*-self-locating credences about which possible world you are in – for example, it may involve changing the credence you assign to the proposition that you are in a world governed by one particular set of laws, or the credence you assign to the occurrence of certain measurement outcomes in parts of your world that you have not yet observed, or the credence you assign to the proposition that your world has certain ontological features, such as being a branching Everettian world. So if the Relevance-Limiting thesis is true, then by admitting that we obtain only self-locating information in an Everettian measurement, the Everettians are effectively conceding that we can't actually learn anything relevant to empirical confirmation from an Everettian measurement, in which case the Everett interpretation will clearly be unable to solve the epistemic version of the measurement problem.

A natural response for the Everettian is to deny the Relevance-Limiting thesis. But there are fairly good reasons to think the Relevance-Limiting thesis is correct – for

[1] The name 'The Relevance-Limiting Thesis' is due to Titelbaum (2008). However, Titelbaum himself does not support this thesis.

as discussed by Adlam (2014), it appears to be implied by a number of very natural schemas one might employ for updating one's beliefs based on new information, including the simplest possible approach which *'simply amounts to setting one's credence to zero on the proposition whose falsity one has just learnt, and renormalizing'* (Greaves, 2006, p. 31–32, online version). Indeed, the same applies to any other schema in which we are supposed to first assign credences to possible worlds and then somehow distribute those credences over the 'centered worlds' corresponding to the possible worlds. The term 'centered world,' which is due to Lewis (1979), refers to the pair of a possible world and a location or perspective within it – for example, the system proposed by Halpern and Tuttle (1989) and the compartmentalised conditionalisation suggested by Meacham (2008) use centered worlds in this way, and thus they uphold the Relevance-Limiting thesis.

Now, there do exist some belief-updating schemas which don't imply the Relevance-Limiting thesis, such as the approach described by Halpern (2006), based on some work due to Elga (2000). However, typically these approaches either don't allow the number of worlds to increase over time, or else they run into trouble when the number of worlds increases over time – for example, Meacham (2008) shows that the approach of Halpern and Elga has the paradoxical consequence that regardless of the evidence, our credence in certain sorts of propositions will increase over time, eventually converging to certainty no matter how low the initial credence assigned to them. And Adlam (2014) argues that the 'Extended Conditionalization' schema proposed by Greaves (2006), an approach specifically designed to make sense of Everettian confirmation, has the same problem. So there are good reasons to think that schemas which don't imply the Relevance-Limiting thesis may be untenable at least in scenarios where the number of worlds can increase over time.

This is problematic for the Everettian because the branching hypothesis central to the Everett interpretation seems to imply that the number of centered worlds does indeed increase over time in an Everettian picture, since more worlds are created every time a measurement is performed. There is a possible alternative view – the 'Diverging' Everettian approach in which the entire set of worlds is present all along, so the number of worlds does not increase over time – but in section 7.3.1 I will argue that this approach has its own epistemic problems. For now it is enough to observe that in the context of the more common 'Overlapping' approach to Everett, in which genuine branching takes place when measurements occur, it seems hard to do without the Relevance-Limiting thesis.

Now, initially the Relevance-Limiting thesis may seem surprising because it's easy to come up examples where it seems as though we do learn something substantive after observing purely self-locating information. However, this usually arises from a failure to distinguish properly between two importantly different kinds of cases – that is, cases in which the information obtained can be redescribed as *non*-self-locating information, and cases where it cannot. To see this, consider the following two examples:

1. You are to undergo a procedure in which seven identical copies of your body and brain will be created, and then each clone will have its hair dyed some colour,

but you don't know what the distribution of colours will be. Your original body will be destroyed.[2]

2. You are to undergo a procedure in which seven identical copies of your body and brain will be created, and then each clone will have its hair dyed a different colour from the seven colours of the rainbow. Your original body will be destroyed.

Suppose that in the first case you wake up and look in the mirror and see green hair. Intuitively, it may seem that you have learned something about the distribution of hair colours across the set of clones. But we can make sense of this intuition in a way that is consistent with the Relevance-Limiting thesis because there is a sense in which this observation actually gives you a piece of *non*-self-locating information. That is, implicitly you are regarding yourself as a random sample from the set of clones, so in the process of learning which clone you are, you are also learning a piece of non-self-locating information about what is seen by a randomly-selected clone. So we can straightforwardly redescribe this scenario from an external point of view, without appealing to self-location or centered worlds: if you were simply given the information that a clone randomly selected from the seven has green hair, you would update your credences in just the same way, so the part of your new information that is relevant to the updating of your non-self-locating beliefs is not really self-locating information at all. And thus even if we make use of a belief-updating schema which *does* uphold the Relevance-Limiting thesis, that schema will still straightforwardly tell us that this new *non*-self-locating information can lead you to update the credences you assign to various possible ways the world could be.

Whereas in the second case, when you wake up and look in the mirror and see green hair, what have you learned? Clearly you have learned some self-locating information: you have learned that you are the green clone. But have you learned any *non*-self-locating information? It seems not: what else could there be to learn? You knew all the facts about the distribution of hair colours before seeing the green hair, so there's no extra information to be gained by seeing what colour of hair you now have. In this case there does not seem to be any possible way of redescribing this scenario such that you learn a piece of non-self-locating information when you see the green hair.

This suggests that one crucial question for the Everettian to address in their account of the physical nature of measurement is to say whether Everettian measurement outcomes are more similar to case one or case two above. That is, in order to offer an account of Everettian measurement which affirms the reliability of our ordinary methods of scientific enquiry, we will likely need to show that observations made by an Everettian observer can be redescribed in such a way that we can think of them as yielding some kind of non-self-locating information, as in case one, rather than yielding *purely* self-locating information, as in case two.

[2] The point of destroying the original body is to make sure that there is no fact of the matter about which clone the original person becomes – each clone has exactly the same claim to being the true successor of the original person, just as in the Everettian case every one of the observers produced in a branching event is a successor of the original pre-branching observer.

Before discussing this possibility, let me offer a caveat: of course, in the real world no observation can possibly give us only non-self-locating information. In case two above, for example, you may believe with 99% confidence that the experiment will be carried out as expected and you will wake up as of your colourful future selves, but you know it is possible that the person performing the cloning will drop dead, or the experiment will be interrupted by a nuclear apocalypse. So when you wake up as one of your future selves, you gain some non-self-locating information to the effect that none of these disasters have occurred. But this doesn't change the fact that *the specific colour of your hair* is purely self-locating information which tells you nothing new about the world. And similarly, in the Everett case clearly some minimal non-self-locating information is gained when the observer sees a measurement result, since she now knows that the measurement was successfully performed and that the observed outcome was a possible outcome to that measurement. But what is relevant here is whether the information about *which particular outcome was obtained* can be treated as non-self-locating information, since that is the information needed to construct relative frequencies in order to gain information about the underlying mod-squared amplitudes and thus empirically confirm the mathematical apparatus of quantum mechanics. So I will henceforth disregard other kinds of non-self-locating information that might be obtained in an Everettian measurement, in order to focus on the question of whether an Everettian observer can obtain any non-self-locating information about the distribution of mod-squared amplitudes from the facts about which particular outcome she observes.

Now, clearly a major reason why it is possible to learn non-self-locating information in case one is simply that there *exists* some non-self-locating information to learn, since in this case you do not already know the distribution of hair colours prior to the experiment. And this point may initially appear to work in the Everettian's favour, because any circumstances in which we are interested in performing empirical confirmation will necessarily be circumstances where there is some piece of non-self-locating information relevant to the experimental setting which we do not yet know with certainty – for example, information about the distribution of mod-squared amplitudes. So we might hope that, as in case one, we can redescribe this scenario from an external point of view by thinking of the observation as simply providing information about what is seen by an observer randomly selected from this distribution, meaning that some *non*-self-locating information is obtained in this observation and therefore it can be used to update non-self-locating beliefs.

However, it's not clear that the Everettian observation can, in fact, be redescribed in this way. For in order to understand the observation as providing information about what is seen by a randomly selected observer, we must be able to give a third-person description in which we *first* specify an observer in a way which doesn't fix which outcome she sees, and *then* ask which outcome she does, in fact, see. For example, if I am interested in knowing whether left or right-handedness is most common in a given crowd of people, it will do me no good to request that a right-handed person steps forward and then ask them if they are left or right-handed: I need to select a sample from the group in a way which is independent of left- or right-handedness.

In case one above, the clones will all have to be in different spacetime locations after the hair-colouration is complete, so we can give a third-person description in which we first point to a clone indexed purely by a spacetime location and then ask what colour hair the clone in this spacetime location has. But it's less clear how we can achieve something similar in the Everettian case. First of all, if we are focusing on the observer before the observation takes place, and we adopt the 'Overlapping' view which entails that there is only one observer present before the branching, then it is clearly senseless to point to such an observer and then ask which outcome she sees – there will be some future version of that observer seeing every possible outcome, so it isn't meaningful to ask which outcome *this* particular observer will see. So we must either move to a 'Diverging' view, in which there are multiple copies of the observer present even before the branching, or we must assume the question is posed in the moments between performing the measurement and observing an outcome, at which point the observer will typically have branched into a set of subjectively identical but physically distinct copies.

Second of all, even once we have made this move, it's still not clear that we can single out just one of these observers *without saying which outcome they see*. For what distinguishes the Diverging continuants, or the various different post-measurement selves, is nothing more or less than the fact that they are found in different branches corresponding to different outcomes of the measurement. After all, post-measurement the observers will have exactly the same spacetime location, and their brains will typically be in macroscopically and perceptually identical states until they actually see their different outcomes, or obtain some other information which is equivalent to learning their outcome. So it seems that in general we will only be able to single out a post-measurement self by specifying its branch, which involves specifying the outcome occurring in its branch or giving information equivalent to that specification. Thus it appears that the Everettian case cannot be redescribed from the third-person point of view as a process of random sampling because there is no way to specify an observer in a way which doesn't already fix which outcome she sees; thus unlike in case one, we cannot think of the Everettian observation as providing some non-self-locating information.

Now, Everettians might perhaps try to refute this argument on the grounds that it is not necessary that we should be able to describe the sampling process coherently from a third-person point of view: it is enough that we can describe it from a *first*-person point of view, appealing to the agent's ability to individuate herself in an indexical, self-referential way which singles out which agent she is without saying anything about which measurement outcome occurs in her branch. However, if the scenario can't be understood from a third-person point of view it is no longer clear that we can meaningfully say that non-self-locating information is obtained in the observation, so the Relevance-Limiting thesis becomes a problem once again. In addition, relying on indexical self-reference here comes at a price – it involves a particular radical kind of indexical self-reference which is *purely* indexical and can't possibly be given any third-person analysis. This is not how indexical self-reference usually works – as van Fraassen (2008, p. 182) puts it, in more ordinary contexts, '*The measurement outcome content typically involves indexical reference to a particular vantage point, but this vantage point is a publicly ascertainable feature of the measurement set-up.*'

Of course, the fact that indexical self-reference of this kind is radically new doesn't necessarily mean we shouldn't accept it, but from the point of view of the overall coherence of our belief system it is not ideal that in order to make sense of the epistemology of the theory we have to introduce a new kind of fact which appears to be relevant only in one specific circumstance and which is postulated purely to solve an otherwise very serious epistemic problem – for as noted in section 4.4, allowing this kind of ad hoc addition makes the bootstrapping involved in progressively building a coherent epistemology of science less meaningful. It also doesn't help that the nature of this fact is quite mysterious: if irreducibly indexical facts are to play the same kind of role as spacetime location in the hair colour case, one might think they need to be a physical fact somewhat similar to spacetime location, but it seems hard to see how something like this *could* be a physical fact. And the coherence problem is multiplied by the fact that irreducibly indexical information seems hard to square with the Everettians' other commitments – for example, many Everettians hold that branching is an approximate, emergent concept which arises in the decoherence limit, but if we are really going to postulate that agents in branching scenarios can use irreducibly indexical reference to single out which particular agent they are, one might then think there must be some fairly definite, precise fact about how to individuate a single agent, which is at odds with the approximate nature of the branching.

So I am fairly sceptical about the 'irreducibly indexical reference' route. However, for now let us just grant for the sake of argument that there is really a meaningful sense in which either a Diverging Everettian observer or a post-measurement Overlapping Everettian observer can individuate herself and pose a meaningful question about what she will observe. Even granting this, the Everettian still is not out of the woods. For even if she is able to regard herself as a randomly sampled observer from the set of post-measurement observers, this will not allow her to obtain distributional information about mod-squared amplitudes unless a case can be made that the mod-squared amplitudes are actually relevant to what is seen by a randomly sampled observer. To see the difficulty, consider adding to the examples above a third case:

3. You are to undergo a procedure in which seven identical copies of your body and brain will be created, and then each clone will have its hair dyed a different colour from the seven colours of the rainbow. Each clone will also be injected with a different quantity of saline, so the clones all have different blood sodium levels (within normal sodium levels, so there is no perceptible effect). The original body will be destroyed.

Now the facts about what observers there are and what hair colours they have no longer exhaust the physical information about the situation: there are also facts about the distribution of sodium levels and about what blood sodium level each individual clone has, so there is indeed some non-self-locating information which could in principle be learned here. Moreover, it seems reasonable to think that the observers in this scenario can indeed regard themselves as random samples from the set of observers, since we can use their different spacetime locations to identify them in a way which is independent of the facts about what hair their colour has been dyed. But nonetheless, there does not seem to be any way for these observers to use an observation of hair colour to learn anything about the distribution of blood sodium levels – if you happen to see green hair, you would surely not automatically update your beliefs to assign higher probabilities to distributions of blood sodium where the green-haired clone has higher blood sodium, because you have no reason to think that blood sodium is relevant in any way to the final hair colour. And in much the same way, what is still missing for a fully satisfactory account of measurement in an Everettian context is some argument to the effect that mod-squared amplitudes are actually relevant to what Everettian observers should expect to observe during measurements. This brings us to the second aspect of the Everettian confirmation problem.

6.2 Task Two: The Everettian Statistical Assumption

The second aspect of the problem is about explaining why it would make sense in the context of an Everettian picture for observers to update their beliefs *in the specific quantitative way* that we actually have done when empirically confirming quantum mechanics.

So what does this actual process of belief-updating entail? For the moment I will suppose, as proponents of the Everett interpretation often do, that the empirical confirmation associated with quantum mechanics is entirely the product of Bayesian updating (though I will question that assumption in chapter 8). The term 'Bayesian updating' refers to a belief-updating strategy in which, upon witnessing a novel piece of evidence E, we update the credence assigned to hypothesis H according to Bayes' rule, which means setting it equal to $\frac{p(H)p(E|H)}{p(E)}$, where $p(H)$ and $p(E)$ are the prior probabilities we assigned to the hypothesis H and evidence E before witnessing E actually occurring, and $p(E|H)$ is the probability of evidence E conditional on the hypothesis H being true.

To see how this works in practice, imagine that we have a set of hypotheses $\{h_i\}$ describing different ways the world could be, and suppose we assign prior credences $c(h_i)$ to each of these hypotheses. Suppose that we are about to perform a measurement, and suppose that for each possible outcome O of that measurement, each of our hypotheses h_i assigns some probability $p_{h_i}(O)$ to that outcome – in the case of quantum mechanics, this probability is given by the mod-squared amplitude associated with the outcome O. Then a straightforward application of the formula for Bayesian updating tells us that on seeing outcome O, we should update the probability assigned to hypothesis h_i to $\frac{p_{h_i}(O)c(h_i)}{\sum_i c(h_i)p_{h_i}(O)}$. Similarly, we can also imagine performing a sequence of measurements, such that our hypotheses assign probabilities $p_{h_i}(S)$ for each possible sequence of outcomes S, and therefore after seeing a sequence of outcomes S we will update the probability assigned to h_i to $\frac{p_{h_i}(S)c(h_i)}{\sum_i c(h_i)p_{h_i}(S)}$. If it is assumed that the outcomes in the sequence are all probabilistically independent, the result of this updating is that we will end up with significantly increased credences for hypotheses which predict probabilities or mod-squared amplitudes that approximately match the relative frequencies in the observed sequence S, and significantly decreased credences for hypotheses whose predictions do not match the observed relative frequencies.

Thus, if we understand the empirical confirmation of quantum mechanics purely in terms of Bayesian updating, we require an account of the physical nature of measurement which explains why it would make sense for Everettian observers to assign high credence to hypotheses or mod-squared amplitudes that approximately match the relative frequencies they have observed in the past. That is, we need an account which is capable of giving some kind of justification for what I have called the Everettian Statistical Assumption (ESA), which stipulates that the relative frequencies observed in experiments are, after a large enough number of experiments, reliably close to the mod-squared-amplitudes assigned by the theory. As Friederich (2021, p. 162) puts it, '*The onus is on the Everettian to demonstrate that we can be reasonably confident to not find ourselves in one of [the low mod-squared amplitude] branches if the [Everett Interpretation] is true.*'

Now, most proposals for solving the Everettian probability problem are concerned with showing that Everettian agents should assign credences matching the mod-squared amplitudes for *individual* measurement outcomes, rather than focusing on sequences – i.e. the emphasis is typically on justifying the Born rule for individual measurements. The reason I am instead emphasising sequences is that we clearly cannot confirm a complex theory like quantum mechanics by just one observation; in fact, quantum mechanics has been confirmed by appeal to relative frequencies observed over large numbers of experiments, so in order to show that the methods we have used to arrive at quantum mechanics are reliable, we need only show that the relative frequencies in sufficiently large numbers of experiments can generally be expected to match the appropriate mod-squared amplitudes – strictly speaking it is not necessary to say anything about individual experiments at all. However, if it can successfully be argued that Everettians should assign credences matching the mod-squared amplitudes for individual measurement outcomes, and if we assume that individual experiments are probabilistically independent, it can then be shown that Everettians should also assign high credence to the proposition that over a sufficiently large number of experiments they will see relative frequencies matching the mod-squared amplitudes. Indeed we can use the law of large numbers (Saunders, 2021) to argue that they should be *certain* that they will see relative frequencies matching the mod-squared amplitudes in the limit of an infinite number of experiments. Thus if any of the approaches I discuss here can give the right kind of justification for assigning credences matching mod-squared amplitudes for individual outcomes, that approach can plausibly also be leveraged to justify the ESA.

Before explaining the available approaches, let me first make some clarifying remarks about what it is that the Everettian needs to do here. First, the mandate is not to show that ESA is true without any shadow of a doubt. Saunders (2021b, p. 6–7) seems to accuse critics of the Everett interpretation of demanding this, suggesting that they would be only satisfied with '*a causal mechanism that will reliably bring about true beliefs about the amplitudes, where "reliable" does not involve probability; that it must be possible, if the unitary evolution is all that there is and if the theory is to be empirically adequate, for ratios in branch weights to be deterministically driven in to the memory of some measurement device.*' And, of course, this would not be a reasonable thing to ask for – after all, even in a collapse approach where there is no doubt that the mod-squared amplitudes correspond to probabilities, there is still some non-zero probability of seeing a branch with low mod-squared weight, so it would make no sense to demand a proof that this cannot possibly occur. Thus, in fact, in the Everettian case the goal is not to have an account of measurement which *guarantees* that observers will end up with true beliefs, but simply to have one which has the consequence that it would be *epistemically rational* for an Everettian observer to assign high credence to the ESA, even while admitting that the ESA could possibly fail to be true – just as we consider it to be epistemically rational in the ordinary probabilistic case to assign high credence to the proposition that the series of outcomes you have seen is one assigned high probability by the underlying probabilistic theory, even while admitting that this could possibly fail to be the case.

Second, the mandate is also not to give an irrefutable proof that the ESA is likely to be true. For as discussed in chapter 2, we should not expect a scientific theory to conclusively prove the correctness of the epistemological assumptions from which it earns its credentials, any more than we should expect a scientific theory to conclusively falsify philosophical scepticism. Rather than conclusive proof, what we are seeking is merely coherence: the task is to show that there is some plausible way about thinking about the Everett interpretation such that, if you believe that the world is indeed like this, then it would be epistemically rational for you to assign high credence to the ESA.

Third, note that even if no reason can be found to affirm the ESA, that doesn't mean *nothing at all* can be learned from Everettian measurement outcomes. For even without any probabilistic interpretation of measurements, an Everettian observer who performs a measurement will at least learn something about which outcomes are possible, and with repetition they can build up a picture of the set of possible outcomes (though, of course, without a probabilistic interpretation which would justify an application of the law of large numbers, they could never be confident they had seen all of the possible outcomes).

However, it seems unlikely that an observer who knows only the *possible* outcomes for all quantum mechanical operations would have enough information to empirically confirm the entire apparatus of quantum mechanics, for a great deal of the most important content of quantum mechanics is to be found in what it says about relative frequencies. For example, an important class of evidence for quantum mechanics comes from experiments making use of 'interference,' in which we can indirectly demonstrate the presence of different branches of the wavefunction by causing them to interfere with each other – that is, they will either cancel each other out or reinforce each other in certain quantifiable ways. But it is not possible to witness the full range of quantum interference effects looking at just the set of possible outcomes. If we only know the possible outcomes, then the only kind of interference we can occur is *complete destructive interference* in which one branch completely cancels out another and thus a certain measurement outcome becomes impossible; but as argued by Catani et al. (2023 a,b), the patterns of complete destructive interference exhibited in quantum mechanics can occur even in a classical model, so the truly quantum aspects of the interference phenomenology are all to be found in the intermediate cases where cancellations are not complete. And those cases can only be properly observed by looking at a pattern '*built up from detections of individual particles that exhibit statistical variation in their location*' (Catani et al., 2023a, p. 3) – i.e. a pattern based on relative frequencies. So it is implausible to think that we would be able to empirically confirm the full mathematical apparatus of quantum mechanics without using relative frequencies, and without the full mathematical structure of quantum mechanics we likely would not end up with any good justification for postulating branching and an Everettian multiverse, so as argued in chapter 3, this kind of position seems self-undermining.

Thus I will henceforth take it that in order to provide a satisfactory account of the physical nature of measurement in an Everettian world, we must be able to provide an account of Everettian measurement which entails that it makes sense for Everettian observers to affirm the ESA. In this section I will discuss a number of different arguments that might be leveraged to derive this conclusion, which I will divide into three categories. The first involves giving a decision-theoretic argument aiming to show that it is rational for Everettian agents to assign credences over measurement outcomes in proportion with mod-squared amplitudes; the second involves using principles governing self-locating belief to argue that it is epistemically rational for Everettian agents to believe that they will see outcomes with high mod-squared amplitudes; the third involves seeking to show that as a matter of objective physical fact, mod-squared amplitudes really do correspond to probabilities for observations.

6.2.1 Arguments of the First Kind: Decision Theory

A well-known approach to the Everettian probability problem, originally due to Deutsch (1999) and subsequently elaborated and defended by Wallace (2010, 2012), involves the use of decision theory. The approach is inspired by 'representation theorems' from classical decision theory, in which it is proved that any agent with a set of preferences obeying certain axioms of rationality can be assigned a quasi-credence function P and a utility function U over possible outcomes $\{O\}$ such that the agent always chooses the action A which maximises expected utility, $\sum_O P(O|A)U(O)$ where $P(O|A)$ is the quasi-credence assigned to outcome O conditional on action A being performed, and $U(O)$ is the utility attached to outcome O. The Everettian version uses a similar decision-theoretic argument to prove the Born rule.

The Everettian proof begins by imagining an agent faced with a choice of actions: these may be quantum measurements, or possibly preparations, and/or transformations followed by measurements, but in the unitary-only Everettian setting all of these possibilities can always be combined into a single unitary transformation associated with the action A. We then assume that an Everettian agent has a set of preferences over these possible actions which obey certain principles of rationality. Some of these principles are quite generic, e.g. we require that the ordering on preferences is transitive, while other principles are specific to the Everettian case, such as 'Branching Indifference,' which requires that '*an agent doesn't care about branching per se: if a certain measurement leaves his future selves in N different macrostates but doesn't change any of their rewards, he is indifferent as to whether or not the measurement is performed*' (Wallace, 2009, p. 12). It is then shown that as in the classical case, the agent's preferences may be represented by a quasi-credence function P and a utility function U such that the agent always chooses the act A which maximises expected utility $\sum_O P(O|A)U(O)$, where $P(O|A)$ is the quasi-credence assigned to outcome O

conditional on performing the action A. But the Everettian representation theorem goes further – it can also be shown that the quasi-credence $P(O|A)$ will always be equal to the mod-squared amplitude assigned by the unitary quantum formalism to the outcome O, given the prior quantum state of the system together with the unitary transformation encoded in the action A. Hence, it is claimed, a rational agent who believes they are in an Everettian universe and knows that the mod-squared amplitude for a certain outcome is p will always act as if the probability of that outcome is p – i.e. rational agents in an Everettian universe will behave as if the Born rule is true.

As noted in chapter 3, the decision-theoretic approach to the Everettian probability problem may well be satisfactory if our main priority is to explain why agents should still behave as if quantum measurements are uncertain events in an Everettian universe, despite knowing that they are not. But will the decision-theoretic approach work to make sense of the specific kind of probabilistic reasoning involved in empirical confirmation? Proponents of the decision-theoretic approach argue that, because they have shown that it is rational for Everettian agents to assign quantitative credences to measurement outcomes in proportion to mod-squared amplitudes, after the measurement has been completed the agents can perform Bayesian updating using these credences in just the same way as non-Everettian agents would perform Bayesian updating using the probabilities they assign to measurement outcomes. That is, in the updating procedure described in section 6.2, the probabilities $p_{h_i}(O)$ and $p_{h_i}(S)$ would, in the case where h_i is Everettian quantum mechanics, simply be replaced with the credences obtained from the decision-theoretic argument – and since these credences are equal to the mod-squared amplitudes, it follows that if we see relative frequencies approximately matching the mod-squared amplitudes predicted by standard quantum mechanics, standard quantum mechanics will be empirically confirmed. Thus this argument seems to show that indeed, it is rational for an Everettian agent to perform Bayesian updating on measurement outcomes in just the same way as a single-world agent, thus affirming the rationality of the means of enquiry that we have used to arrive at quantum mechanics in an Everettian context.

However, this argument depends on the assumption that we can simply replace single-world probabilities with decision-theoretic credences in the process of empirical confirmation, and it is not necessarily obvious that we ought to do so. For the meaning and justification of the decision-theoretic credences is quite different from the meaning and justification of single-world probabilities, so even if you are completely convinced by the decision-theoretic proof as it pertains to practical deliberations, it doesn't necessarily follow that the proof also justifies the use of mod-squared amplitudes in empirical confirmation. And, in fact, there are two key issues with the decision-theoretic proof which raise questions about the applicability of the resulting credences to the process of empirical confirmation.

Past and Future

First of all, note that the decision-theoretic proof is ultimately *forward*-looking: it seeks to tell us how we should weight various possible future branches of the wavefunction when we are making decisions relevant to our personal future. For example, suppose you are offered two bets. Bet B_1 involves a case where you branch into two copies, and in the branch with mod-squared amplitude 0.8 you will receive a million dollars, while you will receive \$0 in the complementary branch. Bet B_2 involves a case where you branch into two copies, and in the branch with mod-squared amplitude 0.2 you will receive a million dollars, while you will receive \$0 in the complementary branch. So which bet should you take? The decision-theoretic proof provides a straightforward answer: you should take bet B_1, since the expected return is \$600,000 greater than the second bet.

On the other hand, the process of empirical confirmation is *backward*-looking: the goal is for us to be able to look at a series of events which *have already occurred* and have some justification for assigning high credence to the proposition that this series of outcomes is assigned high mod-squared amplitude by the underlying theory. The decision-theoretic approach has to insist that this high credence is grounded in principles of rationality governing decision-making in an Everettian world, and yet there seems to be no obvious reason why constraints on our forward-looking decisions should have any consequences at all for credences we assign to propositions about the mod-squared amplitudes associated with events that have already occurred. The key difference is that the credences we assign over the outcomes of events yet to occur can be related in a direct operational way to our decisions, since we can be offered bets with different payoffs depending on these outcomes; whereas there cannot be a similar operationalisation of credences pertaining to the mod-squared amplitudes associated with events in the past, because we already have all the information about which events *did* occur, and there is no further test we can do now to determine what mod-squared amplitude was associated with those events, unless we make some significant additional assumptions about the relationship between past mod-squared amplitudes and future observations. So without such additional assumptions there is no possible way to design a bet which pays out differently depending on the mod-squared amplitude associated with events in the past. Thus the standard decision-theoretic methodology on its own can't be used to constrain backward-looking credences about the mod-squared amplitude assigned to events that have already occurred – to achieve that we must make some further assumption to the effect that rationality demands a systematic link between credences pertaining to propositions about the mod-squared amplitudes associated with events in the past, and credences or utilities pertaining to the future.

One way to achieve this would be to appeal to some kind of principle of diachronic consistency. Indeed, the decision-theoretic proof itself uses 'Diachronic Consistency' as one of its axioms (Wallace, 2010). However, this particular consistency principle will not help with the confirmation problem because it pertains only to an agent's preferences over *future* decisions – e.g. if it is known that after performing action U, an agent's future self will prefer performing action V to action V', then right now the agent should prefer performing U and then V over performing U and then V'. This forward-looking principle does not directly tell us anything about the relationship between forward-looking credences and backwards-looking credences.

Alternatively, we might appeal to the kinds of argument sometimes used in the ordinary classical case to argue that Bayesian updating is a uniquely rational way of responding to evidence. For example, Lewis (1999) offers an argument (originally published by Teller (1976)) to the effect that if you fail to update your beliefs by Bayesian updating you are vulnerable to having a diachronic Dutch Book made against you, i.e. you can be convinced to accept bets in which you are guaranteed to lose money. And Greaves (2006) argues that exactly the same argument can be used in the decision-theoretic Everettian approach because the decision-theoretic proof shows that rational Everettian agents should act as if they assign credences equal to mod-squared amplitudes, and accepting a bet is a kind of action, so a rational Everettian who acts as if they assign credences equal to mod-squared amplitudes must then also act as if they are performing Bayesian updating on those credences in the ordinary way, otherwise they will be vulnerable to a Dutch Book.

However, let us look more closely at the bets involved in Lewis' argument. To apply it to the Everettian case, let Q be the proposition that quantum mechanics is true, and suppose you are about to perform a measurement which has outcome E with probability greater than zero. Before the measurement, the Dutch Bookie sells you two bets for a total price of \$x: one which pays \$1 if Q is true and the measurement has outcome E, and \$0 otherwise, and another which pays \$x if the measurement does not have outcome E, and \$0 otherwise. Here $x = \frac{p(E|Q)p(Q)}{p(E)}$ using your credence function p; this value corresponds to the maximum price that you should pay for this pair of bets if they aim for expected winnings of no less than 0. Then, after the measurement, the Dutch Bookie buys from you a bet which pays \$1 if Q is true and \$0 otherwise, for a price of \$y, where y is set equal to the credence you now assign to Q; again, this value corresponds to the maximum price that you should pay for this bet if you aim for expected winnings of no less than 0. Thus, regardless of whether or not Q is true, if E fails to occur your winnings are \$0, and if E occurs your winnings are $y - x$, so if $y < x$ your total expected winnings will be negative. Moreover, if $y > x$, the Dutch Bookie can simply perform the opposite procedure in which she buys the two bets before the measurement and sells the one after, so again your total expected loss is negative. So, the argument goes, the only rational policy here is to choose $x = y$, which means you must perform Bayesian updating.

However, note that in order for you and the Bookie to perform these transactions, the two of you must come to agreement about the conditions under which the bets will pay out. Of course for a probabilistic theory like quantum mechanics you can presumably never be completely sure whether or not the theory is true, so technically bets on Q should never pay out, but let's suppose the two of you agree that the bet will pay out once enough evidence has accrued for you both to assign credence of at least 0.99 to Q, or for you both to assign credence at least 0.99 to $\sim Q$. Yet if the Bookie is performing Bayesian updating, and you are doing some wildly different kind of updating, there is no reason to think that you will *ever* come to agree in this way – it would be entirely possible for the Bookie to converge towards credence 1 for Q and you to converge towards credence 1 for $\sim Q$ at the same time. Indeed, if you are not performing Bayesian updating – if, for example, you are just leaving your credence in Q unchanged upon seeing the measurement outcome, because you know you are in some kind of branching world and you don't think measurement outcomes provide information relevant to empirical confirmation in a branching world – then you may well be completely certain that regardless of what the Bookie does, you yourself will *never* assign credence of at least 0.99 to either Q or $\sim Q$. And in this case, rather than paying x for the original pair of bets, you should pay nothing at all for the first bet, since you can't win anything from a bet which will never pay out, and you should pay no more than $1 - p(E)$ for the second bet. Thus the Bayesian Dutch Book argument can't be made to work in the Everettian case unless we have already committed to the ESA or some other concrete updating policy in order that bets on theoretical claims like Q will eventually pay out, so it would be circular to use such an argument to support for Bayesian updating in the Everettian context.

It should be noted that, of course, a similar argument could be made with regard to the non-Everettian case; it would similarly be circular to use an argument of this form to insist that we must update our beliefs in the truth of a single-world probabilistic theory by appeal to Bayesian updating alone. This offers some reason to think that the epistemology of science cannot be fully accounted for in terms of only Bayesian updating; I will return to this point in chapter 8, but for now it is enough to observe that Dutch Book arguments do not seem adequate to justify the use of Bayesian updating with respect to Everettian quantum mechanics.

Alternatively, we could arrive at a link between credences pertaining to the past and credences pertaining to the future by simply appealing to something like the principle of induction to argue that we ought to expect that future relative frequencies will be similar to the ones we have observed in the past. I will return to the relation between the ESA and the principle of induction in section 7.4, but for now I will just point out that in the Everettian context we know that all possible sequences of outcomes occur in some branch, so Everettian agents cannot possibly have any reason to think that future relative frequencies will be similar to those they have observed in the past unless they *already* have some reason to believe that the specific relative frequencies that they have personally observed in the past are in some way special. And if we simply assume that the relative frequencies that have been observed in the past are special, then we are already quite close to assuming the ESA, since the whole point of the ESA is to licence the assumption that the relative frequencies we have observed are structurally privileged by the underlying theory, and thus can give us some useful information about the structure of the underlying theory. Therefore again, it would be circular to presuppose something like an Everettian principle of induction as part of a decision-theoretic proof aiming to show that it is rational for Everettian agents to act as if they are doing Bayesian belief-updating on measurement outcomes. So it appears that the forward-looking nature of the decision-theoretic arguments prevents us from using the resulting credences in a backward-looking way to do empirical confirmation, since the only way to make backwards-looking credences relevant to forward-looking decisions is to assume exactly the kind of connection between past and future which we are trying to prove.

Pragmatic versus Epistemic Rationality

But let us for now put aside these concerns about past and future and suppose that we are willing to accept that the forward-looking decision-theoretic argument also constrains past-looking credences, allowing us to argue that it is rational for an Everettian agent to act as if they are performing Bayesian updating on measurement outcomes in just the same way as a single-world agent. Still, there is a further problem – is it the right *kind* of rationality?

The issue is that the decision-theoretic method seems to completely detach rational credence from truth – as David and Thébault (2015, p. 57) put it, '*No deductive step is available that leads from the statement that only betting on a specific statistical data distribution is rational to the statement that this distribution will in fact be found in experiments.*' That is, in an approach based on decision theory, assigning credences is just about deciding how it is rational to behave at a pragmatic level, rather than asking what is true; so at best, the decision-theoretic approach can show that in an Everettian universe it would be *pragmatically* rational to behave as if one is doing Bayesian belief-updating on measurement outcomes.

Yet in order to solve the epistemic version of the measurement problem, we cannot merely pose questions about what agents should do – the goal is to arrive at an account of the physical nature of measurement which retrospectively justifies our epistemic commitment to quantum mechanics by showing how measurements connect agents up with the structures described by quantum mechanics, and therefore we are clearly concerned with *epistemic* rationality rather than pragmatic rationality, where epistemic rationality is understood as '*the kind of rationality which one displays when one believes propositions that are strongly supported by one's evidence and refrains from believing propositions that are improbable given one's evidence*' (Kelly, 2003, p. 612). As argued in chapter 4, one important aspect of epistemic rationality in the context of science is that is not epistemically rational to believe a theory which tells us that the means of enquiry we have used to arrive at that theory are not reliable ways of arriving at true beliefs. And the fact that it is *pragmatically* rational for Everettian agents to behave as if they are doing Bayesian updating on measurement outcomes does not mitigate the fact that it is not *epistemically* rational for Everettian agents to believe that quantum mechanics has been empirically confirmed, unless, of course, the reason why it is rational for agents to behave this way is that measurement outcomes in an Everettian world can be shown to reliably put agents in contact with the kinds of theoretical structures described by quantum mechanics – which is clearly not the kind of reason that the decision-theoretic proof offers.

Indeed, the basic principles of the decision-theoretic proof seem incompatible with the goal of giving an account of the physical nature of measurement in an Everettian world. For example, the proof uses axioms like 'Microstate Indifference' – '*An agent doesn't care what the microstate is provided it's within a particular macrostate*' (Wallace, 2010, p. 12, online version) – which explicitly advise agents to disregard facts that are not relevant to what they will personally experience in the future. Whereas in seeking to account for the physical nature of measurement, the point is to situate the agent within a more general theoretical description in order to show how their experiences are related to important structures in the theory they are seeking to confirm, and providing this more general description is likely to involve making reference to facts which are not directly relevant to the agent's personal future. Similarly, as Price (2010) points out, the decision-theoretic proof assumes that Everettian agents can only have preferences over 'in-branch,' matters, i.e. their utility must be some weighted sum of utilities attached to each of their future branches, so the proof rules out by fiat the possibility that Everettian agents might attach their preferences directly to the overall global structure of the future wavefunction rather than to individual branches. And yet the situation in which you are trying to empirically confirm a scientific description of your reality is precisely the kind of situation in which you *ought* to care about these kinds of global facts, even if they are not relevant to your personal future, because the global structure of the future wavefunction is exactly the kind of thing that is likely to be relevant to that scientific description; thus it wouldn't make sense to begin the process of empirical confirmation by assuming that these kinds of facts are not relevant for your credences. So although the pragmatic underpinnings of the decision-theoretic proof are an ingenious way to allow us to get a normative claim out of the purely descriptive statements of Everettian quantum mechanics, they also seem to make the approach fundamentally unsuited to provide probabilities which can be used to empirically confirm a scientific theory, and thus this approach doesn't seem to have the resources to adequately affirm the reliability of our means of enquiry in an Everettian world.

Now, the Everettian might push back against this argument by questioning whether there is really a meaningful distinction between pragmatic and epistemic rationality. For example, it may seem natural to pair a decision-theoretic approach to empirical confirmation with a broadly behaviourist or functional account of belief, which suggests that to believe something is simply to act as if one believes that thing (Place, 2000; Ryle, 2009). And if something along these lines is the right way of thinking about belief, it would seem to follow that we simply cannot properly distinguish belief from action in the way that would be needed to define a sensible notion of distinctly *epistemic* rationality. For example, one way of modelling behaviourism mathematically would involve a decision-theoretic analysis in which an agent is said to have a certain credence function only if her revealed preferences are consistent with that credence function together with some preferred choice of utility function. Clearly this approach leaves no room for us to distinguish between 'credences' determined purely by epistemic rationality and 'credences' based on other more pragmatic considerations; one could, as in Greaves and Wallace (2006), imagine constructing an 'epistemic utility' function in order to capture purely epistemic credences, but then the problem of identifying credences would become extremely degenerate, since any behaviour would be compatible with an infinite number of possible ways of decomposing the utilities and credences into epistemic and pragmatic parts. So indeed, it seems difficult to sensibly uphold a notion of purely epistemic credence unless it is accepted that belief is not just a matter of behaviour.

However, pragmatic rationality, particularly within the decision-theoretic analysis, is a form of instrumental rationality, i.e. the rationality we exhibit when we take actions which are a good way of achieving our goals, as encoded in our utility function. So epistemic rationality cannot be inextricably bound up with pragmatic rationality in the way that this analysis would suggest unless epistemic rationality is itself a form of instrumental rationality – e.g. Laudan (1990) suggests that epistemic rationality is simply a form of instrumental rationality where the relevant goals are epistemic ones, such as the goal of believing true things. But, in fact, it is controversial whether epistemic rationality *is* just a form of instrumental rationality, for as Kelly (2003) points out, '*What a person has reason to believe does not seem to depend on the content of his or her goals in the way that one would expect if the instrumentalist conception were correct.*' In particular, Kelly notes that epistemic reasons are typically considered to be intersubjective, i.e. if you and I have the same information about some situation, we have exactly the same epistemic reasons for holding beliefs about this situation, regardless of differences in our epistemic goals. Moreover, Kelly notes that this cannot be explained by arguing that we all share the same epistemic goals, because for most of us there are many propositions whose truth or falsity we do not really care about. Hence we have no particular goals with respect to our cognitive attitude to those kinds of propositions, but nonetheless given relevant evidence about these propositions we can still have compelling *epistemic* reasons to believe or disbelieve them. These considerations suggests that epistemic and pragmatic rationality are indeed quite distinct from each other, so it is reasonable to criticise the decision-theoretic approach for focusing on pragmatic rather than epistemic rationality.

Moreover, no matter how plausible the functional analysis of belief and credence may be, I think it is not really a serious objection to the existence of a notion of epistemic rationality which is distinct from pragmatic rationality, for the simple reason that the notion of epistemic rationality need not have anything to do with the definition of 'belief' or 'credence.' For when you are trying to decide which beliefs or credences are epistemically rational given some set of evidence, you are not asking yourself anything about what the word 'belief' or 'credence' means; rather you are asking yourself 'What inferences does this evidence support?' Although proponents of the decision-theoretic approach might try to reconceptualise this as a deliberation about what you should do, in the sense of 'Which of these things should I (act as if I) believe about the world?' this is a misdirection: the question is about the world, not about you. Thus it is perfectly possible to insist that epistemic rationality is distinct from instrumental rationality whilst still accepting the behaviourist analysis of belief – this would simply entail that even when we are in principle fully capable of recognising what it is epistemically rational to believe, we do not always fully succeed in believing those things, since our actions may not always be fully in accord with our epistemically rational conclusions. And yet with epistemic as with pragmatic rationality, limitations in our abilities to behave in rational ways surely do not make questions about rationality meaningless, so regardless of one's preferred analysis of belief I think it is reasonable to aspire to a scientific picture of reality which is epistemically rather than merely pragmatically rational.

6.2.2 Arguments of the Second Kind: Principles for Self-Locating Credences

The problems we have discussed for the decision-theoretic approach ultimately stem from the fact that the proof is motivated by pragmatic rather than epistemic considerations, making it unsuited to address the epistemic version of the measurement problem. But the mathematical structure of the proof, which simply serves to pick out the mod-squared amplitudes as physically privileged, is somewhat independent of the decision-theoretic motivation, and thus one might hope to avoid these problems by giving a similar proof using more explicitly epistemic principles. This is what the more recent approaches of Sebens and Carroll (2018) and McQueen and Vaidman (2018) set out to do; they aim to use basic principles of *epistemic* rationality to show that we should assign credences to branches in accordance with their Born weights.

The proof of Sebens and Carroll is based on one simple principle, the epistemic separability principle (ESP), which Sebens and Carroll (2018, p. 40) explain as follows: '*ESP: The credence one should assign to being any one of several observers having identical experiences is independent of the state of the environment.*' They then argue that the natural generalisation of this to the quantum case is '*ESP-QM: Suppose that an experiment has just measured observable O_b of system S and registered some eigenvalue O_i on each branch of the wave function. The probability that agent A ought to assign to the detector D having registered O_i when the universal wave function is Ψ, $P(O_i|\Psi)$, only depends on the reduced density matrix of A and D, ρ_{AD}: $P(Oi|\Psi) = P(O_i|\rho_{AD})$.*' From ESP-QM, they are then able to derive the Born rule, i.e. they prove that, after the agent has performed the measurement but before she has seen the outcome, it is rational for her to assign probabilities to measurement outcomes equal to the corresponding mod-squared amplitudes. Similarly, the proof of McQueen and Vaidman uses two principles – first, they assume that when an experiment respects a symmetry, it will lead to a symmetry between descendants corresponding to the measurement outcomes, and second, they assume that whatever happens in region A depends only on the quantum description of this region and its immediate vicinity – and again, they are able to derive the Born rule from these principles.

How does this approach fare with respect to empirical confirmation? Both Sebens and Carroll and Vaidman and McQueen argue that once we have derived the Born rule, we can then immediately use these probabilities to do Bayesian updating as described in section 6.2. But it's not completely clear that the self-locating credences can really be used in such a way. One key reason to have doubts about this application of self-locating credences is the Relevance-Limiting thesis, which would immediately imply that self-locating credences should not be used this way. But in addition to these kinds of general worries, there are also difficulties more specific to the arguments used by Sebens and Carroll and Vaidman and McQueen. In particular, the arguments used by Sebens and Carroll to motivate ESP, and more generally the examples used in the literature to motivate other similar principles constraining assignments of self-locating credences, focus on leveraging our intuitions about how we would intuitively be inclined to assign credences while we are in a state of self-locating uncertainty *prior* to observing an outcome. Indeed, both Sebens and Carroll and Vaidman and McQueen specifically understand their proof to apply in the moment *after* performing the measurement but before observing the outcome, so the justification for the recommended assignation of credences seems to rely quite crucially on specific features of the kind of uncertainty that agents feel at this particular time.

However, as noted in section 6.2.1, if we are to use these credences in Bayesian updating, then we must appeal to credences of this kind applied to events that have already occurred. And thus if such an approach is to be used to affirm the reliability of our means of enquiry in an Everettian universe, it cannot be the case that the principles they employ are reasonable only in the specific period between measurement and observation that the authors focus on – we must be entitled to continue invoking these credences *retrospectively* after the observation has been made, at which point we are no longer in a state of self-locating uncertainty. Of course ordinary non-self-locating credences do translate across time in this way, but self-locating-credences are conceptually quite different in many ways from non-self-locating credences, so even if ESP, Symmetry, and Local Supervenience (LS) are reasonable constraints on credences while we are in a state of self-locating uncertainty prior to an observation, it doesn't follow that we can go on to use self-locating credences derived in this way to do empirical confirmation after the observation in the same way that we would use non-self-locating credences.

Indeed, Friederich highlights an important disanalogy between the role of self-locating and non-self-locating credences in the process of confirmation: when we gain new non-self-locating information, we typically respond by performing Bayesian updating of the credences we assign to hypotheses, where each of these hypotheses assigns a different probability distribution over the outcomes of the measurement we have just observed, so we are updating credences we assign to various different probability distributions. But when we gain new self-locating information, '*the impact of self-locating information is adequately modeled only by updating the [self-locating] distribution itself ... rather than any updating any probability distribution over [self-locating] distributions*' Friederich (2021, p. 107). That is, it's unclear that we can sensibly assign 'meta-credences' *in retrospect* over possibilities which themselves correspond to different possible distributions of self-locating credences pertaining to an observation that has already been made. For these self-locating credences represent nothing other than the state of uncertainty that an observer might have been in prior to the observation, so what could it possibly mean to assign credences over different possible past states of uncertainty, after the uncertainty is at an end? In light of this disanalogy, we should be very cautious about using the credences derived by Sebens and Carroll and Vaidman and McQueen in a backwards-looking way.

Another reason why it might be inappropriate to use self-locating credences in this way is linked to the fact that, as described in section 4.2, one important function of a measurement model is to reassure us that the theoretical structures which are supposed to be confirmed by a given observation are in some meaningful sense *responsible* for this observation rather than some other observation being made. And can this really be the case if the relationship between the structure and the observation depends crucially on a principle like ESP, Symmetry, or LS which is not grounded in any physical fact? Looking back at case one from section 6.1, suppose you learn that six out of the seven clones will be given green hair, and the other will be given pink hair. Then it does seem plausible that prior to the cloning you would set your credence for ending up with green hair equal to roughly $\frac{6}{7}$, as ESP would suggest. But when you wake up and see that you have green hair, can you really say that the distribution of hair colours is *the reason why*, or part of the reason why, you have green hair rather than pink hair? There is surely an equally good case to be made that there is just *no reason* why you have green hair rather than pink hair – there were always going to be six clones with green hair and one with pink hair, and you happen to be one of the green-haired ones, but it doesn't make sense to ask why you are this one rather than that one. So it's at best unclear that a principle like ESP can provide the right kind of link between theory and observation for us to be able to say that the observations are really providing meaningful information about the theoretical structures.

It is helpful here to note that ESP, Symmetry, and Local Supervenience are all, in effect, symmetry principles: this is self-evident for Symmetry, and meanwhile ESP and Local Supervenience are both variations on the assertion that the appropriate probability distribution to assign over measurement outcomes should be invariant under transformations of the state of the environment, which simply amounts to saying that the probability distribution obeys a certain symmetry. And this use of symmetry principles has a high degree of intuitive plausibility because there is a long and successful history of using symmetry principles to constrain probability assignations. In particular, it is common in *non*-self-locating cases to assign probabilities using a principle known as the Principle of Sufficient Reason (PSR), which suggests that if there is no known reason to prefer one element of a partition over another, we should assign them all equal probabilities (Schupbach, 2022). The PSR seems superficially very plausible, but it encounters a problem known as 'Bertrand's Paradox': it can produce vastly different predictions depending on the way we specify the elements of the partition to which we will be assigning equal probabilities. A famous example is known as the Buffon's needle problem, where we imagine that there are a large number of parallel lines marked on the floor, and a needle is dropped – what is the probability that the needle cuts one of the lines? We can get two different answers depending on how we apply the principle of indifference here: either we will end up with a uniform distribution over the angle θ that the needle makes with the nearest line, or a uniform distribution over the y coordinate of the needle point relative to the nearest line, and these two choices predict different values for the probability of cutting a line (Hey et al., 2010; van Fraassen, 2012).

Now, one quite successful response to this problem, developed by a number of authors including Jeffreys (1998) and Jaynes and Bretthorst (2003), is to say that the PSR will often produce good results if the partition is chosen in a way which respects the symmetries of the underlying physical situation or process. Now, the PSR is supposed to be applied taking into account everything we know about the situation in question, but the point here is that regardless of what we currently know about the problem, our use of the PSR will typically work well if we should happen to choose a probability distribution which is invariant under all transformations that we naturally understand as producing different formulations of the same problem. For example, in a real implementation of the Buffon's needle experiment, it may be the case that the needle is dropped blindly, i.e. no attempt is made to align its starting position with the lines on the floor. In that case rotating the lines on the floor produces exactly the same problem, and hence, regardless of what the person applying the PSR knows about the way in which the needle is dropped, as a matter of fact their application of the PSR will produce the most accurate predictions if they should happen to choose a probability distribution which is invariant under rotations (van Fraassen, 2012). In fact, it turns out that a uniform distribution over the angle θ is invariant under rotations but a uniform distribution over the y coordinate is not, which suggests that the uniform-angle distribution is the one which will be most successful under most normal circumstances – and indeed, this can in principle be verified in real experiments.

But the key point here is that that this reasoning works precisely because there is an actual physical process by which the outcome is produced, and the rotation symmetry is a symmetry *of that process*. If we instead consider a case where the needle is not dropped blindly, i.e. the person dropping it is deliberately trying to make it fall parallel to the lines, then rotation symmetry is no longer a symmetry of the process by which the outcome is produced, and hence the uniform-angle distribution will no longer be the most successful – which again can in principle be demonstrated in real experiments. So in cases where symmetry principles are useful to address with Bertrand's Paradox, there are often concrete physical facts which explain why some particular probability distribution is, in a sense, objectively right – the use of the PSR is not just an a priori principle of rationality, but rather a methodology we use to try to characterise real physical features of the probabilistic process by which the outcome is selected. As van Fraassen (1989, p. 316) puts it, '*This method always rests on assumptions which may or may not fit the physical situation. Hence it cannot lead to a priori predictions. Success, when achieved, must be attributed to the good fortune that nature fits and continues to fit the general model with which the solution begins.*'

By contrast, in cases where the probabilities in question are purely self-locating, reasoning of this kind doesn't apply: there are no objective *physical* facts about how best to assign credences because there is no objective physical process which selects one outcome rather than another. Of course scenarios involving self-locating credences may certainly have symmetries of various kinds, but there can't possibly be the same kind of rationale for using these symmetries to constrain the credences as there is in the non-self-locating case, because in the self-locating case there is no process by which the outcome is selected, and thus any symmetries present cannot be symmetries *of* such a process.

Thus ultimately the only rationales that Sebens and Carroll and Vaidman and McQueen can offer for their principles are based on intuition or historical precedent – Sebens and Carroll (2018, p. 40) argue for ESP on the basis that '*it seems clear that certain facts are irrelevant*' while Vaidman and McQueen (2018, p. 6) argue for Symmetry on the basis that symmetry in general is '*a strong universal principle of physics*.' Yet these intuitions and precedents have been formed in scenarios of *non*-self-locating uncertainty – very few of us have ever encountered pure self-locating uncertainty, after all, so our intuitions in this matter are presumably largely inherited from the intuitions we have developed around the more ordinary non-self-locating cases. But we have just seen that the rationale for assigning credences cannot be the same in the self-locating and non-self-locating cases, so there's no particular reason to think these intuitions and precedents should be a good guide in the self-locating cases. Thus it's unclear that the symmetry-based arguments offered by Sebens and Carroll and Vaidman and McQueen are physically substantive enough to provide probabilities which can play the necessary epistemic role of reassuring us that our means of enquiry really do provide reliable information about the underlying structural features of the theory, so they too may not suffice to solve the epistemic part of the probability problem.

6.2.3 Arguments of the Third Kind: Branch Counting and Measures of Existence

In both arguments of the first kind and arguments of the second kind, the difficulty we encountered was that in order to use relative frequencies in empirical confirmation, it seems we must be able to make the case that Everettian agents really are, in some quite strong and objective sense, likely to be in branches with a high mod-squared amplitudes – it is not enough to merely say it is rational for them to behave as if they are, or that some particularly intuitive way of assigning credences would give the impression that they are. So in this section I will consider several proposals which aim to give a more 'objective' account of the Everettian probabilities.

The first such approach is branch-counting, which is perhaps the most obvious way of thinking about probabilities in an Everettian context. The idea here is that in a branching event we should simply say the observer has equal probability of ending up in any one of the branches, just as we would naturally be inclined to say that if your brain is split in two and used to create two clones you have equal probability of ending up as either of the clones.

However, this naïve approach will not work to justify the ESA, for the simple reason that it delivers the wrong probabilities: in most cases, quantum mechanics does not assign equal probabilities to all branches, rather it assigns probabilities equal to the mod-squared amplitude of the branches. In addition, in many versions of the Everett approach it is usually suggested that branches are approximate, emergent structures arising out of the decoherence process, which means there is not even a fact of the matter about exactly how many branches there are, so branch-counting cannot even get off the ground (Wallace, 2003, 2011).

A more sophisticated version of branch-counting has been developed by Saunders (2021), who argues that we ought to seek a branch-counting rule which depends non-trivially on the state, is a continuous function of the state, and is free of arbitrary conventions. Saunders thus offers an approach inspired by Boltzmann's rule for counting microstates in classical statistical mechanics, in which we create a partition made up of equi-amplitude branches, such that assigning equal probability to all branches will reproduce the Born rule. Saunders makes a strong case that this is the most reasonable and physically motivated way to offer a count of branches. However the fact that we have a reasonable way to count branches still does not mean that observers are in some sense being randomly sampled from amongst those branches, and therefore the existence of a count in and of itself isn't enough to solve the problem – there are many things in the world which can be counted, but the fact that they're countable doesn't mean it automatically makes sense to assign probabilities over them, unless one is actually making a random selection from amongst them. Furthermore, even when there is a well-defined count of distinct items, it is not automatically the case that we can assign equal probabilities to each distinct item – for example, if you are choosing gifts from a lucky dip, there is a well-defined count of gifts, but the probability for each gift is not equal, since you are more likely to select the larger gifts. So the well-defined branch count won't suffice to affirm the ESA unless we can provide a) a meaningful sense in which a measurement can be understood as a random selection from the set of branches as defined by Saunders, and b) a good reason for thinking that this random selection occurs with equal probabilities for every such branch. In fact, Saunders himself recognises the need to supplement his branch-counting methodology with some ways of answering these questions, and advocates using the decision-theoretic approach to do so; but I have already argued that decision-theoretic probabilities don't give us what we need for the purpose of empirical confirmation, so in order to get any further with the branch-counting approach we'd have to find some other way of answering these questions.

An alternative to the branch counting approach involves calling the mod-squared amplitude a 'measure of existence' and adding an additional postulate sometimes known as the Vaidman rule: '*The quantum probability of a quantum history is equal to its measure of existence in the universal wavefunction*' (Ridley, 2023, p. 502). Evidently this postulate does provide justification for an Everettian agent to believe the ESA, provided we take it that the term 'quantum probability' is supposed to have something to do with the credence that it is epistemically rational for an agent to assign to being in the relevant branch. However, one might worry that this approach just amounts to renaming the mod-squared amplitudes in a way that makes the idea that they have something to do with what agents will experience sound more plausible, without actually offering any explanation as to *why* they have anything to do with what agents will experience. As Friederich (2021, p. 161) puts it, '*The profound-sounding term 'measure of existence' should not distract us from the fact that Vaidman simply postulates the Born rule here as a primitive rationality principle.*' Referring back to case three in section 6.1 above, presumably you would not suddenly start thinking that the hypothesis that green-haired clones have high blood-sodium explains your experience of seeing green hair just because somebody proposed renaming 'blood sodium' as 'measure of existence' – you would want to know what that term is supposed to mean and what it has to do with the fact that you end up as one clone rather than another.

So if the measure of existence approach is to provide a satisfactory justification for ESA, we will need more information about what a measure of existence actually is and why it is epistemically rational for Everettian agents to expect to end up in a branch with a large measure of existence. After all, existence is normally understood as a binary – either something exists or it does not – and though we should of course be open to the possibility of a meaningful non-binary generalisation of the notion, the generalisation is vacuous unless it is possible to explain what sense can be made of fractional existences. And it is not enough to simply say that by definition, the 'measure of existence' reflects the probability of something being experienced, because then we are back to just making an ad hoc stipulation that mod-squared amplitudes correspond to probabilities.

One possible way to give such a generalisation content would be to appeal to the idea that *consciousness* can come in degrees, and then argue that observers in higher mod-squared amplitude branches are in some sense more conscious. However, those philosophers who believe that consciousness can come in degrees have typically made this argument by considering very different kinds of observers – e.g. perhaps a human may be more conscious than a fish – or the same observer in very different states – e.g. in some sense a fully awake person is more conscious than a drowsy person (Lee, 2023). But degrees of consciousness in the Everettian case could not be understood in either of these ways, since the Everettian scenario often postulates branches containing representations of what appear to be almost qualitatively identical observers having almost qualitatively different experiences, and yet those branches may be assigned very different mod-squared amplitudes. So if the observers in these branches somehow have different degrees of consciousness, this can't be cashed out in the usual way in terms of differences in the qualitative content of their conscious experiences; so the proponents of measures of existence would still owe us an explanation of what exactly the degree of consciousness *does* mean in this context.

Indeed, it is not even clear what kind of fact the 'measure of existence' is supposed to be. It can't be a purely normative fact if it has consequences for the conclusions we are to draw about what the actual world is like, so presumably it must be understood as making some kind of descriptive statement about physical reality. But also, according to Vaidman the measure of existence is defined by the stipulation that Everettian observers should assign credences equal to the measure of existence for the corresponding branch, and so whatever descriptive content it has seems to also have normative implications. Thus we appear to run head-first into Hume's Is-Ought dilemma (Black, 1964) – it is impossible to derive normative claims from purely descriptive statements about physical reality, so if the measure of existence does have these normative implications, it can't be a purely descriptive statement about physical reality either.

A similar problem arises in the context of naturalised epistemology: as Laudan (1990, p. 45) puts it, '*How, given the contrast between the descriptive character of science and the prescriptive character of traditional epistemology, can the naturalist plausibly maintain that scientific claims and philosophical ones are woven of the same cloth?*' The answer for Laudan is to understand rules of epistemology as connecting means and ends: that is, the normativity comes from our epistemic goals, and then naturalised epistemology tells us that we should use the descriptive claims we get from scientific enquiry to explain why certain ways of forming beliefs are good ways to achieve these epistemic goals, given the scientific facts about what the world is like. So one might perhaps read Vaidman as making a similar claim – that is, perhaps he is suggesting that the measure of existence is some thing whose physical nature is such that as a matter of fact, believing something like the ESA is typically a good way of achieving our epistemic goals, given the scientific facts about what the world is like.

But if that is the idea, then ultimately we will have to do as Laudan advises and use scientific enquiry to explain *what it is* about the measure of existence which makes it the case that believing that one is in a branch with high measure of existence is a good way of achieving our epistemic goals. And if our epistemic goals include believing true things, as one would usually imagine, then in the Everettian case this cannot be done in the way that Laudan (1990, p. 46) describes, i.e. invoking '*empirical information about the relative frequencies with which various epistemic means are likely to promote sundry epistemic ends*.' For in the Everettian case, 'relative frequencies' defined using a naïve branch counting approach will not give the desired result – counting branches would suggest that believing that one is in a branch with a high measure of existence is *not* likely to promote the epistemic goal of believing the truth – and any other way of calculating the relative frequencies would be in need of justification, which could presumably only be given by appealing to some other solution to the probability problem.

So unless we are willing to simply change our epistemic goals – a topic I will return to in section 7.1 – it seems difficult to see how we could argue in this context that believing one is in a branch with a high measure of existence is a good way of achieving our epistemic goals.

Now, one way for the Everettian to push back against these criticisms would be to appeal to inference to the best explanation (IBE). This would involve suggesting that the purpose of Everettian quantum mechanics is to explain the empirical evidence, but it does *not* explain the empirical evidence unless we are, in fact, seeing outcomes with relative frequencies roughly corresponding to high branch-weights. Thus, by appeal to IBE, we are entitled to conclude that we must be seeing relative frequencies roughly corresponding to branch weights because this is the best available explanation of the evidence – and then one could fill in the details about seeing these relative frequencies either because branch weights are proportional to a count of Saunders-style branches or because they are a 'measure of existence.'

However, what exactly we supposed to be inferring in this inference to the best explanation? If the idea is that we should infer the truth of Everettian quantum mechanics plus the claim that as a matter of fact we ourselves happen to be in a high mod-squared amplitude branch, then it would seem that the infrastructure of quantum mechanics is not doing any work to explain our empirical experiences at all, since the fact that the branch has high mod-squared amplitude is not the reason why the agent is in it; so this putative explanation just becomes the trivial statement that we have such-and-such experiences because we have such-and-such experiences. Or perhaps we are supposed to be inferring the truth of Everettian quantum mechanics plus the claim that Everettian agents are likely to find themselves in a high mod-squared amplitude branch; but without any further physical explanation for the latter statement, it looks not much different to simply assume that you *are*, in fact, in a high mod-squared amplitude branch. Or perhaps we should infer the truth of Everettian quantum mechanics plus a normative claim to the effect that Everettian agents should believe they are likely to be in a high mod-squared amplitude branch; but then we would not be entitled to use inference to the best explanation any longer, since IBE is usually understood as a tool for inferring descriptive facts about physical reality, and therefore it is not a legitimate way of arriving at normative statements. So it's at best very unclear that there's any sensible inference to be made here, and thus we can't remedy the problems with the branch-counting or measure of existence approaches merely by appealing to IBE.

6.3 Finding the Right Assumptions

Throughout this chapter we have seen that there are a series of quite significant obstacles standing in the way of giving a fully satisfactory account of the physical nature of measurement and the way in which it connects observers to the real structure of quantum mechanics in the context of the Everett interpretation. That said, it doesn't seem completely impossible to chart some course through the Everett interpretation which gives rise to a set of beliefs which avoids at least the most obvious kind of inconsistency. For example, suppose we simply stipulate that Everettian observers can individuate themselves using irreducibly indexical self-reference, and we simply assume that branch weights correspond to a 'measure of existence,' and we take it as a brute fact that this measure is somehow related to what observers see in Everettian branching events, even though no one can explain how or why this is the case. Perhaps one might motivate this by saying that there are just some things about how the universe works which are beyond human comprehension, and the relation between mod-squared amplitudes and what observers experience is one of those things. Then indeed we could argue that an Everettian observer can see herself as a random sample from a set of observers and that she can use observed relative frequencies to do empirical confirmation in the usual way.

But obviously this is a fairly weak position involving a highly ad hoc claim – we could reach basically any conclusion we liked by just insisting that the reasons for its truth are beyond human comprehension! So this route may in some technical sense allow the Everettian to give a not completely inconsistent account of the physical nature of measurement, but it is exactly the kind of ad hoc account which, as argued in section 4.4, is not really adequate to solve the epistemic version of the measurement problem, since it has the consequence that all of our empirical knowledge is founded on one ad hoc assumption which is not technically inconsistent with but which also is not suggested or supported by any of our other beliefs about the content of physical reality.

Thus, since the epistemic version of the measurement problem is about giving a consistent epistemology for quantum mechanics which preserves scientific knowledge (or at least empirical scientific knowledge), I think it is fair to say that, even if the Everett interpretation can in some sense solve the measurement problem thus construed, it doesn't solve it *well*. In addition, as noted in chapter 2, one major reason why we have a measurement problem in the first place is because we do not generally think it is plausible that a fundamental theory of physics should contain ineliminable references to vague, agent-centric notions like 'measurement.' So if the Everett interpretation actually cannot be fully formulated without an assumption like 'you should believe you are in a branch with high mod-squared amplitude,' which makes reference to subjective experience, it arguably has not fully solved the measurement problem – it has just shifted the vague, agent-centric notions away from 'measurement' and into this new assumption. So even for those who are willing to accept the Everettian solution if there is really no other option, there are certainly compelling reasons to keep looking for other approaches which might solve the problem better.

7
No Worse Off

There is one Everettian response to the probability problem which we have not yet discussed: the 'No Worse Off' argument. This argument comes in various forms, but the general idea is that even if the Everett interpretation has some epistemic problems, nonetheless we encounter very similar problems in other, less controversial scientific contexts, and therefore it would be unreasonable to hold the Everett approach to a higher standard by demanding that it should solve its epistemic problems, since similar problems are allowed to go unsolved in other contexts.

Now, the No Worse Off argument does admittedly sound like something of a last resort, and one might be inclined to respond to it by saying that if similar problems are encountered in the context of other kinds of hypotheses, then so much the worse for those hypotheses – we should be trying to solve their epistemic problems too! And, in fact, I do think that a number of the examples I will discuss in this chapter exhibit epistemic difficulties which we have some reason to be concerned about, though these difficulties are arguably not as severe as the Everettian ones. Thus this chapter will seek to clarify the extent to which the kind of epistemic difficulties we have examined in the Everettian case extend beyond the context of the Everett interpretation. Some of the No Worse Off arguments I consider below have actually been made explicitly by Everettians, while others have not yet been offered in print, but are still interesting as a way of getting a sense of what a coherent epistemology of science would look like, and what the Everett interpretation might need to do to reach that standard.

Note that throughout this chapter I will be focusing on the case where the Everettian simply assumes the ESA without further justification, perhaps in combination with the use of terminology like 'measure of existence.' This is because my aim is to establish whether this assumption is relevantly similar to other methodological assumptions we must make in the practice of science, or whether it is distinctively problematic in a way that justifies its rejection. Of course, if the decision-theoretic approach or an approach based on principles for self-locating credences could be made to work, then we would not have to simply assume the ESA, and thus in that case the Everett approach might actually be better off than some of the examples I discuss here – but as described in chapter 6, it is not obvious that these approaches are adequate to address the epistemic version of the measurement problem, so it is useful to consider how things would stand for the Everett approach if the ESA does have to be simply a foundational assumption.

7.1 Objective Chance

One version of the No Worse Off argument, laid out explicitly by Papineau (2010), involves the idea that, in fact, nobody really understands what objective chances are,

Saving Science from Quantum Mechanics. Emily Adlam, Oxford University Press. © Oxford University Press (2025).
DOI: 10.1093/9780197808887.003.0007

even in the ordinary classical case, so the Everettians are not worse off in this regard than any other approach to quantum mechanics which includes objective chances.

This argument calls our attention to one important feature of quantum mechanics which is relevant to its epistemic status: it is a probabilistic theory and, perhaps for the first time in modern scientific history, it is a probabilistic theory where it is commonly (albeit not universally) accepted that the probabilities involved are objective, in the sense that they are not derivable from any underlying deterministic theory. Does this make a difference to the process of empirical confirmation? That is, should we consider indeterminism itself to be a form of epistemic weirdness, or is it simply benign weirdness?

As a first pass, one might argue that it surely cannot matter for the purposes of empirical confirmation whether the probabilities in quantum mechanics are objective or not, for all of the macroscopic predictions are completely identical regardless of whether the relevant probabilities are objective or whether they are derived from some underlying deterministic theory. The idea that quantum mechanics must be inherently probabilistic comes from various no-go theorems purporting to rule out hidden variables, but, in fact, these theorems only rule out hidden variables with certain kinds of properties, such as locality – for example, the de Broglie-Bohm approach (Holland, 1995; Esfeld et al., 2014), which I will discuss further in chapter 11, shows clearly that the same empirical predictions can be reproduced in a deterministic way, though at the cost of sacrificing locality.

However, this framing of the situation only serves to highlight the fact that hidden variable models in the style of the de Broglie-Bohm approach are deterministic only in a somewhat superficial sense: as noted by Landsman (2022, p. 2), their alleged determinism is '*parasitical on some external random sampling mechanism ("oracle") the theory has to invoke in order to state the specific value of the hidden variable (i.e. position) in each experiment.*' That is, these models achieve deterministic dynamics simply by putting all of the randomness into the initial distribution of the hidden variables, and this means that any difficulties with respect to empirical confirmation that we may encounter in the intrinsically probabilistic case actually crop up in just the same way in the deterministic hidden variable case. For example, suppose you are worried about appealing to observed relative frequencies to empirically confirm an indeterministic theory, on the grounds that we can never be sure that the series of outcomes we have seen are 'likely' rather than 'unlikely.' Then you should have exactly the same concern in the de Broglie Bohm setting, for in that case we can never be sure that the outcomes we saw resulted from 'likely' rather than 'unlikely' initial conditions. Thus with regard to the process of empirical confirmation, the issues at hand don't substantially change if we shift the objective chances between the dynamics and the initial conditions – what matters is simply the fact that some kind of objective chances play an essential role in the explanation offered by the theory for the empirical results.

So it is certainly possible that indeterminism, either at the level of dynamics or at the level of a probability distribution for initial conditions, could be a form of epistemic weirdness. To make the case that it is, recall that one central epistemic problem for the Everettian agent is that she must justify the ESA – the assumption that the relative frequencies observed in experiments are, after a large enough number of experiments, reliably very close to the mod-squared amplitudes assigned by the theory – even though she must also acknowledge that it is *possible* that she is in a branch in which the relative frequencies are not close to the mod-squared amplitudes assigned by the theory. And if we are trying to confirm a probabilistic theory, we must similarly justify what I will refer to as the Probabilistic Statistical Assumption (PSA) – 'the relative frequencies observed in experiments are, after a large enough number of experiments, reliably very close to the probabilities assigned by the theory' – even though we must also acknowledge that it is *possible* that the relative frequencies are not close to the theoretical probabilities. So a No Worse Off argument may get its purchase here by claiming that the ESA is no more ad hoc than the PSA, so if we believe that probabilistic theories can be empirically confirmed, we should also accept that Everettian quantum mechanics can be empirically confirmed.

In effect, the difficulty for probabilistic theories here is that the conception of scientific epistemology based on 'closing the circle,' as discussed in chapter 4, suggests that we should be seeking to form beliefs using means of enquiry which are good ways of forming true or approximately true beliefs, according to the way in which we model the physical processes associated with these means of enquiry. But once indeterminism enters the picture we encounter an ambiguity – a good way of forming true beliefs *for whom*? In an indeterministic setting it would not be reasonable to demand that our beliefs about the physical nature of our means of enquiry entail that some approach to forming beliefs is a good way of forming true beliefs for *all* physically possible observers – for example, standard Bayesian updating on probabilistic measurement outcomes cannot possibly lead to true beliefs for all physically possible observers, since it will only do so for people who happen to have seen measurement outcomes which are likely, and no one can know for sure whether or not they have seen measurement outcomes which are likely.

So in the indeterministic context, it makes sense to clarify our approach to scientific epistemology so as to stipulate that the goal is to form beliefs in ways which will be good ways of forming true beliefs if you see the outcomes you are most likely to see, so you end up with beliefs which are *likely* to be true. That is, we ought to be open to a probabilistic generalisation of the notion of coherence discussed in chapter 4, with the goal of arriving at a set of credences which are jointly reasonable and which, in particular, affirm that the methods of enquiry by which we have arrived at them are *likely* to lead to true beliefs. And indeed, a closely related generalisation of the notion of epistemic rationality for the indeterministic context is encoded in the Principal Principle (Lewis, 1980), which says that the (epistemically) rational credence to assign to the occurrence of event C, conditional on the knowledge that the objective chance of C is P(C), is simply P(C). The Principal Principle is often regarded as being simply a *definition* of objective chance, which suggests that the idea that it is epistemically rational to form beliefs which are likely to be true can perhaps be understood as being part of the definition of objective chance.

But if this is an acceptable way of securing epistemic rationality in an indeterministic theory, the Everettian might wish to make a similar move, by arguing that in an Everettian universe we should just change the definition of epistemic rationality to say that it is epistemically rational to form beliefs in ways which will be good ways of forming true beliefs if you are one of the people in a high mod-squared amplitude branch. This could be encoded in an Everettian version of the Principal Principle, which might say that the rational credence to assign to the occurrence of event C, conditional on the knowledge that the mod-squared amplitude associated with C is A(C), is A(C).

However, there are two problems with this approach. The first is that amplitudes already have a definition, since they play a substantive dynamical role in quantum mechanics – for example, they determine the way in which quantum interference takes place. Therefore we cannot simply pronounce that mod-squared amplitudes are by definition probabilities, because that definition may not be compatible with the prior definition of amplitudes in terms of their dynamical role. In particular, Franklin (2023, p. 4) notes that it cannot *always* be the case that amplitudes play the role of probabilities: '*We may not regard fundamental (mod-squared) amplitudes as corresponding to probabilities before decoherence because such amplitudes interfere. If one were, for example, to regard the (mod-squared) amplitudes half way between the double slit and the screen as probabilities in the double slit experiment, one would end up with the wrong predictions. It's only once decoherence has ensured effective non-interference that mod-square amplitudes behave as probabilities.*' That is, mod-squared amplitudes do not always act like probabilities and thus it cannot always be rational to set our credences equal to them: only under certain circumstances – i.e. when decoherence has occurred to a sufficient extent – is it reasonable to set credences this way. Thus the rationality of setting credences this way cannot be simply a matter of definition: there must be some higher standard of epistemic rationality that we can appeal to in order to decide whether or not it is appropriate to set our credences equal to mod-squared amplitudes in any particular case.

The second problem is that the epistemic concerns to be addressed here are not at the level of assigning credences to individual events, but rather at the level of assigning credences to entire theories, and/or to accounts of reality which can be associated with those theories. So not only must the Everettian justify the choice to set their credence equal to $A(C)$ conditional on the truth of a version of quantum mechanics which assigns amplitude $A(C)$ to C, but they must also justify a choice about how they will update their credences over whole theories upon seeing event C – that is, they must justify the choice to perform Bayesian updating or something similar using the credence $A(C)$. And the Everettian Principal Principle on its own does not seem to tell us whether or not this is a good way of forming beliefs – in particular, as we saw in section 6.2.1, Dutch Book arguments don't seem to work in this case, since they can only be made if we already have some agreed-upon way of determining the truth of quantum mechanics as a whole or Everettian quantum mechanics in particular.

The problem the Everettian faces is that the Everettian Principal Principle is a standard of epistemic rationality which is *internal* to quantum mechanics – it does not even make sense unless quantum mechanics or some theoretical framework relevantly similar to quantum mechanics is correct. The Everettian Principal Principle therefore cannot be the standard that tells us how to assign credences over a range of different theories including theories which are not relevantly similar to quantum mechanics, because it cannot be meaningfully applied at all in a more general context. So even if the Everettian does believe that the Everettian Principal Principle gives the correct way of assigning credences to individual events conditional on the truth of Everettian quantum mechanics, nonetheless they must still subscribe to some other higher-level notion of epistemic rationality or some suitable set of epistemic goals which determine how they will decide whether or not Everettian quantum mechanics itself is, in fact, true. But if they *do* subscribe to some higher-level notion of epistemic rationality, then they surely ought not to simply stipulate the Everettian Principal Principle – rather they ought to show how the use of Everettian Principal Principle is a good way of achieving the epistemic goals set out by this higher-level notion of epistemic rationality. Thus it seems that the Everettian cannot simply define the Everettian Principal Principle to be true, because they necessarily already subscribe to some *other* standard of epistemic rationality by which the principle can be judged. Therefore they will need to use some other method, such as possibly one of the approaches discussed in chapter 6, to argue that the Everettian Principal Principle and hence the ESA are good ways of forming beliefs according to their own pre-existing standard of epistemic rationality.

This is in contrast with the original Principal Principle, which can be expressed in a theory-independent way, because the notion of objective chance is not internal to any particular mathematical formalism or scientific theory. Of course, some scientific theories are deterministic and thus they admit only objective chances of 1 and 0, but such predictions can still be meaningfully compared with deterministic or probabilistic predictions made by other theories under consideration, and thus the Principal Principle can be used to define the general framework of epistemic rationality that we use to assign credences across theories as wholes. Thus there need not be any higher-level standard of epistemic rationality that we could appeal to in order to judge whether or not the probabilistic Principal Principle is a good way of forming beliefs, and therefore it *does* seem reasonable to simply stipulate the Principal Principle as part of the definition of epistemic rationality. Thus ultimately it doesn't seem right to say the Everett interpretation is No Worse Off than any theory involving objective chances, since the normative role of objective chances can simply be stipulated, whereas it doesn't seem reasonable to just stipulate the normative role of mod-squared amplitudes, given that there necessarily exists a higher-level standard of epistemic rationality relative to which this way of forming beliefs ought to be judged.

With that said, I do agree there is something somewhat unsatisfying about simply invoking objective chance as part of the definition of epistemic rationality – one might still think that in a theory which invokes objective chances, it would be preferable to replace the term 'objective chance' with some analysis which tells us the physical nature of the chances in this setting, and explains why it is epistemically rational to set credences equal to them. Indeed, this is exactly what is achieved in contexts where the probabilities predicted by theories can be derived from an underlying deterministic theory, but it is yet to be seen if something similar could be achieved in contexts where the probabilities are genuinely objective, although approaches which propose defining objective chances in terms of laws about frequencies seem promising in this regard (Roberts, 2009; Barrett and Chen, 2023). So although I do think theories with objective chances are better off than the Everett interpretation, the No Worse Off argument does nonetheless highlight the fact that from the epistemic point of view it would be valuable to have a clearer account of the epistemology of objective chances, at least in the context of our significant scientific theories.

7.2 Boltzmann Brains

No meaningful scientific theory can be confirmed by just a single observation, so in order to engage in empirical confirmation of any theory at all, we must have access to reliable records of the past. Moreover, in order to be able to interpret a putative record as providing reliable information about the past, we need to make some non-trivial assumptions about the state of the world before the record was created. Loewer (2012, p. 17) explains: '*In order for the final state of a measuring device (the photographic images) to be taken to be a record at time t of the prior condition the order of finishing at time t* we need to assume that the camera was in its 'ready to measure' state (film is blank, camera pointed at the finishing line, etc.) at an even earlier time t**. This is completely general and holds for all records whether or not they were produced by measuring devices.*' So in order to justify the use we make of records in real scientific practice, we must make some assumption about the state of the world at a time before the formation of any records that we might wish to consult. And since we may wish to appeal to effective records of even the very early universe, in practice this amounts to making an assumption about the state of the universe at the very beginning of time, which is usually known as the 'Past Hypothesis' (Albert, 2000).

There is still some debate around the exact form that this assumption ought to take, but one common proposal is that it involves assuming that the initial state of the universe at a microscopic level is such that it corresponds to a macroscopic state which we would characterise as having 'low entropy,' which roughly speaking means that it is highly ordered and uniform (Price, 2004). However, based on our current understanding of statistical mechanics and thermodynamics, the set of possible initial microscopic states associated with a 'low entropy' macroscopic state is very small relative to the set of all theoretically possible initial microscopic states, and therefore if we are assuming a uniform measure over initial states, it is surely in some sense fantastically unlikely that the universe really began in such a state (Price, 2004). This raises some awkward questions for the epistemology of science because your memories can be expected to be accurate records of the past only if something like the Past Hypothesis is true, so this argument seems to suggest that it is fantastically unlikely that your memories are, in fact, accurate records of the past. Indeed, it would seem that a far more likely explanation for your current experiences is that the universe has had roughly constant entropy throughout its history, and therefore the memories of the past that you possess are not accurate – they are nothing more than the result of a random thermodynamic fluctuation producing something which persists for only a brief moment but which is akin to a human brain with all of your memories and present sensations all programmed in. Such a thing is known as a 'Boltzmann brain' (Carroll, 2017).

Now if you were really a Boltzmann brain, then almost everything you think you know about science would be worthless, since it would all be based on memories of a completely fictitious past that cannot be expected to contain much useful information about what the world is really like. So this looks like another example of empirical incoherence, similar to the problem encountered in the Everettian case – we have arrived at and empirically confirmed the laws of thermodynamics and statistical mechanics based on the assumption that our records and memories give us accurate information about the past, but those very laws seem to be saying that it is very likely you are just a Boltzmann brain, which would mean that the records and memories available to you do not give accurate information about the past after all. The bootstrapping structure of scientific enquiry seems to have failed here, for the physical account of measurement that we have arrived at – and here I am using 'measurement' in the most general sense, to include the process of consulting records and memories – does not lead to a satisfactory account of how we could have come to obtain reliable knowledge of the theory by the methods of enquiry that we have actually used.

So in order to make sense of the epistemology of real scientific practice, it seems we must make an assumption like the Past Hypothesis, in order to secure the general reliability of memories and records. However, a common objection to the Past Hypothesis is that it is ad hoc; for there is no reason internal to statistical mechanics why such a thing should be true. We simply have to postulate it in order that we can make use of our empirical evidence to confirm the theory. So a proponent of the Everett interpretation might be tempted to argue, again, that the Everett approach is really No Worse Off than ordinary statistical mechanics – for if we are entitled to simply postulate the Past Hypothesis in the case of statistical mechanics, then surely we are equally entitled to just postulate the ESA in the Everettian case.

However, the Past Hypothesis actually seems less problematic than the ESA, for several reasons. First, the Past Hypothesis is not *only* postulated to avoid the Boltzmann brain problem. We require some version of the Past Hypothesis in order to explain how the laws of thermodynamics emerge out of statistical mechanics because thermodynamics is manifestly temporally asymmetric, whereas there are no temporal asymmetries in statistical mechanics unless we add them in the form of an assumption about the initial state (Albert, 2000). Thus the Past Hypothesis plays an important role in unifying two distinct scientific domains, resulting in an account of reality that combines statistical mechanics and thermodynamics in a natural, coherent way, linked together by the Past Hypothesis. So the Past Hypothesis is much better integrated into the overall account of reality that we end up with than the Everettian assumption ESA. We don't complete our account of physical reality, realise something is missing, and then add the Past Hypothesis as an additional hypothesis to save the epistemology; rather we are led naturally to the Past Hypothesis as part of our process of theory-building and unification, and then it turns out that the Past Hypothesis also happens to play an important role in the story that the resulting theory has to tell about the physical nature of measurement, particularly with regard to 'measurements' that involve consulting records, or memories, or reports made by other observers. So, in fact, the bootstrapping process seems to be working as hoped – our methods of enquiry have led us to a unified account of reality which affirms our original assumptions about the reliability of those methods of enquiry.

Relatedly, another important disanalogy is that the content of the Past Hypothesis is relatively clear – there are still debates over what specific properties we should ascribe to the initial state, but whatever the outcome of those debates, in the end the Past Hypothesis is likely to be a relatively well-defined assumption about the state of a physical system at a certain time. And the straightforward physical nature of the Past Hypothesis is precisely why it can be well integrated into and indeed play a unifying role in the account of physical reality offered by the theory of statistical mechanics plus thermodynamics. By contrast, we saw in the previous chapter that the content of the ESA is quite opaque: it is not just an assumption about physical facts, and yet it's unclear that we can see it as purely normative either, so it cannot straightforwardly be integrated into the account of physical reality offered by the Everett interpretation. Furthermore, because the Past Hypothesis has clear physical content, it is arguably open to us to elevate it to the status of a law, as suggested for example by Chen (2022) and Albert (2000), which plausibly makes it less ad hoc; whereas we certainly cannot do this in the case of the ESA, since we know for sure that the ESA is not true for all observers in an Everettian universe, and thus it cannot be a universal law.

Finally, note that the Past Hypothesis is always postulated in the context of the time-evolution paradigm, in which the universe is conceptualised as something like a computer which starts from an initial state and evolves forwards in time (Wharton, 2015; Adlam, 2022). And when we work within the time-evolution paradigm, we are in effect committed to the idea that at the most fundamental level, physical explanations can only proceed by showing how some condition evolves from some earlier condition. Thus a time-evolution approach can't be expected to offer any physical explanation of the initial state of the universe, and therefore it is in effect built into the ontology of a time-evolution theory that we will either have to make some arbitrary choice of initial state, or add a law singling out an initial state. That being so, choosing the Past Hypothesis as our stipulation about the initial state perhaps seems less ad hoc – the theory itself gives us no reason to expect one initial state rather than another, nor any sensible way to assign a measure over initial states, but there must nonetheless be an initial state, so it seems quite reasonable to just stipulate that the initial state is simply whatever we need to it to be to arrive at a coherent account of reality.

By contrast, the basic ontology of the branching-world picture does not seem to entail that there has to be any objective fact of the matter about what observers are more likely to see after branching events, and indeed the difficulties discussed in chapter 6 stem from the fact that it is not easy to see how there *could* be such a fact. So the assumption ESA is not just making some particular choice for an arbitrary specification that would have to be made in any case; it is adding something completely new which apparently can't even be stated in precise physical terms and which is completely uncalled-for other than as a way to solve a serious epistemic problem. That being so, adding the assumption ESA seems significantly more problematic than the Past Hypothesis, since the basic structure of the theory gave us no reason to make any such assumption.

Now, none of these points completely negates the fact that the Past Hypothesis has a somewhat ad hoc nature. And indeed, it does seem fair to say that applying the kind of epistemic analysis that I have advocated in this book to statistical mechanics and thermodynamics yields the conclusion that we should not be completely satisfied with the state of play – perhaps there could be some other way of unifying these domains which is less ad hoc and thus more epistemically robust. For example, Adlam (2023) suggests that this might be possible if we consider moving away from the time evolution paradigm, in which case temporal asymmetries can potentially be attributed to something other than the initial state of the universe. But even so, I think it is clear that overall the epistemology is in better shape in the unified domain of statistical mechanics and thermodynamics compared to the Everett picture, and thus in this instance the Everett approach still seems to be Worse Off.

7.3 Other Multiverse Theories

A different kind of No Worse Off argument involves comparing the Everett interpretation to some other kind of 'many-worlds' hypothesis. A number of 'many-world' proposals exist within physics, often with the aim of addressing fine-tuning problems in cosmology. For example, some approaches postulate an ensemble of spatiotemporally unconnected universes, with laws and/or constants varying across the ensemble, while others postulate '*a single, connected space-time manifold where certain constants – e.g. Newton's constant – vary over large temporal and/or spatial length scales*' (Friederich, 2021, p. 4), which would yield something like a multiverse, with different 'universes' occupying different regions of the same spacetime.

However, in this section I will focus on a different kind of proposal, known as the 'landscape multiverse.' This particular multiverse seems to lead to the strongest version of the No Worse Off argument – first because it has a more concrete theoretical motivation than other multiverse proposals, and second because it involves a branching process, which makes it more closely analogous to the Everett picture than other multiverse proposals in which the worlds are distinct at all times.

The landscape multiverse (Smeenk, 2017; Carroll, 2018; Read and Bihan, 2021) has its origin in the idea that in order to explain various cosmic coincidences, we must postulate a process known as 'cosmic inflation' in the early universe. It turns out that certain models of cosmic inflation entail that inflation creates a multiverse: '*According to many inflation models, notably ones in which the potential of the inflation field depends quadratically on the field strength, island universe formation is globally "eternal". When it comes to an end, it does so only locally, resulting in the formation of a causally isolated space-time region that effectively behaves as an "island universe". This process of continuing island universe formation never stops. As a result of it, a vast (and according to most models, infinite) "multiverse" of island universes is continually being produced*' (Friederich, 2021, p. 8). Now, as Friederich (2021) discusses, there are still a number of questions about the extent to which inflationary cosmology succeeds in its predictive and explanatory goals, so the landscape multiverse cannot yet be regarded as established scientific fact. But for the sake of argument let us consider the case which would be most favourable for the Everett picture, i.e. let us try to decide what would follow if we did come to regard the landscape multiverse as established scientific fact. Should we then also be willing to accept the existence of an Everettian multiverse?

First, it is important to understand what kind of predictions we can extract from the landscape multiverse picture and how those predictions are obtained. Obviously this will vary across models, but some of the most well-understood scenarios combine inflation with string theory, in which case the models suggest that we will get a different string theoretic vacuum within each of the universes, and each of these different vacua will give rise to different high-level physics, i.e. both different laws and different values for the fundamental constants (Carroll, 2018; Read and Bihan, 2021). So we end up with a large ensemble of universes, each exhibiting very different high-level physical laws.

However, although the physical laws vary across universes in this picture, typically they can still be expected to be constant within each individual universe. Thus in the context of the landscape multiverse, once we have established the laws of our universe, we can expect that these laws will continue to apply in the future, with exactly the same justification as in the non-multiverse context (for now I will put aside worries surrounding the rationality of induction within a single universe, although I will shortly return to this point in section 7.4). By contrast, in the Everett picture, even if you could magically come to know for sure the actual laws governing the Everettian multiverse, it's unclear that you would be able to make any useful predictions with these laws, for in order to make such predictions you would still need some justification for assigning credences over future events in proportion to mod-squared amplitudes. That is, in the landscape multiverse case the laws most relevant for most ordinary predictions are associated with an individual universe, and thus once you have established the laws holding in your universe, you can be confident that they will also apply to your personal future, whereas in the Everettian case, there are no laws except at the level of the whole multiverse, so coming to know the laws without a solution to the probability problem tells you almost nothing about your personal future.

Does this distinction mean that agents in an Everettian universe are, indeed, Worse Off with regard to epistemology than agents in a landscape multiverse? Not necessarily, because the distinction will be relevant only if there *is*, in fact, some way for agents in a landscape multiverse picture to reliably come to know what the laws of their universe are. Now, if the laws in a typical universe belonging to the landscape multiverse were deterministic at the level of observables, then it would be quite straightforward for agents in the multiverse to come to know the laws of their individual universe: they could simply observe some exceptionless regularities around them and then infer the existence of deterministic laws mandating those regularities. But what if the laws are indeterministic at the level of observables, as they are likely to be? In that case, we face the following worry. If the number of universes is very large, or indeed infinite, then it is plausible that for any set of experiments we might perform, there will be many other universes in which the same set of experiments is performed, and in some of those universes the relative frequencies exhibited across the relevant experiments will diverge considerably from the probabilities predicted by the underlying laws of the universe, so any inhabitants of those universes will be systematically deceived about what the laws of their universe are. Knowing this, what justification can we have for believing that we are in a universe which exhibits relative frequencies close to the theoretically expected ones? The worry is that if we can't give such a justification, then in the landscape multiverse context we will fall short of being able to give an account of the physical process of measurement which affirms the reliability of our means of enquiry, for essentially similar reasons as in the Everettian case.

A natural way to address this problem is to regard our universe as a random sample from the set of all universes, or perhaps simply from the set of all universes which are friendly to life, and then predict that a certain outcome or series of outcomes will be seen with probability equal to the fraction of all relevant universes which exhibits this outcome. The idea that the predictions of a multiverse theory are simply the observations that would be made by a typical observer has been referred to as the 'principle of mediocrity' (Vilenkin, 1995), and it can be regarded as a consequence of Bostrom's 'self-sampling assumption' (Bostrom, 2002). Principles like this are most frequently used to make predictions for whole-universe features – particularly derivations of the cosmological constant, such as the calculation of Bousso and Freivogel (2007) – and as such they play a major role in arguments to the effect that a multiverse can explain the apparent fine-tuning of certain constants (Friederich, 2021). However, presumably something similar could be done to predict the outcomes of probabilistic events within a single multiverse. In particular, if we assume that all possible relative frequencies for the relevant outcomes are equally compatible with life, then it seems reasonable to expect that a properly-constructed multiverse would have the feature that for a sufficiently long sequence of measurements, a large majority of the (life-friendly) universes in which these measurements occur will exhibit relative frequencies very close to the theoretically expected ones. Indeed, for a properly-constructed multiverse we would probably expect to get some analogue of the law of large numbers, so the fraction of (life-friendly) universes in which the measurements occur which exhibit relative frequencies very close to the theoretically expected ones should tend to 1 as the length of the sequence becomes sufficiently long. And therefore, if we are satisfied with the approach of treating our universe as a random sample from the set of multiverses, we should be able to justify the use of relative frequencies to confirm scientific theories in a multiverse context.

This highlights two further senses in which the Everett approach appears to be Worse Off than the landscape multiverse, relating to the incoherence problem and the quantitative problems, respectively. First, although the landscape multiverse does involve branching, it also locates all of the branching firmly in the distant past. That is, although the multiverse models may postulate eternal branching and thus eternal universe creation, they also tell us that our own universe branched off at a time very early in its history, and it will never branch again. Moreover, although there may be cases in which two or more universes are internally qualitatively identical up until a certain point in their history, they still have external features which unambiguously render them physically distinct – for example, once we have specified a frame of reference we can put the universes in order of creation, so we can index them by the order in which they came into being. So in the landscape multiverse, although it may be true that every possible outcome of a measurement occurs and is observed by a copy of the observer somewhere in the multiverse, any branching involved is purely epistemic: for each of the copies of the observer there is, even prior to the measurement, a well-defined physical fact which can be described from a third-person point of view about which universe they are in and which outcome they will ultimately obtain. Thus it is comparatively straightforward to see how it could be reasonable to regard oneself as a random sample from the set of observers in this context, without appealing to some kind of irreducibly indexical information only expressible from a first-person perspective. By contrast, we saw in chapter 6 that, since Everettian worlds are not physically distinct in this unambiguous way, there are significant conceptual difficulties standing in the way of regarding oneself as a random sample from one's set of post-branching successors.

Second, in a *finite* landscape multiverse, once we have justified the claim that we can treat ourselves as a random sample, there is also a natural way to arrive at probabilities for this random sampling, since the universes become completely physically distinct during the landscape multiverse branching process, and therefore we can simply count universes in order to arrive at a fraction of universes exhibiting a certain feature. Moreover, it seems plausible that fractions of universes will usually give the probabilities we would expect, so at least in the finite case it seems relatively simple to solve the landscape multiverse version of the quantitative problem. By contrast, in the Everett picture we cannot simply count branches, since branches are not completely physically distinct and thus there will not be a well-defined count of them; and we also noted in chapter 6 that the most obvious way of counting branches doesn't give the right probabilities in any case. So solving the quantitative problem definitely seems more complex in the Everett case than in the finite landscape multiverse case.

That said, this counting approach can only be used in the case where we have a finite number of universes – and unfortunately, many current models suggest that the inflationary period which produces the universes is very likely to go on forever (Guth, 2007), which means that the total number of universes will be infinite. In this case we can no longer simply calculate a proportion of universes with some feature because '*ratios between infinite numbers (of the same cardinality) are not unambiguously defined, at least not without specifying a regularization procedure: a prescription for comparing finite numbers and only then taking the limit to infinity in a controlled way*' (Friederich, 2021, p. 119). Thus in the infinite case, we will have to adopt some choice of regularisation procedure in order to arrive at a measure over universes which we can use to calculate ratios.

There exist many proposals for such measures – for example, the proper time measure, defined by starting at some fixed initial spacelike surface and then constructing a measure using the spacetime volume traversed by time-like geodesics leaving the surface towards the future, then taking the limit as the spacelike surface moves towards infinite proper time (Linde, 2007). However, as discussed by Friederich (2021), all existing proposals for cosmic measures lead to some counterintuitive consequences – for example, the proper time measure has the consequence that we are vastly more likely to be in extremely young universes, meaning that we are much more likely to be Boltzmann brains in very young universes than to be observers produced by a long history of evolution, as we typically take ourselves to be (Bousso and Freivogel, 2007; Linde, 2007). And furthermore, even if we *could* find a measure which satisfied all of our intuitive expectations, would that really solve the problem? After all, if there are many possible measures, it's unclear what it would actually mean to say that any one of them is objectively correct – if the measure of choice is not counting multiverses, what *is* it doing, and why should we expect it to have anything to do with probability? One might worry that such a thing is, in fact, no better than a 'measure of existence,' since it is simply postulated ad hoc to arrive at the desired results. And indeed, a number of philosophers do see this issue as reason enough to reject the infinite multiverse: for example, Friederich (2021, p. 121) writes, '*My own view is that this stance is, in the end, the right one to take: the measure problem does seem to make scenarios that involve eternal inflation untestable.*' So perhaps there is a case to be made that the Everettian multiverse is No Worse Off or at least Not Much Worse Off than the infinite cosmic multiverse in this regard, but if this is so then the comparison does not seem to do the Everett approach any favours.

But also, this leads to a further concern about the finite multiverse case. For can it really be true that a very large but finite multiverse is perfectly acceptable from the epistemic point of view, but simply making the multiverse infinite makes the whole thing impossible? After all, even in the finite case we are essentially adopting a choice of 'measure' – in the finite case the measure is a probability distribution over universes, and by adopting the counting approach we are choosing a uniform probability distribution over universes. So in a sense, the only real difference between the finite and the infinite case is that in the finite case there is an intuitively natural way of assigning probabilities, whereas in the infinite case there is no similarly natural way of singling out a unique measure. But does the intuitiveness of the measure mean that it is in some sense objectively correct, or does it just make us feel more comfortable with what is really an entirely alien kind of probabilistic reasoning?

Friederich (2021) argues that in the infinite case, we face the following dilemma: '*Even if some specific measure were established as physically privileged in the context of external inflation, this would not by itself show that this measure should guide our assignment of probabilities.*' But exactly the same point can be made in the finite case as well – just because the probability distribution assigning equal probabilities to all observers seems more natural than other distributions, that does not necessarily entail that it is thereby some kind of objective physical fact about probabilities of observation. This line of argument may give rise to the worry that even in finite cases there may simply be no objectively correct way of assigning probabilities with regard to this kind of self-locating uncertainty – which of course is exactly the worry that I raised in section 6.2.2 in discussing the Sebens-Carroll and Vaidman-McQueen Carroll;? approaches to the Everettian probability problem. And thus, just as there is no clear justification for a probabilistic interpretation of the amplitude in the Everett case, one may worry that there is ultimately no satisfactory justification for a probabilistic interpretation of fractions of the multiverse in the landscape multiverse case, even when the number of universes is finite.

However, there is still an important disanalogy between the landscape multiverse case and the Everettian case: in the landscape multiverse, appealing to fractions or measures over universes is not the only possible way of understanding the epistemology of measurement. For in the landscape multiverse case, instead of appealing to the structure of the whole multiverse to predict the outcome of some measurement, we can simply select one particular universe – for example, our own universe, identified by some kind of indexical individuation, which as noted earlier can be achieved in this kind of case without any particularly exotic kind of indexicality – and then simply make predictions for that universe on the basis of what we might call the 'Universe Statistical Assumption,' or USA: in any given universe, the relative frequencies that the inhabitants of that universe observe in experiments are, after a large enough number of experiments, reliably very close to the probabilities assigned by the laws of that universe. Obviously, the USA will not in general be correct for all observers in a given universe, but it follows from our usual understanding of the meaning of the word 'probability' that it is very likely to be correct for any individual observer, and thus it can be justified in a similar way to the PSA, as discussed in section 7.1. So if we decide that the probabilistic interpretation of fractions of universes won't work, we can simply refrain from making predictions based on the principle of mediocrity or the self-sampling assumption, and instead just use the USA directly to learn about the laws of our own universe, which will then allow us to make predictions about what we will observe in the future in our own universe.

This means that in the landscape multiverse case, we can distinguish a large class of 'ordinary' observations which can still be sensibly predicted and used for empirical confirmation even if there is, in fact, no viable probabilistic interpretation of fractions or measures of the multiverse. We only really need a probabilistic interpretation of fractions or measures in the special case where we wish to make predictions about whole-universe properties which are not assigned probabilities by the laws internal to the individual universes, such as the values of fundamental constants. So doubts around the objectivity of credences in self-locating cases may undermine attempts to use the multiverse to solve fine-tuning problems, as in attempted derivations of the cosmological constant, but these doubts do not undermine most ordinary scientific observations, and hence under these circumstances we would still be able to affirm the reliability of our means of enquiry in most ordinary contexts. On the other hand, no such distinction can be made in the Everettian case because we saw in section 3.4 that in an Everettian universe nearly all uncertain events can potentially involve branching, so it would seem that very few predictions in an Everettian universe will make sense without a viable probabilistic interpretation of the mod-squared amplitudes. Therefore even if it turns out to be impossible to arrive at objectively correct probabilities based on fractions of universes or measures over universes in the landscape multiverse case, nonetheless the Everett approach does still seem to be Worse Off than the landscape multiverse approach.

With that said, there is another kind of problem that may arise in the landscape multiverse approach if we accept that there is no objectively correct probabilistic interpretation of fractions or measures of multiverses. For we have already noted that certain choices of measure over the cosmological multiverse seem to suggest that we are very likely to be Boltzmann brains in a very young universe, rather than evolved observers in a very old universe. As long as there is a probabilistic interpretation of fractions or measures of multiverses, we can hold out hope that the numbers can be made to work out in such a way that we are relatively unlikely to be Boltzmann brains in a very young universe; but if there is no probabilistic interpretation of fractions or measures of multiverses, then we can have no principled reason for assuming that we are an evolved observer in an old universe rather than a Boltzmann brain, and therefore we can have no principled reasons for assuming that our memories reflect real features of the past. And, of course, we need to assume our memories reflect real features of the past before we can assume that the outcomes we remember are the ones assigned high probability by the laws of our universe, so the possibility that we could be Boltzmann brains would seem to block the straightforward path I have just sketched out from observations to probabilistic laws.

As we saw in section 7.2, this is indeed a serious epistemic problem. But on the other hand, if again we take as our starting point our own indexically individuated universe rather than the whole multiverse, then all we need to do to avoid the Boltzmann brain problem is postulate some analogue of the Past Hypothesis for this specific universe – perhaps not at the start of time, but at the moment of branching, or early in the post-branching history. This would then change the balance of probabilities in such a way as to make it unlikely that you are a Boltzmann brain, in just the same way as in the single-universe case. So it may be that if one is willing to simply postulate the Past Hypothesis in the single-world case for the reasons sketched in section 7.2, one can just do the same in the multiverse case, and hence avoid the Boltzmann brain problem, in which case it again seems fair to say that the Everettian multiverse really is Worse Off than the landscape multiverse.

7.3.1 Diverging Everettian Worlds

We have just noted that one key feature of the landscape multiverse which makes its epistemology more tractable than in the Everettian multiverse is the fact that branching is finished during the early universe, so by the time observers come along they are each unambiguously located in a physically distinct universe where every measurement has exactly one outcome. On this basis, the Everettian sympathiser might be tempted to adopt a different approach to the Everettian probability problem. That is, we could adopt a *Diverging* rather than *Overlapping* Everettian metaphysics, in which we say that Everettian worlds are actually completely distinct at all moments of time (Wilson, 2012). In the Diverging picture, rather than saying that worlds genuinely split or branch at various points in the course of history, we simply have a set of worlds which are distinct at all times, but which are qualitatively identical up until a certain moment in time, and thereafter no longer identical. Adopting a Diverging picture would then allow us to say that, as in the landscape multiverse case, Everettian observers are at all times located in physically distinct universes where every measurement has exactly one outcome.

Various different Diverging Everettian proposals have been made, falling into two main categories: proposals in which the distinction between Overlapping and Diverging is understood as a semantic claim about the identity conditions for agents, and proposals in which the distinction is understood as an assertion about the real physical existence of distinct universes. Saunders and Wallace (2008) offer a proposal of the first kind, taking as their starting pointLewis' argument that in cases of personal branching, such as surgeries dividing a brain in two, we ought to say that there were always two persons present all along – that is, we should take 'person' to refer to a '*unique cradle-to-grave continuant*' rather than an instantaneous time-slice. Building on this, Saunders and Wallace (2008) propose that in order to make sense of ordinary talk about personal identity and uncertainty in an Everettian universe, we ought to adopt a semantics which identifies observers in terms of cradle-to-grave continuants, and which assigns thoughts and utterances to whole continuants rather than time-slices. As Saunders and Wallace argue, it's plausible that this semantic move can solve some aspects of the Everettian probability problem, such as the difficulty of giving meaning to talk about 'uncertainty' in an Everettian world: for according to their proposed semantics, it comes out as true to say that all post-measurement observers are present even before the measurement, and therefore even before the measurement an observer can meaningfully say they are uncertain about which one of these observers they are.

However, if the Overlapping rather than Diverging nature of the worlds merely pertains to semantics, it's unclear that it can help with the *epistemic* aspects of the problem, for intuitively it seems that our beliefs about whether or not a theory is empirically confirmed shouldn't depend in any crucial way on our choice of semantics. Now, admittedly it is quite non-trivial to make this intuitive notion precise – the logical positivists might have wished to apply the label 'just semantics' to all sentences which have no verification conditions, but as we saw in chapter 4, it has subsequently been recognised that sentences in isolation can't be mapped directly to verification conditions. Thus the failure of the logical positivist criterion of significance may also have the consequence that there is no straightforward way to draw a line between assertions which are 'just semantics' and assertions which have factual content. However, if the argument given for replacing one kind of proposition with another is purely in terms of perspicuity and convenience, it seems reasonable to conclude that in this particular context, these propositions are intended to be construed in such a way that the difference between them is purely semantic, for after all if there were any factual difference between them, then an appeal to linguistic convenience would not be an adequate argument – some considerations of factual accuracy would have to be adduced. And indeed, Saunders and Wallace (2008) are explicit here about their intentions: '*We are not proposing a metaphysical theory, but a manual of translation, in something like Quine's sense: the standard of correctness is set by fluency of discourse and the principle of charity, not by metaphysical principles.*' Thus, if the argument of Saunders and Wallace is the right kind of argument for the claim that we should speak in terms of diverging rather than overlapping Everettian continuants, the content of that claim *must* be 'purely semantic,' and therefore the choice between Overlapping and Diverging worlds presumably can't change the degree of confirmation of the theory.

On the other hand, proposals of the second kind treat the claim that Everettian worlds diverge rather than overlap as a substantive claim about physics or metaphysics, rather than a semantic matter; and thus it is reasonable to think that in this kind of proposal the degree of confirmation attached to the Overlapping and Diverging pictures could be different. However, there are several obstacles standing in the way of using an approach of this kind to solve the Everettian epistemic problem.

First, even if the worlds *are* distinct, in order to provide an affirmation of the reliability of our means of enquiry in this context we still have to find some way of justifying the claim that the relative frequencies observed within some specific world provide meaningful information about laws or the future. In the landscape multiverse case, each universe has its own laws which ex hypothesi are the same throughout the whole history of the universe, so we are entitled to suppose that our observations of the past are likely to give reliable information about the laws of that universe and hence also about the future of that universe. But in the Everettian case, even if the worlds are Diverging rather than Overlapping, nonetheless laws are defined at the level of the whole multiverse at once, not individually within the worlds; so if we don't already have a probabilistic interpretation for the mod-squared amplitudes, then we just end up with a set of worlds containing almost every possible course of history, and therefore the fact that a world has exhibited certain relative frequencies up to a point provides no reason at all to think that those relative frequencies correctly reflect the underlying laws, or that the relative frequencies in that world will look similar in the future. Thus although we can plausibly make many kinds of predictions in the landscape multiverse case without a probabilistic interpretation for the fractions/measures, we can't similarly make predictions in the Everett case without a probabilistic interpretation for the mod-squared amplitudes, even if we do have worlds Diverging rather than Overlapping. Moving to Diverging worlds therefore doesn't fully resolve the epistemology of the Everett picture.

Second, there is a technical challenge regarding how this set of distinct diverging worlds is to be defined. Some early Everettian proposals can arguably be understood as proposing a set of diverging worlds, including versions proposed by Deutsch (1985), and Barbour (1994), but these approaches necessarily add something to the ontology in order to define a set of exact distinct worlds, rather than using decoherence to define worlds approximately as in most modern versions of the Everett approach. And we have already noted that one key motivation for the Everettian picture is the idea that nothing needs to be added to the mathematical formalism of unitary quantum mechanics, so these approaches seem to undercut one of the central motivations for the theory. In particular, we will see in chapter 11 that putative solutions to the measurement problem which add something significant to the formalism often run into trouble with quantum field theory, and thus it seems natural to worry that Diverging approaches with additional ontology may ultimately have similar problems.

Alternatively, Wilson (2012) offers a diverging picture defined within the modern decoherence-only version of Everett, using the consistent histories formalism. Wilson's approach is thus able to maintain that nothing needs to be added to the mathematical formalism of unitary quantum mechanics – but on the other hand, as we saw in section 3.3, there will always be many consistent sets which contain maverick histories in which records of the past are completely inaccurate, and since the formalism doesn't assign probabilities or mod-squared amplitudes for the sets themselves, there would seem to be no way of arguing that these histories are 'unlikely,' even if we *do* have some justification for interpreting mod-squared amplitudes as probabilities; so this approach would potentially end up just replacing one set of epistemic difficulties with another. Thus it's unclear that there's any satisfactory way of defining a unique set of Diverging histories with the properties we require for an adequate solution to the epistemic version of the measurement problem, using only the standard unitary formalism of quantum mechanics.

7.4 Induction

To finish, let us try to imagine an Everettian No Worse Off argument based on the problem of induction (Hume, 2000; Lange, 2011; Henderson, 2022). This problem arises from the fact that in order to carry out the process of scientific enquiry, we must assume that the future will resemble the past in some way, since otherwise we could not possibly learn anything about the future by studying past measurement results. I will refer to this as the 'induction assumption.' But the induction assumption clearly cannot be verified empirically, for even if it has been correct to assume the future will resemble the past up until the present moment, this doesn't entail that the future will *continue* to resemble the past. Additionally, with a few notable exceptions (Kant, 1890), most philosophers don't think it is knowable a priori, logically necessary or analytically true that the future must resemble the past. So the induction assumption is in some sense unjustified and ad hoc. And thus an Everettian might be tempted to argue that the ESA is not any worse than the induction assumption – and indeed in some ways it is very similar to the induction assumption, since it is ultimately unfounded but nonetheless a necessary presupposition for the possibility of doing science in an Everettian context.

As a first response, note that the induction assumption is not actually as crucial as the ESA, for even if one does not accept the induction assumption, one can still maintain that we have some empirical scientific knowledge in the form of knowledge about the *past*. After all, even if the regularities we have identified in our observations do not extend any further than the patch of the universe we have so far observed, we have still learned a highly non-trivial fact about reality – at least in one particular region of spacetime, it exhibits very strict regularities which can be described mathematically. Of course many of us will feel an almost overwhelming urge to argue that the success of this mathematical description is good reason to suppose that the regularities can be extrapolated beyond our observations, but even without that extrapolation it is clear that we have obtained some substantial knowledge about reality by systematising it in this way.

On the other hand, in the Everett interpretation we do not have recourse to this way of narrowing scientific knowledge, because if we are not entitled to assume that we are in a high mod-squared amplitude branch, then we cannot say that the regularities we remember are in any sense representative of some kind of substantive regularity, even one that holds just in some region – *all* of the possible regularities will appear in some branch, and without a solution to the probability problem we have no reason to think there is anything special about our own branch. So if the Everettian probability problem can't be solved, we apparently have to say that in an Everettian context, empirical scientific knowledge is really knowledge about *nothing*: any systematisation we manage to achieve encodes nothing other than random features of how things happen to be in the branch we happen to be on. Thus the failure of the ESA in an Everettian context would undermine scientific knowledge in a much more serious way than the failure of the induction assumption, and therefore we should be commensurately more wary about adopting a picture in which scientific knowledge depends on the ESA.

Second, note that the ESA does not *replace* the induction assumption – any Everettian who wants to include knowledge about the future in their accounting of scientific knowledge will have to make the induction assumption *in addition* to any other assumptions they need to make sense of Everettian probabilities. And it doesn't follow from the fact that we have to accept one unjustified assumption in our accounting of scientific knowledge that we should be happy to accept arbitrarily many more such assumptions. I noted in section 4.4 that ad hoc assumptions decrease the coherence of the belief set, and therefore from the point of view of overall coherence we surely ought to be seeking to minimise our reliance on unjustified assumptions. So the Everettian is indeed strictly Worse Off than the non-Everettian in this regard, since Everettian requires strictly more ad hoc assumptions.

Third, although the induction assumption may be in some sense unfounded, it is at least clearly defined and comprehensible: we are simply assuming that the world (the empirically accessible, well-understood classical world of our experience) exhibits similar regularities in the past and future. Maybe that is true, maybe it is false, but either way there is nothing mysterious or ill-defined about it. Whereas we noted in section 7.2 that the assumption ESA is quite opaque and seems to float in an ill-defined way between being descriptive and being normative. So even if the induction assumption and the ESA are equally unfounded, one might think that from the epistemic point of view it is preferable to use assumptions which are clearly stated and comprehensible.

Fourth, although ultimately there is no hope of proving that the induction assumption is correct, that doesn't mean it is not subject to any kind of empirical test. As argued by Adlam (2022), although we can't prove empirically that regularities will always persist across time, it's nonetheless the case that once we have assumed that *some* regularities persist across time, we can use empirical enquiries to help set the parameters of that assumption. For in the practice of science we don't just decide at random which regularities will persist across time: we go out and do experiments to find out which ones generally persist, we test different regimes to see if regularities break down under certain circumstances, we look for explanations of regularities in terms of deeper underlying regularities, and so on. So the assumption itself may be unfounded, but it can then be refined by the methods of science to arrive at a version of the assumption which is well-supported by the empirical evidence and which leads to a better degree of coherence. Indeed, this looks like a good example of Chang's progressive coherentism in action: we accept the induction assumption without ultimate justification, but then we use it to pursue enquiries which allow us to refine and correct the initial assumption, leading to a mutually coherent and self-supporting set of beliefs which affirm the assumptions we started out with.

By contrast, the Everettian assumption does not seem to be subject to any empirical test at all – we never have any direct access to mod-squared amplitudes, so we cannot have any independent reason to think that the assumption ESA is ever true under *any* circumstances, and nor can we investigate in more detail the circumstances in which it fails to be true, or the frequency with which it fails. One might argue that we have *indirectly* arrived at and refined the ESA by obtaining evidence for quantum mechanics itself, but, in fact, we arrived at and empirically confirmed quantum mechanics based on a single-world picture in which measurements were understood to provide ordinary non-self-locating information, and as discussed in chapter 6, this confirmation does not by default transfer across to Everettian quantum mechanics. Thus if we decline to accept the illegitimate transfer of empirical confirmation based on one set of background beliefs to a completely different set of background beliefs, we can't give an account of the role of the ESA in progressive coherentist terms because we have no way of pursuing any lines of enquiry which allow us to refine or correct our initial assumption. Thus the Everettian assumption ultimately seems more unscientific than the induction assumption, since the progressive development of science cannot refine or correct it.

7.4.1 Induction II

A different kind of No Worse Off argument here would involve sidestepping the ESA completely and simply focusing directly on beliefs about the relationship between past and future. For example, one might suggest that Everettian agents should simply adopt a policy of believing that relative frequencies in the future will generally resemble whatever relative frequencies they have seen in the past. One might then argue that this strategy is not any less epistemically rational than our usual approach to forming beliefs about the future based on the past. For after all, the problem of induction makes it clear that we can never have any conclusive empirical proof the future will resemble the past: we have just collectively decided as a scientific community to assume that it will. And if we are entitled to collectively make that decision in the ordinary classical case, what is stopping us from just assuming the same thing in the Everettian case?

However, this route to making sense of Everettian probabilities would only allow the Everettian agents to form beliefs about their own in-branch future; it gives no route to say anything whatsoever about any other branches. And thus what Everettian observers adopting this dictum would arrive at via their empirical investigation could not possibly be the Everett interpretation, since the Everett interpretation makes many claims about the overall structure of the universal wavefunction and hence also about goings-on in branches other than one's own. An agent following this prescription would, if they happened to be in a branch where the relative frequencies match the predictions of quantum mechanics, presumably end up with something like an operationalised version of textbook quantum mechanics, but they would not arrive at Everettian quantum mechanics. So even if we are willing to ignore the situation of all of the agents in branches with the 'wrong' relative frequencies, this approach is still self-undermining because it entails that rational belief updating in an Everettian world cannot possibly lead us to believe the Everett interpretation!

This also raises some questions about how we should think about induction in the ordinary non-branching case. One possibility is that the inferences we make about the future based on records of the past need not be grounded on any particular beliefs: that is, regardless of what you believe about the nature of measurement and its connection to reality, you should simply take it as a methodological principle that your personal future will be similar in some ways to the past as you remember it. If this is the view one takes of induction in the ordinary case, indeed in some sense it wouldn't be unreasonable to go on applying induction in the context of the Everett approach or indeed in any other context whatsoever. But as we have just seen, the resulting beliefs will necessarily be purely operational in character, and thus will not lead to any substantive account of physical reality which could reassure us as to the reliability of our means of enquiry.

The other possibility is to say that the inferences we make about the future are, in fact, grounded on some specific beliefs we hold about reality. At the most general level, a Humean, who believes that laws of nature are nothing more than the best possible systematisations of all actual happenings (Lewis, 1994, 2013), might say that these inferences are grounded on the belief that our world exhibits certain universal regularities, so we can form beliefs about what these regularities might be based on memories and records of past observations and then use what we have learned about the universal regularities to make inferences about the future. Or a non-Humean might say that these inferences are grounded on the belief that our world is governed by laws which enforce the persistence of regularities – for example, Armstrong (1983) and BonJour (1998) offer a response to the problem of induction along these lines. And at a more granular level, one might think that inferences about the future are partly grounded on measurement models (Tal, 2012) which allow us to refine our understanding of what it is that a measurement is actually measuring, and thus provide grounds for distinguishing between regularities which are generalisable and regularities which are not, allowing us to extrapolate into the future in an informed way.

Note that none of these kinds of grounding amount to *proofs* that the future will resemble the past, but they do entail that we are not just blindly assuming that the future will resemble the past for no reason at all – we do have reasons, albeit defeasible ones, for this belief. And in particular, this grounding of the induction assumption in some kind of physical facts is what makes it susceptible to empirical enquiries – if the induction assumption were not linked in any way to any beliefs about physical reality, we could not hope to refine the parameters of that assumption through the methods of science. So from the point of view of the bootstrapping account of scientific epistemology as set out in chapter 4, this seems like the right way to think about the induction assumption.

Additionally, this approach also seems consistent with our everyday understanding of the relationship between past and future. After all, in ordinary life we do not unilaterally accept that memories and records can always be used to make inferences about the future, regardless of what we know about their reliability – if you read some experimental results reported by a colleague, and you happen to know that this colleague habitually fabricates results, you will surely not just proceed to update your beliefs about the possible outcomes of future measurement outcomes on the basis of these results despite knowing they are probably not accurate. And there seems to be no reason why this should apply only for individual instances of evidence and not for general classes of evidence – so, for example, Friederich (2021, p. 172) says of Lewis' modal realism that '*Lewis's claim that acknowledging the reality of other possible worlds does not change the status of induction is unconvincing. ... [Suppose] you learn about the existence of a number of people with exactly the same psychological states and memories as you yourself, where for only some of those people their memories more or less faithfully reflect their personal histories whereas for the others their memories are misleading and "fake" as they are for a Boltzmann brain. Prime facie it would then not be reasonable for you to trust your memories – unless you additionally learned that the number of copies of observers with fake memories who are copies of you is vanishingly small compared with that of observers with reliable memories who are copies of you.*' Here Friederich seems to be endorsing the view that the induction assumption must be grounded in beliefs about reality, rather than merely adopted by fiat as a methodological presupposition.

This point of view on the induction assumption suggests that it is not the case that we should just blindly carry on making inferences about the future based on records of the past regardless of other beliefs we hold which might cause us to question the connection between our current records of the past and the relevant universal regularities or fundamental laws; and thus it is not reasonable to simply carry on doing induction in an Everettian universe without regard to the Everettian epistemic problem. Therefore I think that an attempt to sidestep the Everettian epistemic problem by a blanket application of the induction assumption would not ultimately be successful, and thus this argument, like the others considered in this chapter, ultimately does not succeed in showing that the Everett interpretation is No Worse Off.

8
Bayesianism

Although we have discussed the problem of empirical confirmation in a number of ways thus far, we have not yet addressed a central pillar of modern confirmation theory: Bayesianism, which is the study of degrees of belief or 'credences,' and how they ought to be updated in response to evidence. In this chapter I will explore further the special role played by Bayesianism in two popular proposals for solutions to the measurement problem – the Everett interpretation, and QBism.

Each of these approaches leans heavily on one particular Bayesian idea. First, we saw in chapter 6 that proposed solutions to the Everettian epistemic problem often rely crucially on characterising empirical confirmation in terms of Bayesian updating – that is, it is implicitly assumed that to empirically confirm a scientific theory is to make a series of observations, and after each observation to update the credence one attaches to the theory in accordance with Bayes' rule. In chapter 6 I gave a number of arguments against the idea that credences over Everettian branches can be used in Bayesian updating in just the same way as credences over probabilistic outcomes in single-world contexts. But in this chapter I will make a different kind of point: even if it *were* possible to show that agents in an Everettian universe ought to use Bayesian updating in just the same way as non-Everettian agents, that would not be enough to solve the Everettian epistemic problem because Bayesian updating does not exhaust the epistemology of science. Specifically, I will argue that the kind of problems that crop up in the epistemology of Everettian quantum mechanics are exactly the kind of problems that pure Bayesian updating is ill-equipped to solve.

Meanwhile, although QBism is in many regards very different from the Everett interpretation, I will suggest in this chapter that the two approaches share a common problem – in both cases the focus on Bayesian thinking leads to an overly narrow conception of scientific epistemology. For the QBists the problem is that the approach is heavily inspired by subjective Bayesianism, and in particular by the view, often associated with subjective Bayesianism, that the *only* standard by which we can assess a set of credences is probabilistic consistency. And this central notion of probabilistic consistency requires no more than that a set of credences should have the property that their formalisations can be jointly entailed by at least one probability distribution, or equivalently they obey the axioms of probability (Schupbach, 2022). So taking probabilistic consistency as the only possible standard for the assessment of scientific beliefs disregards both the role of empirical evidence and also stronger kinds of coherence; we will see that this leads the QBist to a set of beliefs which are probabilistically consistent in the Bayesian sense but which nonetheless seem epistemically irrational.

Thus, although the Bayesian toolkit is certainly a very useful means of analysing certain facets of the epistemic difficulties we face in quantum mechanics, I think it is unlikely that we can solve the epistemic version of the measurement problem by appeal to Bayesian ideas alone. We need to remain aware of the bigger picture in which

Saving Science from Quantum Mechanics. Emily Adlam, Oxford University Press. © Oxford University Press (2025).
DOI: 10.1093/9780197808887.003.0008

the goal is to arrive at a coherent system of beliefs which includes a satisfactory explanation of how we could possibly have come to know the things we take ourselves to know. For as noted in chapter 2, it is precisely in the context of fundamental physics that we should expect our theories to provide a satisfactory account of their own epistemology, so in this context the circle of coherentist justification becomes particularly tight, and the Bayesian methods that work straightforwardly in many other areas of science are no longer completely adequate.

8.1 Bayesian Updating and Everett

As noted in chapter 6, the term 'Bayesian updating' refers to a belief-updating strategy in which, upon witnessing a novel piece of evidence E, we update the credence assigned to hypothesis H according to Bayes' rule. That means we set our new credence equal to $\frac{p(H)p(E|H)}{p(E)}$, where $p(H)$ and $p(E)$ are the prior probabilities we assigned to the hypothesis H and evidence E before witnessing E actually occurring, and $p(E|H)$ is the probability of evidence E conditional on the hypothesis H being true.

We have seen that both the decision-theoretic approach and the Sebens-Carroll and Vaidman-McQueen approaches ultimately aim to address the Everettian problem of confirmation by showing that ordinary Bayesian updating is still valid in an Everettian world, and thus they seem to be implicitly assuming that empirical confirmation can be entirely characterised in terms of Bayesian updating. As Greaves (2006, p. 17) puts it in her defence of the decision-theoretic approach, '*I therefore take it that, if the Bayesian account yields the result that [observations of relative frequencies matching the single-case chances predicted by quantum mechanics] confirms Everettian quantum mechanics just as much as it confirms stochastic quantum mechanics, then the Epistemic Problem will be solved.*' Moreover, Greaves summarises the Bayesian account as the idea that '*empirical confirmation of a theory consists in the increase of degree of belief in that theory as a result of conditionalizing on evidence*' (here, 'conditionalizing' refers to Bayesian updating) and hence she does indeed seem to assume that the epistemic problem can be entirely solved by simply explicating Bayesian updating.

Now, Bayesian updating is closely related to Bayes' theorem, a theorem of the probability calculus which says that the conditional probability of H relative to E is given by $p(H|E) = \frac{p(H)p(E|H)}{p(E)}$. But as Schupbach (2022, p. 37) notes, '*It is crucial to distinguish Bayes' rule, the epistemological postulate, from Bayes' theorem, the theorem of probability.*' Bayes' theorem is, obviously, a theorem, and thus we know that under circumstances in which the probability calculus is applicable it will always give the unique probabilistically consistent conditional probability for $p(H|E)$ given the values for $p(H)$, $p(E|H)$, $p(E)$; but this does not necessarily entail that Bayes' rule is always the right way of updating one's beliefs in response to empirical evidence. The claim that empirical confirmation can be fully characterised in terms of Bayesian updating involves substantive epistemic and theoretical commitments which extrapolate significantly beyond the simple theorem of the probability calculus expressed in the above formula.

Note that one could understand the claim that Bayes' theorem is fully sufficient to characterise empirical confirmation in either a descriptive or normative way – the descriptive version would involve arguing that Bayesian updating is the way in which scientists actually do update their beliefs, whereas the normative version would involve arguing that this is the *right* way to update scientific beliefs. And, in fact, I think both the descriptive and normative versions of these positions do successfully capture some elements of scientific practice. There are undoubtedly cases in which scientists update their beliefs using Bayes' rule – indeed, it is not hard to find examples of them doing so explicitly, as in Navarro-Boullosa et al. (2023) and Jiang et al. (2023) – and it also seems true that there are many circumstances in which Bayesian updating is indeed the right way to update our beliefs. However, I think neither the descriptive nor normative versions of this position are completely correct – there are many circumstances in which scientists do more than merely Bayesian updating when assessing hypotheses, and moreover there are many circumstances in which epistemic rationality does, in fact, *require* that we should take factors other than basic Bayesian updating into account.

Now, debate over the use of Bayesian updating in empirical confirmation has often taken the form of discussions about whether various historical episodes in which scientists changed their beliefs can be understood in terms of Bayesian updating. Many examples are controversial, but it does seem fair to say that at least some important cases are difficult to conceptualise in Bayesian terms, particularly cases involving significant conceptual revisions – indeed Miller (1987) claims that '*Bayesian inference to the preferred alternative has not resolved, even temporarily, a single fundamental scientific dispute.*' However, even if true, this historical observation is not the end of the story because, of course, it is still open to proponents of Bayesian updating to adopt the normative version of the view and argue that the only *rational* belief-updating is Bayesian updating, which would mean that episodes in which scientists updated their beliefs in some other way are just examples of irrationality and error. So in this section, rather than enumerating historical examples, I will instead argue that the structure of scientific knowledge makes it clear that Bayesian updating is the rational way of updating one's beliefs only in an idealised special case where one's set of background beliefs is completely fixed. While there are certainly episodes in scientific practice which can reasonably be modelled in this way, many scenarios have a more complex epistemic structure – including, in particular, exactly the kinds of scenarios that are most relevant to the Everett interpretation.

To begin, note that Bayesian updating is defined for instances of 'incremental confirmation,' where we say '*an hypothesis h is incrementally confirmed by some evidence e, in the light of an agent's background k, exactly when e increases h's probability, conditional on k*' (Schupbach, 2022). And thus, as Ramsey (1926, p. 21) points out, Bayes' rule is straightforwardly applicable only if we assume that learning that the evidence occurs doesn't change the rest of our system of beliefs: '*The degree of [an agent's] belief in p given q ... is not the same as the degree to which he would believe p, if he believed q for certain; for knowledge of q might for psychological reasons profoundly alter his whole system of beliefs.*' Moreover, although Ramsay's use of the phrase '*for psychological reasons*' may suggest there is something subjective or irrational about altering one's whole system of beliefs in response to new evidence, this need not be the case:

changing one's background beliefs rather than updating a specific hypothesis can be a perfectly rational response to a piece of evidence. The reasons for this go back to the Quinean observation that in the face of new empirical evidence which may appear to refute some belief, we always have freedom to keep that belief fixed by adjusting other beliefs instead. For it follows from this that although changing one's credences for a particular belief P may be a rational response to new evidence, it cannot be the *uniquely* rational response to such evidence, since as Quine demonstrated, there is no unique mapping from propositions to empirical verification criteria.

In fact, there are a number of ways in which background beliefs may bear on Bayesian updating. Most obviously, our set of background beliefs usually determine the priors $p(H)$ and $p(E)$ appearing in the formula for Bayesian updating. Of course, this may not ultimately be very important, as in many circumstances the specific values of the initial priors actually have little effect on the beliefs we ultimately end up with – various convergence theorems (Blackwell and Dubins, 1962; Gaifman and Snir, 1982; Savage, 2012) show that '*no matter the extent to which subjects disagree in their initial confidences, so long as they all follow Bayes's Rule, a sufficient amount of common evidence will inevitably lead to a convergence in their opinions*' (Schupbach, 2022, p. 45). But Lipton (1991) notes that the conditional probabilities $p(E|H)$ are also partly determined by our background beliefs – in idealised examples of Bayesian updating it is usually assumed that the conditional probabilities depend only on the hypothesis to be tested and nothing else, but in real-world situations that cannot actually be the case. For example, as emphasized by Tal (2019), we need a measurement model in order to convert the predictions of the hypothesis to 'indications' that can actually be read off the instrument, and that measurement model will typically be based on a number of different background beliefs about the structure, composition, and functioning of the instrument. Moreover, as emphasized by Curiel (2020), to properly model a measuring device and observer we will typically have to appeal to several different scientific theories covering different parts of the measurement process as well as external factors which must be accounted for, so we will need background beliefs from theories beyond the one we are currently testing. And it certainly is not the case that agents who employ Bayesian updating with significantly different *conditional probabilities* will still converge to the same final credences given common evidence, so in general the results of Bayesian updating do depend quite sensitively on the background beliefs determining the conditional probabilities. Thus Bayesian updating can be applied straightforwardly only when all of the relevant background beliefs and hence conditional probabilities are held constant.

Now, Bayesians often offer arguments to the effect that Bayesian updating is the uniquely rational way to update one's belief in the face of new evidence, which may seem to be in conflict with the observation that Bayesian updating is not suitable for all contexts. But, in fact, there is no conflict here. As we saw in section 6.2.1, claims that Bayesian updating is uniquely rational are often based on Dutch Book arguments as presented by Lewis (1999) and Teller (1976). However, in order to make such arguments, we must first assume that all the prior probabilities and conditional probabilities used in Bayes' formula are fixed, which is only true if all of the background beliefs used to arrive at those conditional probabilities are considered to be immune to revision in the face of new empirical evidence. If we allow that the evidence might lead us to change our assignation of conditional probabilities, there is no way to say what bets the agent would make over time unless we have some prior knowledge about the 'correct' conditional probabilities are, and hence there is no *internal* irrationality involved in any particular updating policy. So the argument for the rationality of Bayesian updating presumes that we have already decided to keep all of our background beliefs fixed and to change just one credence in some proposition *P* in response to the new evidence, and indeed it seems quite plausible that Bayesian updating is the uniquely rational way of updating our beliefs under these circumstances; but the Quinean point reminds us that we are not necessarily obliged to change the credence for proposition *P* – there is nothing irrational about choosing to alter our background beliefs instead, thus changing the conditional probabilities in such a way that the credence for *P* remains the same or changes in a way different to what Bayes' rule would prescribe.

Van Fraassen makes a similar argument, noting that Dutch Book arguments only work against agents who are committed to some rule other than Bayesian updating for belief-updating; it's not possible to make a Dutch Book against an agent who simply refrains from committing to any particular rule for belief-updating. This, van Fraassen argues, leaves room for creativity and innovation in the formulation of new beliefs, which can't be adequately captured by any fixed rule: '*We are creatively inventing new hypotheses and theories, and ... we are not only prone but rational to embrace them, after viewing favourable evidence, while they still go greatly beyond any accommodated evidence-cum-previous-opinion we could ever have hoped to assemble. This view of rational free enterprise of the spirit ... was fittingly advocated by the American pragmatists*' (van Fraassen, 1989, p. 172). And we can understand how there could be room for creativity in this process when we recall that rationality does not oblige us to update any particular belief in response to evidence, so the unique rationality of Bayesian updating in the context of updating a fixed belief does not preclude the possibility that rationality gives us significant latitude in choosing how to respond to a given piece of empirical evidence. So although it does seem reasonable to think that Bayesian updating is typically a good description of the process of doing 'ordinary science' – i.e. cumulatively adding hypotheses on top of an existing heap of background beliefs, for which the credences are not altered in any way during this process – there is always the option to instead '*leaven this process with theoretical innovation and courageous embrace of new hypotheses that have gained by admiration*' (van Fraassen, 1989, p. 174), so an adequate account of the epistemology of science probably cannot be given in terms of Bayesian updating alone.

These features of Bayesian updating are nicely illustrated by the long-standing debate over the significance of Bell's theorem. For following the discussion in section 4.0.1, if we have decided on a fixed set of background beliefs including all the assumptions that go into Bell's theorem – including 'no-retrocausality,' 'no-superdeterminism,' 'there exists only one world,' and so on – then it follows from Bell's argument that conditional on the hypothesis of locality, the probability of seeing violations of Bell's inequality becomes increasingly small as the number of experiments gets larger. Thus it could be argued that the 'uniquely rational' response to violations of Bell's inequality is to perform Bayesian updating in such a way as to assign significantly lower credence to the hypothesis of locality. But non-locality's detractors have argued that we should keep our credence for locality relatively high and instead adjust some background beliefs in such a way as to change the *conditional probabilities*. For example, violations of Bell's inequality can be produced reliably by an entirely local mechanism if we allow superdeterminism, and thus if we admit superdeterminism as a viable possibility, it is no longer necessarily the case that conditional on the hypothesis of locality the probability of seeing violations of Bell's inequality becomes very small as the number of experiments becomes large. So once we allow this possibility, decreasing one's credence in the hypothesis of locality is no longer the uniquely rational response to such evidence. That is, because the result of the most straightforward application of Bayesian updating in this situation is unpalatable to some physicists, many of them have elected to go outside of the framework envisioned by Bell by changing a background belief and thus altering some of the conditional probabilities, instead of updating their credence in the hypothesis which the experiments were originally designed to test.

8.1.1 Beyond Bayesian Updating

Now, at this point the Everettian might perhaps argue that although we do in principle have the option of responding to new evidence by doing something other than Bayesian updating, nonetheless Bayesian updating is always a valid option, so as long as the Everettian agent can make sense of Bayesian updating then they can still make sense of scientific epistemology.

However, it is not clear that we *do* always have the option of Bayesian updating – there are a number of situations where we may be rationally compelled to use other modes of belief updating, and I will now argue that several of these situations are relevant to the epistemology of measurement in an Everettian context.

Self-Undermining Cases

The first problem case for Bayesian updating is a version of the kind of self-undermining situation we've already discussed in some detail in chapter 3 – i.e. situations in which ordinary Bayesian updating leads you to a conclusion which is in some way inconsistent with, or at least not very coherent with, the background beliefs you originally used to determine your priors and/or conditional probabilities. In such cases, you cannot just do Bayesian updating and then leave it at that – in order to restore the coherence of your system of beliefs, you will have to go back and reassess some background beliefs, which may lead to a variety of other changes throughout your belief system.

In chapter 4 we looked at the holistic aspects of the epistemology of science in a somewhat simplified context, implicitly assuming that for any proposition it is either believed or not believed, i.e. it is assigned credence either 0 or 1 with no intermediate possibilities. But, of course, in most realistic scientific situations it is more realistic to allow credences between 0 and 1. Still, the considerations introduced in chapter 4 largely continue to apply – scientific enquiry involves progressively working towards improving the coherence of our credences, with the aim that our credences are appropriately supported by the empirical evidence, relative to the background credences we attach to other relevant propositions. One way of defining a system of mutually supporting credences of this kind takes the form of a Bayesian network, i.e. a set of propositions together with conditional probabilities specifying the probabilistic relations between the propositions. There have been many proposals for different ways of quantifying the coherence of such a structure – for example, Fitelson (2003) imagines a probabilistic measure of coherence in which a set E is said to be positively dependent iff for any element E_i of the set, and for any subset S of E not including E_i, E is positively supported by S according to a particular Bayesian measure of degree of support.

Examining the structure of such Bayesian networks, we can clearly see that although Bayesian updating is typically a good way of updating our credences in the face of new evidence in such a way as to maintain or even increase the coherence of the network, Bayesian updating cannot be the whole story. For example, Fitelson's measure involves not just pairwise comparisons of elements of the set but rather comparisons of each element E_i to *every* subset not including E_i, so this is a thoroughly holistic notion of coherence – and indeed, it can be shown that a sensible measure of coherence can only be assessed in a holistic way. To see this, consider the following set of beliefs, as described by Shogenji (1999): '*Exactly one of the Beatles is dead; John is alive; Paul is alive; George is alive; Ringo is alive.*' Clearly all pairs of beliefs in this set are perfectly consistent with each other, but the set as a whole is inconsistent, thus demonstrating that the coherence of a whole belief set cannot be a function of the pairwise coherences. Indeed, Shogenji (1999) notes that much the same argument would also rule out any other approach which proceeds by assigning a measurement of confirmation to individual subsets and then taking some kind of average: a plausible measure of probabilistic coherence measure must take the whole system of beliefs into account at once. This means that in a Bayesian net, just as in the systems of beliefs discussed in chapter 4, changing the credence associated with even just one proposition in the set of 'background beliefs'can have a very significant impact of the coherence of the whole belief set, and we may therefore be required us to change a large number of other credences in order to restore coherence. Whereas Bayesian updating only explicitly considers the 'coherence' between two propositions – the one encoding the evidence and the one encoding the hypothesis – and therefore if our goal is to maintain or improve overall coherence, Bayesian updating is not always going to be an appropriate way of proceeding, since we may sometimes have to take more holistic considerations into account.

And, of course, this kind of case is very relevant to the Everett interpretation. For insofar as our predecessors can be modelled as arriving at quantum mechanics using Bayesian updating, they did so based on fixed likelihoods which were shaped by their pre-existing background belief that there exists only one world and measurements have unique, probabilistically determined outcomes. But if the Everett interpretation is indeed a plausible reading of the quantum formalism, it would seem that confidence in the correctness of quantum mechanics should also increase our confidence that *Everettian* quantum mechanics is true. And Everettian quantum mechanics contradicts the background beliefs about the uniqueness of measurement outcomes employed in the first updating step, so if we increase our confidence that Everettian quantum mechanics is right, then we ought to go back and reassess our priors and conditional probabilities, which may then change the credence we attach to quantum mechanics, and so on. Clearly there is the possibility of a vicious circle in this vicinity, and even if it is not, in fact, vicious, nonetheless whatever ultimately ends up happening here will not just be straightforward Bayesian updating.

This means that arguments aiming to show that Bayesian updating works in an Everettian context just as it would in a non-Everettian context will not, even if correct, necessarily help all that much with the Everettian epistemic problem; for this is precisely the kind of scenario in which we need to use something other than merely Bayesian updating to evaluate our beliefs. In fact, under these circumstances we probably need to perform some kind of general redistribution of credences across our whole belief set which takes into account the holistic nature of scientific knowledge, and it's at best unclear that this kind of holistic reassessment of our entire belief system would ultimately favour an Everettian picture. Moreover, it seems quite unlikely that a simple behavioural criterion, as employed the decision-theoretic approach to deriving Everettian probabilities, could capture this complex redistribution of beliefs across a whole belief system; and therefore there are reasons to doubt that any approach of this kind could work to make sense of the reality of Everettian epistemology, as opposed to an idealised case of sequential belief-updating which does not reflect the real epistemic situation.

The Problem of Old Evidence

One common problem case for Bayesian updating involves situations in which someone comes up with a new theory which is supported in large part by evidence that was already available to us before the theory was formulated (Earman, 1992). Glymour (1981) gives the example of the perihelion of Mercury being used to support Einstein's theory of general relativity, even though the data about the perihelion of Mercury was available long before Einstein came up with his theory. The difficulty is that Bayesian updating is essentially a description of the dynamics of belief-updating as new evidence comes in, so it is complicated to apply Bayesian updating when the relevant evidence is not new. In particular, we have previously noted that the probabilities and conditional probabilities in the Bayesian formula $p(H|E) = \frac{p(H)p(E|H)}{p(E)}$ are typically fixed by the agent's existing background beliefs, so if the agent already knows that E has definitely occurred, we will get $p(E|H) = p(E) = 1$ and hence $p(H|E) = p(H)$, so no updating can occur.

Now, of course, one may simply try to subtract the relevant evidence from one's set of background beliefs before deciding the priors and conditional probabilities to be used in the formula, but as Chihara (1987) points out, given the holistic, mutually supporting nature of our beliefs we cannot always simply subtract some propositions and still be left with a sensible set of beliefs which would lead to a plausible set of conditional probabilities. Alternatively one could try to construct some kind of counterfactual story involving 'Ur-priors' which are supposed to reflect what our beliefs would have been if we had never obtained the relevant evidence, as suggested by Howson (1991). But as Glymour (1981) shows, this will require us to make a lot of non-trivial assumptions about our counterfactual belief system, so it will seldom yield unambiguous answers. Similarly, Monton (2004, p. 5) notes that '*it is not always clear what values the ur-probabilities should take, especially when one has to make extreme modifications to one's opinion, by, for example, supposing that one does not know that one exists.*' So, in fact, it seems likely that in reality what we do in this kind of case is not just simple Bayesian updating, but some potentially quite complex redistribution of credences across our whole system of beliefs, aiming to produce a network which is now more coherent and better confirmed by the evidence.

Moreover, the old evidence problem is very relevant to the Everett interpretation – for the Everett interpretation was first introduced in 1957 (Everett, 1957), at which point there was already a significant amount of empirical evidence for quantum mechanics. Now, the Everettians who have presented arguments to the effect that it is rational for an Everettian agent to use Bayesian updating in exactly the same way as non-Everettian agents will presumably want to suggest that we should just look back and re-interpret the process of Bayesian updating that took place *before* 1957 in terms of the Everettian version of belief-updating, rather than imagining some new belief-updating being done after 1957. However, one might worry that it is not really consistent to re-interpret the process of belief-updating in this way while leaving all the priors and conditional probabilities the same because the beliefs that physicists actually held prior to quantum mechanics were themselves tied to a single-world perspective – for after all, we noted in section 3.4 that the multiple-world hypothesis has the potential to alter a large number of ordinary beliefs about parts of the world not obviously related to quantum mechanics. So if we're going to use some re-interpretation of Bayesian updating to justify believing in Everettian quantum mechanics, we will potentially have to adopt some kind of 'Ur-priors' appropriate to a set of observers who do not yet have evidence for quantum mechanics but who do believe that they live in a branching universe – or at least, who believe that there is some non-trivial probability that they live in a branching universe, and are therefore deliberating between various different branching and non-branching accounts of reality, as envisioned by Greaves and Myrvold (2010).

Yet it is not obvious why one would be inclined to believe in a branching universe of this kind if not because of quantum mechanics. In particular, the idea of a 'mod-squared amplitude' and the justification for treating such a thing as a probability seems closely tied to the theoretical structure of quantum mechanics, so there is no straightforward way to arrive at a plausible hypothetical belief system which assigns non-trivial credences to the possibility of branching processes weighted by something like a mod-squared amplitude, without yet assigning non-trivial credence to a theory like quantum mechanics.

Perhaps the Everettian might hope to avoid the old evidence problem by appealing to convergence theorems (Blackwell and Dubins, 1962; Gaifman and Snir, 1982; Savage, 2012) to argue that for almost any choice of priors or Ur-priors, after a long enough series of experiments the credences of an Everettian agent who updates according to Bayes' rule will converge to a distribution assigning high credence to Everettian quantum mechanics. However, as previously noted, although this may work for priors, it will not work for conditional probabilities – observers with very different conditional probabilities will seldom converge to the same final beliefs. And although discussions of the problem of old evidence have often focused on priors, exactly the same issue arises for the conditional probabilities – they, too, are significantly responsive to the system of beliefs we hold before updating begins, and thus they too must be evaluated using some suitable set of counterfactual beliefs that we might plausibly have held if we had arrived at the theory before the evidence. So convergence theorems alone cannot resolve the problem of old evidence, since they do not help with the problem of the conditional probabilities.

So again, it isn't obvious that we can give a full justification for believing in Everettian quantum mechanics based purely on Bayesian updating. Instead we will need an approach which more closely mirrors the actual course of events – i.e. an approach which allows us to move from a large body of existing evidence for single-world quantum mechanics to a re-interpretation of this entire body of evidence in terms of the possibility of a branching universe scenario. And this retrospective shift will presumably involve something more complex than Bayesian updating one belief at a time – rather we will have to re-distribute our credences across our whole belief set in a holistic way, properly taking account of the crucial role that the single-world assumption has played in the interpretation of many scientific results throughout history.

Undefined Priors

The third kind of problem case for Bayesian updating involves cases where the priors are just completely undefined. To be clear, I am not referring here to cases where priors are fuzzy because people do not think in completely precise ways – the Bayesians have strategies to deal with that problem (Viertl, 2008). Rather I am referring to cases where we require priors for propositions which are so outside of our usual frame of reasoning that there seems to be neither any rational ground for assigning any particular prior credences to them, nor indeed any psychological evidence that actual people *do* assign even very approximate prior credences to them. Evidently we can't appeal to convergence arguments in these kinds of cases, since starting with undefined priors will never lead to any well-defined final credences.

This problem is most likely to occur for 'sceptical' hypotheses which significantly disrupt the kinds of relations we usually expect to hold between evidence and reality, such as the hypothesis that I am a brain in a vat. I myself do not know how to attach even an extremely vague credence to this hypothesis, since I have no idea what kind of framework I could use to arrive at such a credence, nor what kind of evidence could be relevant to it. I would not say that my credence is merely fuzzy or approximate; I just don't have any defined credence for it at all, and I simply disregard it on methodological grounds in most of my ordinary reasoning, so no amount of Bayesian updating is going to lead me to assign a finite, well-defined credence to this hypothesis. This is not to say that there could not be evidence which would lead me to take the hypothesis more seriously, but such a change could not be described as simply a process of Bayesian updating – it would be more like a kind of paradigm shift which would radically change my whole system of beliefs.

This kind of problem is again relevant to the Everettian situation, since the hypothesis that we are in a branching multiverse is another kind of 'sceptical' hypothesis which disrupts the relations we usually expect to hold between evidence and reality. Indeed, prior to having access to the theoretical structure of quantum mechanics there would seem to be almost no possible way to say anything in particular about what we should expect to experience if we are, in fact, living in a multiverse, nor what kind of framework we could possibly use to assess the probability that we are. So it's very plausible that the appropriate credences to assign to branching universes, when we are trying to imagine appropriate pre-quantum Ur-priors, are simply not well-defined. And without well-defined priors we will never be able to converge on any final credences for various properties that the multiverse might have, such as the property of being the specific quantum-mechanical multiverse described by Everettian quantum mechanics.

So again, simply giving an argument that Bayesian updating is still rational in an Everettian universe does not fully account for the epistemology of the Everett interpretation; we are going to have to employ some more holistic belief-updating process which can be applied in retrospect. Thus simply providing a justification of Bayesian updating in an Everettian context does not suffice to fully affirm the reliability of our means of enquiry, since our means of enquiry outstrip simple Bayesian updating in exactly this context; an adequate account of the epistemology of the Everett interpretation must do justice to the full structure of scientific knowledge, rather than just the special cases in which Bayesian updating is appropriate.

8.2 Probabilistic Consistency and QBism

Having discussed some of the limitations of Bayesian epistemology as applied to the Everett interpretation, let us now move on to consider the use of Bayesian epistemology in QBism. Both Bayesianism and subjective Bayesianism come in many different versions (Joyce, 2004; Sprenger, 2018; Wallmann and Williamson, 2019), but the version of subjective Bayesianism espoused by the founders of QBism is a philosophy of probability which suggests that probabilities are nothing more than personal degrees of belief associated with some agent or other, and that these degrees of belief can be characterised by the Bayesian probability calculus. That is, this is a version of subjective Bayesian which involves the claim that probabilities don't reflect any objective feature of physical reality: they are *purely* subjective degrees of belief (Savage, 1954; Jeffrey, 1965).

Now, on the face of it quantum mechanics seems to pose a serious problem for this kind of subjective Bayesianism. For if we reject deterministic hidden-variable and many-worlds approaches, then we are apparently obliged to accept that quantum mechanics postulates intrinsic, irreducible probabilities, and yet the subjective Bayesian must say that these probabilities are merely subjective degrees of belief. This has quite significant implications – in particular, since the main function of quantum states is to encode probabilities, then as Fuchs (2010, p. 3) puts it, '*If probabilities are personal in the Bayesian sense, then so too must be quantum states.*' Moreover, if it is true that the probabilities in quantum mechanics are ex hypothesi intrinsic and irreducible, they cannot be understood in the way we often understand subjective degrees of belief, i.e. as encoding uncertainty about some underlying objective reality – they are simply degrees of belief which are not grounded in any objective feature of reality. Thus the subjective Bayesian appears to be committed to the view that in the context of quantum mechanics we hold a range of precise, quantitative, and detailed beliefs for no reason at all.

QBism, as originally proposed by Caves et al. (2002), seeks to make a virtue out of necessity by arguing that we *should* understand quantum mechanics as simply encoding the personal degrees of beliefs of users of the theory. The founders of QBism contend that quantum states are nothing more than characterisations of an agent's degrees of belief, and therefore there is no 'objectively correct' quantum state to assign in a given scenario. This entails that quantum mechanics has no descriptive content, rather it is purely normative – it is '*a tool decision-making agents are advised to adopt in light of the peculiar uncertainties we find in our world*' Fuchs, 2023, p. 5). QBists further argue that a range of conceptual problems can thereby be defused – for example, Extended Wigner's Friend paradoxes (see section 2.2.2) will no longer have any bite because if systems do not have objectively correct quantum states, it follows that different observers can assign different quantum states to the same system without either of them being wrong.

Now, some of the literature on QBism may give the impression that it is intended only as an answer to the problem of interpretation discussed in section 5.2, rather than a solution to what I have called the epistemic version of the measurement problem. For example, (Fuchs (2012, p. 8) states that '*quantum states are not real things ... a quantum state is a set of numbers an agent uses to guide the gambles he might take on the consequences of his potential interactions with a quantum system. It has no more substantiality than that.*' There is clearly something right about this if it is intended merely as an account of the nature of the quantum formalism – the specific mathematical formalism of quantum mechanics is, indeed, a tool developed by humans to predict the outcomes of interactions with certain kinds of physical systems. But if this is the total content of QBism then clearly it is not solving or even seeking to solve the epistemic version of the measurement problem, since no account of the physical nature of these interactions is on offer.

On the other hand, at times QBists also seem to suggest that QBism *is* a solution to the measurement problem, or rather that it reveals the measurement problem to be ill-posed: '*The "measurement problem" is purely an artefact of a wrong-headed view of what quantum states and/or quantum probabilities ought to be*' (Fuchs, 2012, p. 12). And in light of the epistemic issues discussed in chapters 2 and 4, I think it is reasonable to dismiss the measurement problem on the basis of QBism only if QBism has some coherent story to offer about the epistemology of quantum mechanics, so in this section I will try to understand what that story might look like.

Some QBist writings give the impression that QBism falls in the class of observer-relative interpretations, as defined in section 3.2. For example, Fuchs (2023, p. 13) offers as one of the eight tenets of QBism that '*a measurement outcome is personal to the agent doing the measurement*' which looks similar to the kinds claims often made by proponents of observer-relative interpretations. So this suggests that QBism may be vulnerable to the same kind of objections as the other observer-relative approaches. Indeed, French (2023, p. 187) seems to advocate a version of QBism of this kind, noting that intersubjective agreement may appear to be threatened in QBism, since it '*allows that, strictly speaking, two agents observing the same measurement may not have the same experience.*' French (p. 187) contends that this problem is solved by noting that '*updating different prior probabilities in the light of new but common information will lead to convergence*' but it is not clear how this is supposed to work – for if two agents observing the same measurement may not have the same experience, then surely the relevant information is *not* guaranteed to be common. The convergence argument seems to presuppose that the experiences of the agents differ only in virtue of their different priors and not in virtue of what information they obtain from witnessing a given measurement outcome, and yet the problem that we face in an observer-relative interpretation is precisely that it does *not* appear to be the case that agents will in general acquire the same information about a given measurement outcome. Clearly then no convergence theorems can be applied without some prior specification of a systematic relation between the information obtained by different observers about measurement outcomes, so there is still a question mark over intersubjectivity in this context.

With that said, it is not entirely clear that QBism should be interpreted as an observer-relative interpretation because QBists do not necessarily think that unitary quantum mechanics is universal or complete – at various points they seem to be positing some kind of mind-independent external reality which is separate from quantum mechanics. As Fuchs and Stacey (2019, p. 9) put it, '*The world is filled with all the same things it was before quantum theory came along, like each of our experiences, that rock and that tree, and all the other things under the sun; it is just that quantum theory provides a calculus for gambling on each agent's own experiences – it doesn't give anything else than that.*' This kind of statement appears to deny the first postulate of the observer-relative interpretations as set out in section 3.2, which suggests that QBism may not after all be an observer-relative view. And indeed, while explaining the idea that measurement outcomes are personal, DeBrota et al. (2020, p. 5) also argues that '*different agents may inform each other of their outcomes and thus agree upon the consequences of a measurement, but a measurement outcome should not be viewed as an agent-independent fact which is available for anyone to see*,' and this seems to amount to a denial that QBism is an observer-relative view, for we have seen that in an observer-relative picture agents cannot reliably come to agree upon the outcome of a measurement.

Now, one might feel there is some tension between the various claims that QBists make here – for after all, if any observer can, in principle, find out what my measurement outcome was by asking me about it, doesn't that mean my measurement outcome *is* indeed available for anyone to see, albeit via interaction with me rather than via interaction with my measuring device? Perhaps the idea is that quantum measuring instruments and measurement outcomes do not belong to the mind-independent external world posited by QBism, so other observers cannot directly interact with them, whereas human bodies do inhabit the mind-independent external world posited by QBism, so other observers can learn about my measurement outcome by interacting with my body. On a physicalist view it seems a little odd to treat measuring instruments and human bodies as radically different ontological categories in this way, but perhaps QBism is not intended to be a physicalist view. But in any case, since we will be looking at the problems of observer-relative interpretations in some detail in chapter 9, let us try to pin down the content of the version of QBism that appears to be suggested here, in which it *is* possible for observers to share the outcomes of their measurements with one another. I will refer to this as 'Intersubjective QBism.'

Here it is helpful to consider the following question: does Intersubjective QBism say that the probabilities encoded in quantum states will generally be reflected in observed relative frequencies, or does it not? Beginning with the first horn of this dilemma, if probabilities *are* generally reflected in observed relative frequencies, then the central claim of QBism appears to be false – quantum states are not just normative, they are also descriptive, since they describe actual relative frequencies. Moreover, one of the key motivations for QBism is the idea that different observers should be able to assign different quantum states to the same system without either of them being wrong – this is the crux of the solution offered by QBism to the Wigner's Friend paradoxes. But if observers are able to communicate with each other to share the outcomes they have observed, then for a given state preparation we can have a group of observers perform a series of different measurements on identically prepared systems and then share their outcomes with one another, so that there are agent-independent facts agreed upon by everyone concerned about the relative frequencies observed in measurements on systems prepared in this way. These relative frequencies will be consistent with certain quantum state assignations and not others, and therefore if the probabilities encoded in quantum states are reflected in observed relative frequencies, it would seem that there *is* after all a fact of the matter about which quantum states we ought to assign to various kinds of systems, in contradiction with one of the key tenets of QBism.

Now, there is a possible compromise here: we could imagine a theory which permits observers to share their outcomes, but which has a peculiar dynamics such that the relative frequencies seen by an observer Alice performing measurements on a system prepared in a certain way are different from the relative frequencies that would be seen by an observer Bob performing the same measurements on that system, so S can be regarded as having different quantum states ψ_A, ψ_B relative to Alice and Bob, respectively, in virtue of the fact that it would as a matter of objective fact produce different measurement outcomes for each of them. Indeed, we will encounter a view of this kind in chapter 13. However, this kind of approach does not really seem to have anything to do with subjective Bayesianism, for in this picture the probabilities encoded in quantum state are completely objective; it just happens to be the case that to fully specify these objective probabilities we must take into account not only facts about the system itself, but also facts about the observer making the observation. That is, although Alice and Bob might describe the system using different states ψ_A and ψ_B, nonetheless if the two of them properly understand the dynamics of this theory then Alice should agree that ψ_B is the correct state to describe Bob's outcomes even though is it not the correct state to describe *her* outcomes, and likewise mutatis mutandis for Bob – so in this picture, quantum states and the probabilities encoded in them are relativised to a context, but not in any way subjective.

It seems, therefore, that the Intersubjective QBist has little choice but to say that the probabilities encoded in quantum states will not generally be reflected in observed relative frequencies. And yet saying that quantum states have *nothing* to do with observed relative frequencies seems disastrous for the epistemology of quantum mechanics – for as we saw in chapter 6, a very significant amount of the empirical content of quantum mechanics is in the observed relative frequencies, so if we decide that actually quantum mechanics has nothing to do with observed relative frequencies, then it will suddenly have much less empirical content, and will therefore become commensurately less susceptible to empirical confirmation. Indeed, Intersubjective QBism looks even worse off than Everett in this regard, because Everettians who have not solved the probability problem can at least still predict that certain experimental outcomes are possible or impossible, but as Fuchs and Schack (2014) emphasise, the subjective Bayesian idea that probabilities are just degrees of belief applies to probabilities of 0 or 1 as well, and thus the QBist cannot even predict that certain experimental outcomes are possible or impossible. Thus it seems that according to Intersubjective QBism, quantum mechanics effectively has no empirical content at all, and thus there is no possibility of obtaining any empirical evidence for it, so it does about as badly as one could possibly imagine at solving the epistemic version of the measurement problem!

Of course, proponents of Intersubjective QBism will presumably accept that as a matter of historical fact we *did* arrive at quantum mechanics by appeal to empirical evidence, but they will then have to deny that this evidence really provides any meaningful support for the theory. So empirical science must be regarded as only the context of discovery for quantum mechanics, rather than the context of justification – the Intersubjective QBist must argue that we should accept the normative guidance offered by the theory for some reason that has nothing to do with empirical evidence.

What could that reason be? One might optimistically hope for some kind of a priori reason, akin to a Dutch Book argument like those described in section 6.2.1, to think that the quantum mechanical formalism encodes a uniquely rational way to set one's credences. However, it seems unlikely that such an argument is possible – after all, prior to the discovery of quantum mechanics we reasoned in other ways, and it doesn't seem to be the case that there was anything probabilistically inconsistent or irrational about that reasoning. And indeed, Fuchs (2023, p. 10) appears to agree that in QBism, the advisability of following the normative guidance of quantum mechanics is a contingent fact which depends on properties of the actual world: '*Quantum theory on the QBist view, however, is an addition to probability theory which very much takes into account the unique characteristics of our given world. If our world were a different world, agents would not be well advised to use the quantum formalism.*' Yet given that the QBists believe quantum mechanics makes no descriptive claims, they cannot claim that there is anything we could observe which would tell us that this is a world in which we are well-advised to use the quantum formalism rather than some other formalism, for otherwise quantum mechanics would in effect be giving a description of that observable thing.

For example, it has been suggested that QBism can be grounded in phenomenological philosophy, but surely phenomenology cannot by itself lead us to *quantum mechanics* in particular, rather than some other mathematical formalism – the claim is that '*quantum mechanics has the potential to live up to the ideal of a fully rationalized, critical, and ultimately* ***phenomenological*** *physics* (Berghofer and Wiltsche, 2020, p. 37),' not that proper consideration of phenomenology will lead us *uniquely* to quantum mechanics. Thus proponents of the phenomenological approach to QBism must still hold that it is some specific property of our actual lived experience which singles out quantum mechanics – as Bitbol (2021) puts it in his phenomenological account, '*The basis of scientific knowledge is personal lived experience and verbal* *communication between subjects of experience.*' Yet if we take it that at least some of the structure of quantum mechanics is instantiated in some way in our actual lived experience such that those experiences can specifically lead us to quantum mechanics, and if we further assume that verbal communication between subjects of experience can meaningfully be achieved, then we are no longer really doing QBism. In particular, once we allow all this we can no longer avoid the problems posed by the Extended Wigner's Friend theorems in the way that QBism aims to do, and thus if the phenomenological approach is to count as a form of QBism, it must maintain that quantum mechanics does not have implications for any observed relative frequencies.

So although the Intersubjective QBist does not consider the appropriateness of the quantum formalism to be a priori, nonetheless it is apparently not something we can learn from experience either. The only other possibility that comes to mind would involve the Intersubjective QBist arguing that we just come to know the correctness of quantum mechanics by some sort of divine inspiration, akin to an article of religious faith[1]. I confess to being baffled by this position, but in any case, if that is the move the Intersubjective QBists want to make, then we may reasonably say that on their construal, quantum mechanics is not scientific knowledge any more, and therefore this approach to quantum mechanics undermines a significant amount of our empirical scientific knowledge pertaining to the regularities encoded in quantum mechanics. Therefore if this is the intended route, it is fair to say that Intersubjective QBism does not solve the epistemic version of the measurement problem, since it does a poor job at preserving our empirical scientific knowledge.

Now, we can see why these difficulties arise if we put them in the broader context of subjective Bayesianism. More or less all Bayesian views are ultimately founded on the idea that an agent's credences at a given time should always be probabilistically consistent, in the sense that they can be jointly entailed by at least one single probability distribution (Schupbach, 2022). The idea that a rational agent's credences should take this form is known as 'Probablism,' and there are certainly compelling reasons to believe it – for example, it can be shown that agents whose credences fail to satisfy these axioms are vulnerable to Dutch Books (Vineberg, 2022), and we can also invoke 'Expected Utility' arguments (Rosenkrantz, 1981; de Finetti, 2017) '*showing that non-probabilistic confidences are necessarily dominated in terms of expected accuracy by probabilistically explicable confidences*' (Schupbach, 2022).

However, some versions of subjective Bayesianism go beyond Probabilism by suggesting that consistency in this sense is the *only* possible standard to which credences can be held, at least with respect to beliefs that are about probabilities or expressed probabilistically. While not all subjective Bayesians necessarily adopt this view, the proponents of QBism do seem to subscribe to it: as Fuchs (2010, p. 4) puts it, '*Probability theory can only say if various degrees of belief are consistent or inconsistent with each other. The actual beliefs come from another source, and there is nowhere to pin their responsibility but on the agent who holds them.*' And thus, since QBists understand the probabilities in quantum mechanics in these terms, they have little choice but to say that probabilistic consistency is the only standard relevant to assessing beliefs we may hold about any phenomena falling under the domain of quantum

[1] Thanks to Wayne Myrvold for this suggestion.

mechanics – after all, if beliefs about facts in the realm of quantum mechanics are pure subjective degrees of belief which are not grounded in any objective feature of reality, what possible standard could we hold these beliefs to, other than probabilistic consistency?

Thus it is in some ways not surprising that QBism ultimately leads to a system of beliefs with very low coherence. For in real scientific scenarios, probabilistic consistency is not the sole criterion to which we appeal when assessing a set of beliefs, even if those beliefs pertain to probabilities. The problem is that the formal nature of the probabilistic consistency criterion makes it limited in its power: it legislates credences without any attention to what exactly the credences are *about*, so it is blind to various kinds of semantic incoherence. For example, the probabilistic consistency criterion would not recognise anything wrong with a set of beliefs which assigned high credence both to *R*: 'the book is red all over' and *B*: 'the book is blue all over,' even though most of us would regard this set of beliefs as inconsistent or incoherent in some sense. Pre-Quine, one might have hoped to rectify this by adding to our beliefs a set of supposedly analytic truths, all assigned credence 1, which would include something like *A*: 'nothing is both red and blue all over.' And then since *A* presumably entails ~ (*R*&*B*), it is clear that a set of beliefs assigning credence 1 to *A* and also high credence to both *R* and *B* would not be probabilistically consistent. But after Quine's work it is no longer clear that propositions cannot be so neatly divided into analytic and synthetic, and certainly it would not be straightforward to simply enumerate a well-defined list of all the analytic truths to be added to our belief system in this way, which means there is likely no simple method which could turn the assessment of coherence in a general sense back into a mere question of probabilistic consistency.

These limitations lead directly to the problem we have already identified for QBism; for as part of its inability to capture semantic incoherence, probabilistic consistency is also blind to various kinds of *epistemic* incoherence. In particular, we typically expect that beliefs about the physical world – especially very precise quantitative beliefs, such as those associated with using quantum states to make inferences about the future – are arrived at and justified by appeal to empirical evidence. And yet because the QBist epistemology appeals only to probabilistic consistency, it does not give any particular role to empirical evidence. Of course, it is true that if we observe some event and subsequently assign probability 1 to the proposition that this event occurred, the subjective Bayesian approach requires us to update our beliefs in a way which maintains the probabilistic consistency of the whole set, so it is not impossible for new evidence to lead to changes in our beliefs in this context; however, there is also no requirement that any particular kind of belief should be supported by evidence, and thus we can have all sorts of completely unmotivated beliefs about physical reality as long as they are not probabilistically inconsistent with our other beliefs. Yet in scientific contexts – and indeed even in most everyday contexts – we typically think it is reasonable to criticise people for holding substantive empirical beliefs for no reason, or on the basis of poor evidence, or on the basis of evidence which is undermined by that very belief. So assessing belief sets purely by the metric of probabilistic consistency fails to capture the complexities of the judgements involved in scientific practice, and leads inevitably to the kind of empirically incoherent position involved in Intersubjective QBism, where we are advised to hold a lot of substantive quantitative empirical beliefs for no reason at all.

That said, Intersubjective QBism may at least be better off than some of the other putative solutions to the measurement problem that I have discussed here because, arguably, the epistemic danger is confined to quantum mechanics. This is because Intersubjective QBism is not committed to the view that unitary quantum mechanics is complete, so it can still postulate a realm of 'experience' which is distinct from the quantum world: as Fuchs (2010, p. 24) puts it, '*Quantum mechanics is something put on top of raw, unreflected experience.*' That is to say, in Intersubjective QBism the ordinary everyday world of our experience seems to be treated as separate from and prior to any of our beliefs about quantum mechanics, so whatever we believe about quantum mechanics cannot undermine the knowledge that we have about the ordinary everyday world of our experience. Indeed, if quantum mechanics is purely a description of personal probabilities with no descriptive content and no objective physical grounds, it can't possibly have any influence on our understanding of physical reality – true, it is supposed to determine some of our credences about the future, but it also tells us that we shouldn't think of those beliefs as actually describing anything, so they can't play any part in our attempts at giving an account of reality.

However, Intersubjective QBism then faces a different kind of problem: if quantum mechanics has no descriptive content whatsoever, but we still believe that there does exist an objective external world, it follows that quantum mechanics can't possibly play any role in helping resolve puzzles or conundrums about that objective external world. And this is awkward because we arrived at quantum mechanics in the first place partly due to the need to fix inconsistencies in the picture of reality available to physicists immediately prior to the invention of quantum mechanics, such as the ultraviolet catastrophe (Kragh, 2000). The QBist version of quantum mechanics can't be used to address these inconsistencies, since it lacks any propositional content which could be added to our beliefs about external reality to make them consistent or coherent, and thus what we are left with appears to be just the old, inconsistent picture of physical reality that was available to us prior to the advent of quantum mechanics, together with a normative prescription about how to form beliefs which is apparently supposed to be rationally compelling despite having no empirical justification. Thus paradoxically, Intersubjective QBism appears to lead to the conclusion that despite all the scientific progress that has been made since the advent of quantum mechanics, our knowledge and understanding of physical reality is not actually any better than it was in pre-quantum days. So although Intersubjective QBism may at least limit the threat to scientific knowledge outside of quantum mechanics, nonetheless its limited view of scientific epistemology makes it clearly incapable of solving the epistemic version of the measurement problem – not only is it unable to offer a physical account of the measurement process which could reassure us as to the reliability of our means of enquiry, in fact, it ends up denying that quantum mechanics is empirically confirmed by measurement outcomes at all, and indeed possibly undermines the idea that quantum mechanics represents any kind of scientific progress.

9

Observer-Relative Interpretations

In chapter 3 I set out an epistemic problem faced by the general class of observer-relative interpretations – they fail to allow the sharing of information between different observers, and even different versions of the same observer at different times. I will now examine this problem in greater detail. In the following chapter I will describe some more general reasons why intersubjective sharing of information is vital to the methodology of science, but in this chapter I will focus on trying to understand what kind of story we might be able to tell about scientific epistemology in the context of an observer-relative interpretation.

Now, in order to solve the epistemic version of the measurement problem, the goal for the proponents of observer-relative interpretations must be to show that in the kind of world they postulate, our methods of enquiry are still by and large reliable, i.e. they can still deliver some kind of meaningful scientific knowledge. The first hurdle to be overcome is that it is not obvious what scientific knowledge is supposed to be knowledge *about* in the context of an observer-relative approach. I will focus primarily on empirical scientific knowledge here, since that is the easiest case – if we can't even make sense of *empirical* scientific knowledge in this context, there's little hope of having any theoretical scientific knowledge! And note that traditionally empirical scientific knowledge is understood to be knowledge about universal or at least somewhat general regularities, with an implicit understanding that these regularities exist in some mind-independent, external reality. Evidently this kind of empirical knowledge is impossible in the context of an observer-relative approach, since such an approach denies the existence of any mind-independent external reality which we could have knowledge about. So if we consider it essential that empirical scientific knowledge should pertain to a mind-independent external reality, then observer-relative approaches are immediately a non-starter for empirical scientific knowledge; there just cannot be any such thing.

However, this seems a little too quick. We should, surely, allow that empirical scientific knowledge still counts as meaningful knowledge even if it is not instantiated in quite the way we usually expect. After all, it is possible (indeed, likely!) that we are incorrect in some of our most foundational assumptions about what reality is like, but this should not automatically mean that we don't have any empirical scientific knowledge. So let us now consider a few ways in which one might try to reconceptualise the notion of empirical scientific knowledge in the context of an observer-relative interpretation, in order to argue that our means of enquiry are still giving us meaningful knowledge about *something* in this context.

Saving Science from Quantum Mechanics. Emily Adlam, Oxford University Press. © Oxford University Press (2025).
DOI: 10.1093/9780197808887.003.0009

9.1 Many Perspectives

First, one might think that in an observer-relative view, empirical scientific knowledge pertains to universal regularities that apply within every individual perspective, rather than regularities holding in some external reality. And indeed, it seems clear that the proponents of observer-relative interpretations believe that quantum mechanics applies not just within one perspective but across all of the perspectives – for after all, the motivation for adopting an observer-relative approach, as described in section 3.2, only really makes sense if unitary quantum mechanics is supposed to apply generically across all perspectives, since the whole reason for relativising states and outcomes in the first place was to allow us to maintain that unitary quantum mechanics is complete and universal for both Alice and Bob.

But the problem is that it seems very hard to justify the belief that quantum mechanics applies across all perspectives in the context of an observer-relative view. The problem is that in an observer-relative picture we are, by definition, confined to our own perspective: the theory provides no mechanism which could grant us access to even one perspective other than our own. And as noted above, this applies even in cases where commonsense strongly suggests that we do, in fact, share information, such as cases where you and I have a conversation in which I ask you about your measurement outcomes – for in an observer-relative interpretation that conversation must be modelled as a quantum measurement performed by me upon you, and thus as argued in section 3.2, its outcome cannot reliably reflect your actual experience. Thus Type-III disaccord – that is, interactions in which the words heard by one observer fail to reflect what the other believes herself to have said – can be expected to be generic in the context of an observer-relative interpretation (Adlam, 2024c). And therefore, although observers in such a context can observe that events within their own perspective exhibit the regularities predicted by quantum mechanics, they cannot possibly have any empirical evidence for the claim that these regularities also appear in other perspectives. So if we interpret observer-relative approaches as intending to describe universal regularities that apply within every individual perspective, there is an immediate empirical incoherence – such theories imply that we cannot possibly get access to exactly the kind of information that would be needed to confirm a theory of this kind.

Now, one might be tempted to argue that we are entitled to simply make an ad hoc assumption that regularities persist across perspectives, in much the same way as we commonly make the ad hoc assumption that regularities persist across time in order to permit inductive inference. But as discussed in section 7.4, the induction assumption is not really so ad hoc, because although we have no initial justification for it, we are able to shape and refine it by means of empirical enquiries, in the spirit of progressive coherentism. That is, we don't just pick some regularities and decide by fiat that they persist across time; rather we perform experiments to find out which regularities persist across time, check whether the persistence breaks down in various regimes, perform experiments refining our understanding of the mechanisms responsible for the regularities, and so on. Whereas in an observer-relative picture, nobody can have even a single piece of empirical evidence whatsoever about whether any regularity *ever* persists across different perspectives, or about the kinds of regularities that typically persist across different perspectives, or the circumstances in which the persistence breaks down. So in the observer-relative context there is no hope of refining the originally unjustified assumption by means of empirical enquiries – we are left to simply make a guess.

In practice, of course, proponents of observer-relative approaches think the guess we ought to make is that the regularities holding in other perspectives are exactly those predicted by quantum mechanics. But this is surely motivated in large part by the high degree of empirical confirmation attached to quantum mechanics when we interpret measurement outcomes in the ordinary non-observer-relative context, and as we saw in chapter 4, confirmation accrued relative to one set of background beliefs cannot simply be transferred to a very different set of background beliefs in the way that the proponents of the observer-relative interpretations would like to do. When we make a very significant change to our background beliefs, we must make a holistic update to our belief system which will involve assessing the significance of the existing empirical results relative to the new context rather than the old one – and as we have just noted, in the observer-relative context no empirical result can ever be understood as providing information about the content of another perspective, so the existing confirmation attached to the theory in the non-observer-relative context does not appear to furnish any good reason to think that quantum mechanics applies inside other perspectives in the observer-relative context.

Of course, if one finds that one's own perspective exhibits regularities which can all be well-accommodated within a mathematically well-defined physical theory such as quantum mechanics, it is obviously tempting to infer that other perspectives are also described by the same theory – that it is a *universal* scientific theory, as we typically imagine theories to be. But our idea that scientific theories are generalisable to different times and places is grounded in empirical observations of various kinds of theories working in a variety of different times and places and for different observers; whereas in the context of an observer-relative interpretation we cannot ever have any evidence at all to suggest that scientific theories are typically correct across a variety of different perspectives, so we have no real reason to think they are generalisable in this particular way. Moreover, this line of argument will only work if we can, at least, get empirical evidence about the persistence of regularities across time within a single perspective – and as we will shortly see, even that is dubious in the context of the observer-relative interpretations.

In conclusion, in the context of observer-relative interpretations, it seems very hard to argue that we can empirically confirm quantum mechanics if it is to be understood as a description of regularities that apply across many different perspectives. This suggests that the observer-relative interpretations are indeed empirically incoherent, since they appear to tell us that we should not believe the very theory which is the motivation for postulating them in the first place. Approaches of this kind do not seem to be capable of giving a physical account of measurement (which, recall, includes 'measurements' that we perform on other observers when we exchange information with them) in such a way as to affirm the reliability of our means of enquiry, and thus they cannot give rise to a coherent scientific epistemology.

9.2 Just One Perspective

In light of these difficulties, there is one obvious route open to the proponent of the observer-relative interpretations: they could significantly narrow the scope of our empirical scientific knowledge, in order to say that rather than describing universal regularities across all perspectives, empirical scientific knowledge actually pertains only to the regularities exhibited within a single perspective, i.e. the perspective of the observer to whom the knowledge belongs. This would allow the proponents of observer-relative interpretations to say that although their account of reality has the consequence that no measurement or interaction can reveal reliable information about other perspectives, it does reassure us that measurements provide reliable information about everything that lies in the scope of a given observer's empirical scientific knowledge, i.e. regularities exhibited within their own perspective. So in that sense the bootstrapping approach would have succeeded in reassuring us that measurement has the properties it needs in order to stabilise the kind of knowledge the theory attributes to us.

Now, clearly this move would amount to a severe weakening of empirical scientific knowledge, since it requires us to accept that the laws and hypotheses we come up with in the process of doing science are not universal or even generalisable in any significant way. But suppose for the sake of argument that we are able to make our peace with this weakening – can this route then be made to work? Can we arrive at a coherent package of beliefs in an observer-relative picture in which we take ourselves to have empirical scientific knowledge about the regularities occurring within a single perspective across time?

Well, an immediate problem is that, as noted in the previous section, it was the hypothesis that unitary quantum mechanics applies universally for all observers which led to the idea of relativising observations to observers in the first place. If we are no longer insisting that quantum mechanics applies universally for all observers, it's unclear why we should still adopt a view in which everything is relativised to an observer; and yet if we don't adopt that view then we can no longer sensibly say that unitary quantum mechanics and other scientific regularities apply only within a single perspective. So immediately there seems to be an underlying incoherence in this approach.

And even if we can find some way to square that circle, more problems arise from the fact that most of the observer-relative interpretations offer no well-defined way of saying what counts as the same observer at different times. This is not a trivial matter because the long history of the philosophy of personal identity has made it clear that it is very difficult to come up with a precise, unambiguous physical criterion which identifies the same observer across time (Olson, 2003, 2023). And yet if we want to say that observers in an observer-relative world can have empirical scientific knowledge about regularities occurring within their own perspective across time, we must insist that an observer at a given time has access to the same set of relative facts as previous versions of herself, and therefore we must be able to say that there is a precise, unambiguous fact of the matter about whether or not she is the same observer at these different times – there doesn't seem to be room for her to be just *approximately* the same observer, because either she has access to the same set of relative facts, or she doesn't.

This leads to the suspicion that we will ultimately have to say that in observer-relative interpretations different versions of the same person at different times are simply distinct observers, associated with distinct perspectives containing entirely unrelated relative facts. And this would be a disaster for the epistemology of these approaches because if we can't even trust that *our own memories* are giving us reliable information about past measurement outcomes, we will have no meaningful information at all which could be used for empirical confirmation. Clearly such an approach will not be able to solve the epistemic version of the measurement problem, since there is no hope that it could affirm the reliability of any of our means of enquiry.

Perhaps the most die-hard proponents of observer-relative interpretations would try narrowing the scope of empirical scientific knowledge still further – one could perhaps say that that our empirical knowledge is nothing more than knowledge about whatever regularities happen to be displayed within a single perspective at a single moment in time. But this move reduces the scope of empirical scientific knowledge to the extent that it is basically worthless. Moreover, if we can't trust that our minds are stable over time it's unclear that we can even meaningfully carry out an extended process of reasoning, for we would need at least to trust our memories of the conclusions of previous steps of reasoning; and evidently we aren't going to be able to empirically confirm anything or indeed do science at all if we can't even do any temporally extended reasoning!

9.3 Observers Across Time

At this point it will be helpful to consider what a few specific observer-relative interpretations have to say about persistence over time for observers. First consider relational quantum mechanics (RQM). Here I will focus on the original version of RQM, which is an observer-relative approach – in chapter 13 I will consider a recent proposal for an alteration to RQM which has the consequence that it is no longer an observer-relative approach in the sense in which I have used that term, meaning that it no longer faces the problems described here.

Now, in RQM every physical system counts as an observer, so there are probably at least some 'observers' for whom we can give a precise, unambiguous physical criterion of identity over time. For example, in many physical regimes we can straightforwardly track the same particle across time, so the relative facts associated with each of these particles could plausibly remain stable across time. However, these kinds of observers are not much use to us with regard to the epistemology of the theory, because in that setting what we need is to be able to say that observers *like ourselves* are able to persist across time, in order that we can empirically confirm a scientific theory based on information accessible to us in our memories and records. So identity conditions for individual fundamental particles do not solve the epistemological problem: what we really need is to be able to say that *conscious, human observers* persist across time.

Indeed, before we even start to consider the question of persistence over time, we first need to figure out how we could obtain any perspective at all for a conscious human observer within RQM. For in this picture every one of the particles making up a human body, or even just a human brain, will have its own perspective on reality, and moreover we have seen that in the observer-relative picture we cannot expect any particular relations to hold between all of these perspectives, so it's unclear how we could ever get some kind of unified higher-level human perspective arising from all of these individual perspectives: since the individual particles can't share information, they can't reach agreement or form higher-level structures that encode shared facts. Perhaps we should simply say the human perspective arises from averaging over the individual perspectives of all of the particles in the brain, or something of that kind, but it's unclear that we would get anything other than a big mess out of that procedure. Adlam (2024b) notes that this problem is structurally similar to the 'combination problem,' arising in panpsychism (Lockwood, 1993; Rosenberg, 1998; Goff, 2009, 2011), which is the view that fundamental physical entities can have conscious experiences: as Chalmers puts it, panpsychism leads to the question of '*how do the experiences of fundamental physical entities such as quarks and photons combine to yield the familiar sort of human conscious experience that we know and love?*' (Chalmers, 2016, p. 1) However, the problem faced by RQM is arguably more difficult because in ordinary panpsychism we are able to say that different physical systems can share information by means of interactions and thus, as Chalmers notes, we can imagine that higher-level experiences may be '*partly grounded in causal or structural relations among the microexperiences*' (Chalmers, 2016, p. 6), whereas in RQM we do not have the same resources, since by definition there cannot be any causal or structural relations among the experiences of different observers.

In any case, even if we do find a sensible way of getting a human perspective out of the particles in the brain, it's certainly not obvious that this perspective would have any substantial continuity over time. After all, the particles in the brain are constantly being replaced with new particles which will necessarily have a different and unrelated perspective, so even if we're willing to allow the relation of identity between different perspectives to be a matter of degree rather than a binary, such that one's perspective may persist for as long as one's brain is made up of substantially the same particles, clearly this will not permit very long-term knowledge. So maybe there could be some hope that observers could have knowledge of regularities that have held in the sufficiently recent past in this picture, but as we look further back in time this knowledge will become much less reliable, so in this picture empirical scientific knowledge will, at the least, be very severely diminished; exactly how diminished would depend on the nature of the story that is adopted about how conscious perspectives emerge from particles, and how long they persist.

So alternatively, consider an interpretation like the neo-Copenhagen approach of Brukner (2017). Brukner opposes the route taken by RQM of treating all physical systems as observers, noting that '*we are ... encountering the limits of meaningful language when we associate the terms "knowledge" or "taken" to single electrons*' (Brukner, 2021, p. 23). Thus it seems he wishes to consider as 'observers' only conscious observers, or some such category. Brukner does not offer any precise specification of what kinds of things he counts as observers; however, he does not mention anything like a disembodied soul, so it is likely that he would ultimately want to say that consciousness, and hence the perspective associated with a consciousness, is something which arises naturally in certain kinds of physical systems. But if consciousness is just emergent in this way then it seems there cannot be any particularly deep connection between the consciousness of 'the same observer' at different times: there is some causal and spatiotemporal continuity, but no precisely defined thing which accompanies the observer at all stages of their existence, so in this kind of view it seems quite hard to insist that observers do indeed persist across time.

Alternatively, perhaps Brukner's view *should* be understood as postulating primitive facts about a persisting self which is like an essence or soul. Then we could straightforwardly say that there are well-defined facts about what counts as the same observer over time, since that is defined by the facts about where the essence or soul goes. However, making scientific rationality dependent on an invisible, non-physical essence or soul seems somewhat problematic, given the lack of any concrete scientific evidence for anything like a soul. Furthermore, this approach would leave us with no principled way to decide which of our memories we are able to rely on. For it would entail that my memories of the past are meaningful only if they are associated with an 'old' soul which existed in the past and connects me to past versions of myself who had those experiences; but I cannot tell purely from the character of my memories whether they reflect a real history associated with an old soul or whether my soul has just now come into being with brand new set of facts which don't reflect anything that actually happened in the past. Nor would there be any hope of using empirical enquiry to learn anything about where souls are and where they are not, since there is no measurement we can make to check whether or not a soul is present, or whether a soul at one time is the same as a soul at another time. So this approach may allow us to tentatively hypothesise that our memories may provide real information about the past, but not in any stable or reliable way, so this kind of view still severely limits our ability to use empirical evidence to learn about stable regularities persisting across time.

At this juncture one might object that surely every approach to the measurement problem has the same problem, because after all no other approach offers an unambiguous criterion of personal identity either – even classical physics doesn't provide such a thing. But this would be missing the point. Other approaches to the measurement problem don't *need* to postulate a criterion of personal identity because in those contexts nothing of importance hangs on questions about personal identity. If we start from a picture of the world which includes a unique, observer-independent, stable macroscopic reality, then memories and records can simply live in that macroscopic reality, accessible in principle to all observers for as long as they survive, regardless of whether the observers existing now are identical to other observers that existed in the past or not. So in that kind of context observers can in general expect that the records or memories they consult at some time will be at least approximately continuous with records and memories created at some earlier time: they never have to worry about *who* is doing the consulting, since anybody who consults these records will see the same outcome. Thus in most ordinary approaches personal identity will never appear anywhere in the physics, so there is no epistemological mandate to give a precise account of it. And this, surely, is as it should be: as Hedden (2015, p. 452) puts it, it is natural to think that '*determining what an agent ought to believe does not require first figuring out the correct theory of personal identity over time.*' Whereas in the observer-relative interpretations we must appeal to some notion of personal identity before we can say precisely what account of reality the theory offers, and in particular personal identity appears to play a crucial role in the epistemology, so these approaches cannot be complete or coherent unless they are able to provide a sensible criterion of personal identity over time.

9.4 Higher-Level Scientific Knowledge

We have seen that in the context of an observer-relative interpretation it is hard to justify the belief that the regularities described by quantum mechanics apply across all perspectives, or even across the perspectives of different temporal parts of the same observer. But suppose for now that we are willing to just assume without any evidence that indeed quantum mechanics does apply in all perspectives. Even then the observer-relative approach would still pose a number of serious problems for the epistemology of science, for we still have to think about the status of higher-level scientific knowledge – and it seems unlikely that merely assuming the applicability of quantum mechanics across all perspectives is enough to ensure that higher-level scientific regularities are also the same across all perspectives. Higher-level scientific knowledge would be secure in this context if *every* higher-level scientific regularity could be derived entirely from the laws of quantum mechanics, but this does not actually appear to be the case – in fact, initial conditions also play a role in establishing macroscopic regularities, and given that the observer-relative interpretations explicitly deny the existence of a shared external reality, there would seem to be no reason to expect that all the different perspectives will share the same initial conditions.

For example, consider the second law of thermodynamics, which is a time-asymmetric law characterising the increase of entropy over time. Because neither classical nor (unitary) quantum statistical mechanics contains any temporal asymmetries which could give rise to this universal increase in entropy, it is currently thought that the second law can only be derived with the help of the Past Hypothesis, which as discussed in section 7.2, involves postulating that the initial state of the universe was special in some way, for example, postulating that it had low entropy (Albert, 2000, 2014). And it's unclear how this approach could carry over to an observer-relative picture because in that picture there cannot be any universally shared initial state – to get thermodynamic effects we would apparently have to postulate some kind of local Past Hypothesis for *every individual observer*, located at whatever time the perspective associated with that observer comes into being. Even in the usual context where we have just one initial state there have been criticisms of the Past Hypothesis on the basis that it is unlikely or ad hoc (Price, 1996), but being forced to postulate a separate and independent local Past Hypothesis for every individual observer in the universe would certainly seem to make the problem much worse!

More generally, it seems likely that much of the empirical scientific knowledge associated with the special sciences would be at risk in such a context. For example, an ordinary accounting of empirical scientific knowledge would presumably include knowledge about the various different species of birds native to New Zealand, but the facts about birds in New Zealand are certainly not derivable from quantum mechanics alone – many contingent facts about initial conditions will be needed here. So even if you and I both have perspectives which obey the same fundamental quantum-mechanical laws, this does not imply that they also contain all the same facts about the kinds of birds that live in New Zealand. Thus in such a context, the 'knowledge' that I take myself to have about New Zealand bird species will end up just being an observation about how things are in the random fluctuation corresponding to my perspective, rather than representing meaningful knowledge of well-established empirical scientific facts. So even if we can somehow find a justification for thinking that all perspectives must obey the same fundamental quantum-mechanical laws in an observer-relative picture, a great deal of our higher-level scientific knowledge will still be irretrievably lost.

That is, if we are able to justify the claim that quantum-mechanical regularities hold across all perspectives in an observer-relative view, we are left in the odd position of saying that quantum mechanics may have been empirically confirmed, but most of our older, less fundamental scientific theories have *not* been empirically confirmed! So evidently this kind of approach does not offer a very good solution to the epistemic version of the measurement problem, since even if we make some assumption which affirms that we have meaningful knowledge of quantum mechanics as a universal scientific theory, it will still dramatically undermine our means of enquiry with respect to many other kinds of empirical scientific knowledge.

Indeed, it's not clear that this is really a coherent position – for as Curiel (2020) emphasises, empirically confirming a new scientific theory typically requires us to rely on the correctness of older theories, since those theories are needed to describe the functioning of measuring devices, our perceptual apparatus, the stability of the macroscopic world, and so on. In particular, empirically confirming quantum mechanics certainly requires us to rely on the correctness of some parts of thermodynamics – for as discussed in section 7.2, the existence of reliable records of the past seems to depend on the Past Hypothesis, and we can't empirically confirm quantum mechanics or indeed any other theory without reliable records of the past. So ultimately, the progressive, cumulative nature of scientific enquiry means we probably cannot accept an account of reality which affirms the correctness of quantum mechanics whilst undermining all of our other scientific knowledge, and thus solving the 'measurement problem' cannot stop at quantum mechanics – we need to solve it in a way which also upholds the pre-existing scientific knowledge on which the empirical confirmation of quantum mechanics depends.

9.5 Adding Intersubjectivity

The preceding discussion hopefully makes it clear that the possibility of communicating information between observers should be an important consideration in any putative solution to the measurement problem, and doubly so in the context of an observer-relative approach in which we cannot simply lean on the existence of a shared observer-independent reality to ground relations between different observers. Yet in modern discussions of the measurement problem, intersubjectivity, when mentioned it all, is often not given a particularly precise or careful treatment. Shimony (1963, p. 19) was one of the first to criticise the treatment of this issue in work by authors including London, Bauer, Bohm, Feyeraband, and von Neumann; Shimony argued that simply the fact that observers using quantum mechanics will make the same statistical predictions for a given ensemble is not adequate to secure meaningful sharing of measurement outcomes because we also require '*the agreement of the two observers in a **specific** reading of the apparatus*' and this '*would be a coincidence unless the examination of the first observer ... [has] effected a change in [the system] which is not negligible from the standpoint of subsequently observers.*' That is, Shimony emphasises that the standard formalism of quantum mechanics does not provide for intersubjective agreement on the outcomes of *specific* measurements, so something must be added to the formalism to achieve this, otherwise agreement would require '*a kind of implausible pre-established harmony!*' (paraphrase by French (2023, p. 75)).

However, presentations of observer-relative approaches often do not take this mandate seriously. When the issue is mentioned at all, it is common to simply insist that intersubjective agreement must obtain, even if it is not represented in the structure of the theory. For example, Dascal (2020, p. 315) presents a neo-Copenhagen interpretation which appears to be an observer-relative interpretation, and then writes, '*Were Bob to measure in a basis that instead corresponded to Alice's coin toss (i.e. a basis whose vectors corresponded to 'Alice having observed heads' and 'Alice having observed tails'), then the theorist may explain, perhaps, that his information frame grows or changes to encompass Alice's as well.*' But this 'encompassing' is not represented explicitly anywhere in the models Brukner offers, since they are simply the models of standard unitary quantum mechanics. There is something of a double standard here: advocates of observer-relative views certainly expect that physical theories should predict the probability distributions over outcomes for individual observers in quantitative detail, but then for some reason they do not demand that theories should also predict the correlations between different observers in any kind of quantitative detail. Yet there seems to be no good reason for this difference in attitude: agreement between different observers plays a vital role in the empirical confirmation of any physical theory, and thus no physical theory can be empirically adequate if it is incapable of affirming the possibility of such agreement, so correlations between observers ought to be considered as a part of the body of empirical data that such theories must predict.

Similarly, French (2023) (in an exegesis of work by London and Bauer (1939)) discusses intersubjectivity in the context of a phenomenologist interpretation of quantum mechanics. Now, phenomenologists have always placed importance on intersubjectivity as a feature of the 'life-world' or world of common sense – for example, Husserl (1970) emphasised that in his phenomenological picture, intersubjectivity is established via empathy, which involves '*co-attending to the other's experiences in the sense that when comprehending their experiencing, ours passes through that experiencing and reaches all the way through to what they experience*' (paraphrase by French (2023, p. 156)). French thus maintains that a phenomenologist interpretation of quantum mechanics necessarily upholds intersubjectivity. However, no alteration is proposed to the theory to actually *represent* this intersubjectivity. And yet, as French (p. 117) puts it, for the phenomenologist '*what the quantum mechanical formalism captures and represents is precisely that correlation that holds between the observer and the system observed*' – and therefore if the system observed is *another observer*, with whom we are supposed to be able to achieve intersubjective agreement by means of 'empathy,' then the formalism surely ought to represent that kind of correlation as well. Granted, for the phenomenologist the 'life-world' and its evident intersubjectivity is supposed to be primary, but again there seems to be a double standard in action: the phenomenologist still expects that the theory will predict individual events occurring in the life-world in quantitative detail, and yet it is somehow regarded as unnecessary for the theory to say anything about the relationship between different observers, even though the phenomenologist is committed to the existence of such relationships.

The general point here is that a theorist who expects that intersubjective agreement is possible – either by 'empathy' or by any other route – ought to demand that the physics we ultimately arrive at is capable of explicitly affirming that initial expectation. Intersubjectivity cannot be a hand-wavy assumption tacked on after the rest of the theory is already complete – it is central to a theory's empirical content and thus it should be built properly into the theory, along with all of the other quantitative predictions. Prior to the various Extended Wigner's Friend theorems it perhaps seemed adequate to simply postulate that observers can share their measurement outcomes and expect the physics to fall into line, but these recent results have made it clear that intersubjectivity is not a trivial matter in the context of quantum mechanics, so it is essential for a putative solution to the measurement problem to have something concrete to say about when and how intersubjective agreement comes about.

In light of these difficulties, proponents of observer-relative interpretations might perhaps be tempted to relax postulate one slightly, permitting us to add something minimal to unitary quantum mechanics in order to arrive at a view which no longer quite upholds the universality and completeness of quantum mechanics, but which does permit intersubjective agreement. I will discuss what this might look like in chapter 13, but for now let me simply note that once an addition of this kind is made, the result is not an observer-relative interpretation any more, and perhaps more importantly it is no longer unitary-only. So even if the addition needed seems quite small, it would be a significant step – many proponents of observer-relative interpretations have a very deep commitment to the idea that unitary quantum mechanics is universal and complete, and tend to take a dismissive attitude to any approach which postulates anything akin to a 'hidden variable,' so it would be a significant retreat to allow any kind of addition to unitary quantum mechanics.

9.6 No Worse Off Again

In response to the concerns raised in this chapter, one might perhaps object that ordinary scientific realists are similarly unable to prove that intersubjective communication is possible, since we can never know for sure what is going on inside another's consciousness. So this seems to point to another kind of No Worse Off argument, akin to those discussed in chapter 7: perhaps observer-relative approaches are actually not any worse off than standard forms of scientific realism, since either way we are going to have to make some ultimately unjustified assumption about the possibility of communication with other observers.

Now, as a first response, it should be emphasized that the ordinary realist is in some ways less reliant on assumptions about intersubjectivity than the proponent of an observer-relative approach. For in a non-observer-relative realist approach, most scientific knowledge is not considered to be knowledge *about* other perspectives; rather it is understood as pertaining to regularities occurring in some mind-independent external world, to which each observer has at least limited access. So in this context, we can have scientific knowledge about universal regularities instantiated in the mind-independent external world even if we don't take ourselves to know anything about what other observers are experiencing. But things are very different in the context of the observer-relative interpretations, because those approaches specifically deny that there is any mind-independent external reality, so scientific knowledge cannot be about anything other than the content of perspectives. So in *this* context, the conclusion that we cannot actually know anything about the content of any perspective other than our own amounts to a very significant weakening of what we take ourselves to know about the world, because there is really nothing in this picture to have knowledge about other than the perspectives of other observers.

Moreover, although realists cannot prove that intersubjective communication is possible, nonetheless this assumption is not unfounded or ad hoc – it is coherent with and affirmed by the rest of the realist's belief system. For believing in an external, observer-independent reality provides some justification for supposing that other observers are having experiences that are related in systematic ways to one's own, because we can say that this shared external world is the ultimate source of the various experiences that different observers are having. Moreover, in the realist context, the scientific theories we ultimately arrive at will necessarily have something to say about the way in which interactions with the external world give rise to experiences, so they will therefore also say something about how the experiences of different observers are related, and this makes it possible to progressively update and refine the original assumption about intersubjectivity on the basis of empirical evidence and theoretical developments. Thus in the setting of traditional realism, we are able to arrive at a stable position in which our knowledge about our own experiences, our beliefs about the content of external reality, and our beliefs about the experiences of others all fit together and mutually support each other. That is, we need not adopt a foundationalist picture in which the traditional realist must first commit to the possibility of intersubjective communication and then build a system of beliefs on that unfounded assumption – instead they can retrospectively justify their belief in the possibility of intersubjective sharing of information by means of a progressive bootstrapping approach.

By contrast, in the context of an observer-relative approach there is no external world within which this coherent picture could be sited, so an observer-relative picture seems to offer no reason at all to think that the experiences of different observers are related in any way, and it also cannot tell us anything about the specific way in which those experiences are related. Thus in this context, if we nonetheless assume that intersubjective communication is possible, we are assuming something which is quite incoherent with the rest of our belief system, and which is not susceptible to any refinement by means of empirical enquiries or theoretical developments. So in this kind of case the progressive development of science can never possibly justify or improve the original assumptions about intersubjectivity, so there is a clear sense in which the observer-relative interpretations are really Worse Off than ordinary realism.

Considerations of this kind demonstrate that the account of scientific epistemology as a form of bootstrapping, as originally sketched out in chapter 4, must be expanded somewhat. At that point I focused on the problem of giving an account of the physical nature of measurement which affirms our assumptions about the reliability of the measurement process, but we can now see that this is insufficient; in addition to giving an account of the physical nature of measurement, we also need our account to tell us how to model other observers and the interactions between them, in order that the resulting system of beliefs can reassure us that we are in meaningful contact with our wider epistemic community in the way that is needed to support the use of shared information in the process of scientific enquiry. I will return to this point in chapter 10.

9.7 Non-Absoluteness Theorems

Despite the difficulties posed by observer-relative approaches, there has been significant interest in them throughout the history of quantum mechanics, at times bolstered by various arguments to the effect that some quantum phenomenon proves that there cannot be any observer-independent third-person reality. In this section I will examine one specific theorem that has sometimes been employed to support this claim, but the primary objection against such claims can be stated quite succinctly: it cannot be the case that some quantum phenomenon proves that there does not exist any third-person objective reality because no scientific result could *ever* prove such a thing. For doing away with a third-person objective reality which can ground connections between different observers and different temporal parts of the same observer leads to an extremely impoverished view of scientific knowledge as nothing more than a description of the transient experiences of a single observer at a single time, and scientific knowledge so limited certainly cannot lead to a sweeping, universal conclusion like 'there is no objective reality,' so such an argument can never really get off the ground.

With that said, as a matter of fact the various 'Non-Absoluteness' theorems typically invoked in these kinds of arguments do not really say anything about the existence of a third-person reality, and nor is it the case that the physicists who developed these theorems have generally suggested that they do. As the originators of the Non-Absoluteness theorems themselves emphasize, such theorems simply show that some set of assumptions are mutually inconsistent, meaning that one of the assumptions must be rejected. The connection to 'objectivity,' comes from the fact that one of these assumptions is an assumption to the effect that the outcomes of quantum measurements are 'absolute,' so one possible option for responding to such a theorem is to reject this assumption and thus conclude that the outcomes of quantum measurements are not 'absolute,' which is typically taken to mean that they are not observer-independent objective facts. But even if we do take this route, clearly the claim that the outcomes of quantum measurements are not absolute does not entail that *nothing* is absolute – perhaps such a claim can be arrived at if one also assumes that quantum mechanics is a complete description of reality, but that involves extrapolating significantly beyond what is proved by the actual theorems.

So let us now put aside the issue of objective reality as a whole and simply consider what these theorems tell us about the absoluteness or otherwise of quantum measurement outcomes. I will use the theorem of Bong et al. (2020) as a representative example of these approaches. To state this theorem, it is helpful to imagine a hypothetical solution to the measurement problem which I will refer to as 'the Magic Bullet.' This solution is supposed to satisfy the following four assumptions: a) Locality, b) No-Superdeterminism, c) Absoluteness of Observed Events (AOE), and d) Universality of unitary quantum mechanics. Bong et al. construct an extended Wigner's Friend scenario and use it to show that the Magic Bullet cannot exist – that is, at least one of the four properties ascribed to it must fail in the scenario they describe.

As stated by Bong et al. (2020), assumption a), Locality, is similar to the assumption used in Bell's theorem, although slightly weaker. Assumption b), No-Superdeterminism, requires that the result of a measurement at a given time must be probabilistically independent from choices made in the future of that measurement; this assumption as applied by Bong et al. rules out both superdeterminism in the ordinary sense (see chapter 12) and also retrocausality, since it requires us to assume that the later choice of measurement does not have a backwards-in-time influence on the outcome of the earlier measurement. Assumption c), Universality of unitary quantum mechanics, is understood by Bong et al. to mean that the predictions of unitary quantum mechanics *as applied by any individual observer* are always correct. Finally, the assumption d), Absoluteness of Observed Events, has several distinct parts: Adlam (2024c) suggests decomposing it into AOE1, which requires that a measurement has only one outcome relative to the observer who performs that measurement, and AOE2, which requires that if one observer performs a measurement on another observer aiming to establish the outcome of their prior measurement, the value obtained by the first observer will be the same as the value that the second observer actually witnessed in the earlier measurement. Thus understood, AOE1 serves to rule out the Everett interpretation and other similar approaches which allow more than outcome per observer, and AOE2 serves to insist on the possibility of intersubjective communication between observers about measurement outcomes. In particular, the 'measurement' that one observer performs on another could involve simply asking them about their measurement outcome, so the assumption AOE2 rules out Type-III disaccord, at least in this particular kind of scenario. Since AOE1 and AOE2 are together sufficient to derive the conclusions that Bong et al. derive from their combined assumption AOE, it follows that anyone who wishes to respond to the Bong et al. theorem by denying AOE must ultimately be denying either AOE1 or AOE2.

Now let us describe the extended Wigner's Friend scenario postulated by Bong et al. Imagine an experiment where we have two agents, Chidi and Divya, in closed laboratories, each in possession of one particle from a pair of entangled particles. Chidi and Divya perform certain fixed measurements on their particles, obtaining measurements results C and D. Then we have another observer Alice who performs a 'supermeasurement' of a fixed observable on the whole system of Chidi's closed lab, obtaining an outcome A, and we have an observer Bob who performs a similar supermeasurement on the whole system of Divya's closed lab, obtaining an outcome B. Finally, at the end of the experiment Bob tells Alice the result of his measurement. Note that the term 'supermeasurement' means that Alice performs her measurement in a basis which does not commute with the basis in which Chidi's memories of his measurement outcome are stored. The failure to commute simply means that the order in which the corresponding operators are applied changes the probability distribution over the outcomes, and for present purposes what is important about this is that it means Alice's measurement has the effect of erasing Chidi's memories, i.e. once Alice has performed her supermeasurement, no future measurement that she can perform on Chidi will be able to recover the full information about Chidi's measurement outcome. Note that Alice can only perform such a measurement if she maintains complete coherent control over all the degrees of freedom in both Chidi and his lab, which in practice is extremely technologically challenging.

Now, what does the Magic Bullet say about this scenario? First, since the Magic Bullet obeys AOE1, it tells us that no more than four measurement outcomes are observed in this scenario – and since the Magic Bullet also upholds the universality of unitary quantum mechanics for each individual observer, it also tells us that each observer does observe an outcome, so it must be the case that *exactly* four measurement outcomes are observed. Second, since the Magic Bullet obeys AOE2, it must predict that if Alice, instead of measuring the observable corresponding to A, instead goes into Chidi's laboratory and asks him what result he got, then what she hears will match Chidi's observed outcome C. The same is true for Bob and Divya. In addition, AOE2 also ensures that when Bob tells Alice his measurement outcome, what she hears will match his actual outcome. So together these assumptions entail that Alice could in principle find out the values of any one of the four pairs of outcomes {*AB*, *AD*, *CB*, *CD*}, though she can't access them all in a single experiment. This means that the four outcomes A, B, C, and D must all have unique, well-defined values relative to Alice, since she could in principle find out the value of any one of them; and thus the Magic Bullet must be able to assign a joint probability distribution over the values of A, B, C, D. Moreover, the Locality and No-Superdeterminism assumptions together entail that the results of the earlier measurements, C and D, can't depend on choices made later by Alice and Bob about whether to measure A and B or to ask Chidi and Divya about their results, and nor can Alice's result depend on what Bob chooses to do or Bob's result depend on what Alice chooses to do. Hence the Magic Bullet must assign the same probability distributions over A, B, C, and D regardless of which of these things Alice and Bob choose to do. Moreover, because the Magic Bullet obeys the universality of unitary quantum mechanics, it must predict that the pair of results Alice ends up seeing will always match the predictions of unitary quantum mechanics. Thus the probability distribution assigned by the Magic Bullet must reproduce the predictions of quantum mechanics for all four of the pairs AB, AC, CB, CD, without allowing the probability distribution over individual variables depends on which of these pairs Alice ultimately sees.

We can now state straightforwardly what is proved by the theorem of Bong et al.: they show that if we do not allow that the probability distribution over individual variables may depend on which pair Alice sees, then there is no mathematically possible probability distribution over the variables A, B, C, and D' which reproduces the predictions of unitary quantum mechanics for *all* of the pairs AB, AC, CB, CD. This means that the Magic Bullet cannot exist – any mathematically possible model must violate one of the four assumptions a) – d).

Now, one obvious way to respond to this theorem is to simply deny assumption c), universality of unitary quantum mechanics – for example, by adopting a spontaneous collapse approach such as the GRW model. (Ghirardi et al., 1986). In that case, we no longer need to ensure that the probability distribution over A, B, C, D always reproduces the unitary predictions for the pairs AB, AC, CB, CD. However, as noted in chapter 2, many non-unitary approaches have difficulties accommodating QFT, so there are some good reasons to prefer an interpretation in which unitary quantum mechanics is universal, in which case we will have to deny one of the other three assumptions instead. And while any of the three assumptions a), b), and d) could in principle be rejected, the recent resurgence in interest in observer-relative approaches has been driven by the idea that we should respond to this dilemma by rejecting assumption d), AOE, which is then supposed to lead to the conclusion that quantum measurement outcomes are not observer-independent and objective.

However, as argued by Adlam (2024c), it's unclear that rejecting AOE really does provide good motivation for the claim that measurement outcomes are not observer-independent and objective. For denying AOE necessarily amounts to denying either AOE1 or AOE2. And if we deny AOE1, what we are most naturally led to is not a picture in which measurement outcomes fail to be objective, but rather the Everett interpretation, or some similar branching model – in such a model the outcomes all exist in an observer-independent, objective fashion, it just so happens that more than one such outcome occurs. And if we deny AOE2, what we get is merely Type-III disaccord: in the case where Alice asks Chidi about his measurement outcome, what Alice hears Chidi say is not the same as what Chidi experiences himself as reporting. Of course this disaccord *could* occur in a scenario in which outcomes are not observer-independent and objective, but we can still have Type-III disaccord in a context where all relevant events exist in an absolute, observer-independent sense – we need only define the dynamics of the Alice-Chidi interaction such that certain facts about Chidi's intrinsic condition are not dynamically relevant to Alice in this interaction, meaning that the outcome she obtains in her 'measurement' of Chidi does not reliably reflect such facts. So there appears to be no particular reason to interpret the failure of AOE2 as demonstrating that measurement outcomes are not observer-independent and objective, unless one is already predisposed in favour of such views.

Moreover, I have argued throughout this chapter that *generic* Type-III disaccord would be fatal to the epistemology of science. We can perhaps allow that Type-III disaccord occurs in some very rare situations which are not relevant to our ordinary processes of empirical confirmation, but in order to make sense of scientific practice we must insist that it does not happen in most ordinary conversations about measurement outcomes. And since the interaction between Alice and Chidi in this situation is indeed just an ordinary conversation about a measurement outcome, if we allow that Type-III disaccord occurs in this situation, it seems that Type-III disaccord could occur in basically any such interaction between observers, so once we allow this it seems hard to resist the existence of generic Type-III disaccord.

So, in fact, insofar as our goal is to arrive at an account of reality which has a coherent story to tell about the epistemology of science, we likely cannot accept the rejection of AOE2 as a reasonable response to the Bong et al. theorem. And indeed, we arguably can't accept the rejection of AOE1 either, since that will lead to something like the Everett interpretation, and we saw in chapter 6 that such approaches have severe epistemic problems of their own. So once we analyse this experiment with the epistemic version of the measurement problem in mind, it seems likely that, if we are not willing to give up on assumption c), the universality of unitary quantum mechanics, the only possible response to the dilemma it poses is to reject either a) Locality or b) No-Superdeterminism. As a matter of fact, we will see in chapter 12 that superdeterminism in the traditional sense also leads to serious epistemic problems but rejecting retrocausality remains a possibility. So in conclusion, it's not clear that the Bong et al. result offers any compelling reason to take observer-relative approaches seriously, but it does appear to point us towards approaches exhibiting either retrocausality or non-locality or both.

There also exist other recent 'non-absoluteness theorems' in the literature, which use slightly different assumptions but arrive at similar conclusions. For example, Leegwater (2018) and Ormrod and Barrett (2022) offer theorems in which a set of 'inner' observers in laboratories make certain fixed measurements, and then a set of 'outer' observers make certain supermeasurements on the laboratories, and we then demand that any collection of measurements occurring on the same spacelike hyperplane should obey the predictions of unitary quantum mechanics. Here, requiring that the measurements occur on the same 'spacelike hyperplane' just means that there is some frame of reference in which they all occur at the same time. It can then be shown that there is no possible set of outcomes for all of these measurements which does, in fact, obey quantum mechanics on all spacelike hyperplanes, so either we must deny that unitary quantum mechanics is correct on all spacelike hyperplanes, or we must accept that some of the measurement outcomes are relativised to observers rather than being absolute.

Now, these theorems do not employ an assumption of non-locality, no-retrocausality, or no-superdeterminism, and nor do they assume anything about whether or not observers are able to share their observations with one another. However, they do involve a stronger notion of universality than in the Bong et al. result – Bong et al. are only concerned with ensuring that the actual observations made by the single observer Alice will always agree with quantum mechanics, whereas the Leegwater and Ormrod and Barrett results are concerned with ensuring that any collection of quantum-mechanical measurement outcomes which all take place on the same spacelike hyperplane will exhibit the correlations predicted by unitary quantum mechanics *even if no single observer could ever see all of these results.* For it should be emphasised that in the scenarios employed in these theorems, the supermeasurements made by the outer observers will erase the results of the measurements made by the inner observers, so nobody will ever be able to compare the outer and inner results – and yet these theorems insist that the outer and inner results should be correlated in the way predicted by quantum mechanics, even though no individual person could ever observe that they are so correlated.

And this observation indicates that there is a subtle sense in which these theorems *do*, in fact, rely on a no-retrocausality assumption. For if the inner observers make their measurements and then the outer observers do not subsequently perform supermeasurements, then a single person can collect and compare the results obtained by the inner observers, and we have very good empirical evidence to support the hypothesis that in this case the results will match the predictions of quantum mechanics. But we do not, of course, have any direct empirical evidence that the results will match the predictions of quantum mechanics even if they are subsequently erased, since we cannot consult any evidence after it has been erased! So the demand that these results should match the predictions of quantum mechanics *even if the results are subsequently erased* is in some sense predicated on the assumption that the results obtained in the earlier measurements shouldn't depend on whether or not those results are later erased. Whereas if we allow the possibility of retrocausality, then it is open to us to say that in the case where some of the results are later erased, they need not obey the predictions of quantum mechanics, since it is guaranteed that in this scenario no empirical violation of quantum mechanics will ever be observed. So arguably these theorems too can be thought of as pointing us toward accepting retrocausality, even though technically they do not employ a no-retrocausality assumption; for allowing retrocausality can offer a reasonable explanation for why the predictions of unitary quantum mechanics might not be upheld on all spacelike hyperplanes in this particular scenario.

An interesting extension of the Bong et al. theorem, set out by Ying et al. (2023), provides a further example of the way in which taking epistemology seriously can make a significant difference to the analysis of various foundational results. The idea behind this theorem is that we can actually use an operational procedure to verify the No-Superdeterminism assumption used by Bong et al., which would seem to rule out the possibility of rejecting that assumption and would thus force us to reject some other assumption instead.

To perform this operational procedure, we create many copies of the original Bong et al. experiment which will be performed in parallel, with many copies of Chidi but just one external observer Alice who will measure all of these Chidis. Then we add a new observer, Veronika, who talks to all of the copies of Chidi and to Alice in the period between Chidi's measurement and Alice's measurement, thus collecting information on the correlation between the measurement outcomes obtained by the Chidis and on Alice's chosen measurement directions. Veronika can then look at the statistics she has collected and check that Chidi's outcome is probabilistically independent of Alice's later choice, as required by the No-Superdeterminism assumption.

Now, Alice can only perform a supermeasurement of the variable A on one of the Chidis if the measurement fully erases all information about his outcome, so in this case the specific information gathered by Veronika about the outcome must also be erased in this measurement, meaning Alice must exert coherent quantum control over Veronika as well as all the Chidis. However, we can still allow Veronika to send a message out of the experiment, confirming that she found the outcomes and measurement choices to be probabilistically independent, while giving no information about the specific outcomes obtained by any of the Chidis – Ying et al. (2023)

show that unitary quantum mechanics predicts that this message can be sent without preventing the supermeasurement from being performed. Thus we can check the message sent by Veronika to make sure that Chidi's measurement outcome is independent of Alice's later choice of measurement, as the No-Superdeterminism assumption requires.

Note that in order to say that the procedure really succeeds in its verification, we must make some assumptions relating to intersubjectivity; we must assume that i) each Chidi always has a measurement outcome, ii) when Veronika talks to any of the Chidis, what she hears is the same as what Chidi says, iii) when Veronika talks to Alice, what she hears is the same as what Alice says, and iv) when Veronika prepares her message, the content of the message as perceived by another observer after the erasure always correctly reflects what Veronika has actually experienced before the erasure.

Now, these assumptions seem very natural, and indeed, the issues discussed in this chapter suggest that in order to make sense of the epistemology of science we must suppose that these assumptions are correct in most ordinary situations. However, to argue that the verification procedure succeeds, we must assume that i), ii), iii), and iv) all hold *even in the case where the experiences of Veronika and Chidi are subsequently erased*. So the question is, do we still have good grounds to make these assumptions in the special circumstances in which the experiences themselves are later erased?

Now, one might initially argue that since the conversation between Veronika and Chidi and the preparation of the message all occur *before* Alice decides whether or not to perform a super-measurement, then these processes can't depend on whether or not a supermeasurement is later performed. This would suggest that even if the experiences of Veronika and Chidi are later erased, events at the time of their conversation must occur in exactly the way they would if nothing were subsequently erased; and thus, since we require that assumptions i), ii), iii), and iv) should be true in ordinary cases, we have good reason to insist that they must be true in this case also. However, this defence works only if we rule out retrocausality prima facie – and yet one of the things that the operational procedure is seeking to establish is that no retrocausality is exhibited in this situation, since that is one major part of the No-Superdeterminism assumption. So it would be circular for the justification of the verification procedure to assume from the outset that the interactions between the Chidis and Veronika and the content of Veronika's message can't depend in any way on whether or not those experiences are later erased.

One might instead argue that i), ii), iii), and iv) must hold simply because we need to assume that conversations between observers usually succeed in achieving meaningful communication. However, the point of invoking retrocausality here is that it would to allow us to say that communication fails *only* under the very specific circumstances where the experiences are later erased by a supermeasurement. And note that it is for all practical purposes impossible that any of our current experiences, or the experiences we have used to arrive at quantum mechanics, will later be erased by a supermeasurement – that would require collecting together and implementing fine control over a vast number of degrees of freedom which have been spreading across the universe at roughly light-speed for more than a hundred years. So a model in which communication fails in the specific, clearly specified circumstances of this highly unusual procedure need not undermine our confidence that communication is usually successful in ordinary cases, and thus such a model would not seem to pose any threat to the empirical evidence for quantum mechanics or the epistemology of science more generally. Certainly the failure of assumptions i), ii), iii), or iv) in circumstances where a supermeasurement is later performed would be contrary to our usual way of thinking about macroscopic interactions, but it appears to be benign weirdness rather than epistemic weirdness, so we are not rationally compelled to reject it.

This example illustrates the importance of understanding *why* we are making certain assumptions, rather than merely appealing to intuition. Ying et al. (2023) seem to take the view that a), b), c), and d) must hold even in the case of erasure simply because their failure is counterintuitive in light of our standard classical picture of the world – but, of course, we already know that all models of the reality underlying quantum mechanics are likely to be in tension with our standard classical picture in one way or another, so this is not a very strong argument for assumptions i), ii), iii), and iv). Whereas considering these assumptions in the light of the demands of scientific epistemology makes it clear that we do indeed need these assumptions to hold *in ordinary cases*, but if retrocausality is possible, we need not insist they still hold in the case of erasure. Thus we can avoid inconclusive debates about how strongly intuition points in one direction or another by distinguishing between benign weirdness and epistemic weirdness in situations of this kind.

To conclude, it certainly seems true to say that the various extended Wigner's Friend theorems demonstrate that coming up with a viable account of a third-person objective reality is going to be more complex in the context of quantum mechanics than classical physics, and arguably we do not yet know exactly how to do it. But the right response to this is not to just give up and retreat to an observer-relative view. For really the difficulty is the whole point – the fact that it is hard to tell such a story makes it clear that there is epistemic risk in *attempting* to tell the story, and that risk is needed to make the bootstrapping structure of empirical confirmation meaningful. Indeed, part of the reason why it is important to seek an objective third-person account is that that is precisely where the demand for coherence has real bite: in a 'theory' which does nothing other than provide a statistical characterisation of the experiences of a single observer, there's nothing for coherence to get its teeth into, because all the difficult questions appear precisely when we try to describe how the experiences of the observer are situated within a more general picture. So in order for the progress of science to meaningfully affirm, correct, or refine our assumptions about the reliability of our means of enquiry, it is vital that our theories should explicitly model observers and the process of observation.

9.8 The Copenhagen Interpretation

Many observer-relative interpretations can be regarded, or at least are regarded by their proponents, as spiritual heirs of the Copenhagen interpretation, the approach to quantum mechanics particularly associated with Bohr and Heisenberg. So let me close this discussion of the observer-relative interpretations by considering whether or not the epistemic difficulties discussed in this chapter may apply to the Copenhagen interpretation as well.

Now, Bohr's own views vary or are expressed differently throughout his life, and therefore 'the Copenhagen interpretation' is arguably not a single well-defined position. Indeed, it has been argued that what is now called the 'Copenhagen interpretation' may not really be a faithful representation of Bohr's views at all (Howard, 2004). But nonetheless, let us try to see if the Copenhagen interpretation as it is currently understood obeys the three postulates for observer-relative interpretations set out in section 3.2.

First, most versions of the Copenhagen interpretation do seem to posit that quantum mechanics is universal and complete – certainly, all self-professed proponents of the Copenhagen interpretation are in opposition to the idea of adding to quantum mechanics anything like a wavefunction collapse or hidden variables. So the Copenhagen interpretation seems to obey the first postulate. In addition, it seems fairly clear that the Copenhagen interpretation does not intend to posit multiple worlds, which suggests it also obeys the second postulate. There is perhaps more ambiguity around the third postulate, but at least some of Bohr's comments could be regarded as supporting a relational or perspectival approach. For example, Bohr (1939, p. 269) argues that quantum mechanics reveals '*an absolute limit to the possibility of speaking of a behavior of atomic objects which is independent of the means of observation*'. Here he seems to be articulating what we would now consider to be an observer-relative interpretation, in which the properties of quantum systems are relativised to a means of observation and thus perhaps to a specific observer.

But on the other hand, perhaps the most prominent feature of Bohr's views was his principle of complementarity, which involves '*a renunciation of a single picture of the microphysical object in favor of a set of mutually exclusive descriptions appropriate in different circumstances*' (paraphrase by Shimony (1963, p. 23)). That is, Bohr insists that quantum phenomena cannot be given classical explanations, but they must still be '*described in terms of classical physics concepts, because these concepts are indispensable in characterizing the measuring instruments*' (paraphrase by Shimony (1963, p. 24)). So Bohr consistently emphasised the fact that quantum measurement outcomes ultimately live in our familiar classical world, and he noted that '*our task must be to account for such experience in a manner independent of individual subjective judgement and therefore objective in the sense that it can be unambiguously communicated in ordinary human language*' (Bohr, 1958). And when Bohr speaks of 'communication,' it seems natural to think he means not only that one observer can individually articulate their experiences in ordinary human language, but that also other observers can *learn* about their experiences by means of this shared language – so, for example, Shimony (1963, p. 24) attributes to him the view that '*whenever everyday and classical concepts can be unambiguously applied, it is possible to distinguish sharply between the objects observed and the observer ... Intersubjective agreement is thus assured in the reading of the measuring apparatus.*' The observer-relative interpretations may allow observers to formulate their experiences in ordinary language, but then they deny the possibility of using that language to actually share such information with other observers. Thus one suspects that Bohr would likely not have been willing to accept an observer-relative approach which denies the possibility of intersubjectivity, and therefore the Copenhagen interpretation as Bohr understood it is probably not an observer-relative interpretation in the sense in which I have used that term. From this point of view, the modern neo-Copenhagen and perspectival interpretations may be less similar to Bohr's vision than they initially appear, at least until they are supplemented with some structure which secures intersubjectivity.

One conclusion we can draw from this is that the Copenhagen interpretation, if understood as upholding intersubjectivity in the way that Bohr would presumably have wanted, likely cannot be 'unitary-only' in the strongest sense of that term. For we have seen that in an interpretation of unitary quantum mechanics which upholds the possibility of intersubjective communication but which does not postulate multiple worlds, there will necessarily have to be some additional structure which tells us how the measurement results obtained by distinct observers relate to each other. Perhaps Bohr and his contemporaries assumed that this would be trivial and that the outcomes obtained by different macroscopic observers could always be accommodated consistently in a single shared classical reality. But the extended Wigner's friend theorems in particular have drawn attention to the fact that intersubjectivity is not so straightforwardly obtained in the context of quantum mechanics: there is a substantive question about how to implement it, and the particular circumstances in which it breaks down. It is therefore difficult to judge the extent to which '*the* Copenhagen interpretation' succeeds at solving the epistemic version of the measurement problem – we would first need an explicit statement of whether and how the approach is supposed to ensure intersubjectivity. I think it is likely that such an augmented Copenhagen interpretation would end up looking like one of the modified unitary-only interpretations which I will discuss in section 13.1.2, so let us defer further discussion until that point.

10
First and Third Person Views of Science

In this chapter I will consider in greater detail the role of intersubjective sharing of information in the epistemology of science. My central point will be that scientific knowledge is not and has never been something which can be produced by one lone genius: scientific knowledge is jointly created by an entire epistemic community, and thus any reasonable account of the epistemology of science must attach that knowledge to the community rather than to an individual. And therefore, just as it is important for scientific progress to affirm and refine our assumptions about the reliability of the measurements that we use to confirm scientific theories, it is also important for scientific theories to affirm and refine our assumptions about the ways in which we are in meaningful contact with our epistemic community, in order to do justice to the fact that scientific knowledge is built jointly and collaboratively.

Now, we noted in section 9.6 that knowledge of the perspectives of other observers is in some sense more important in the context of the observer-relative interpretations than in the non-observer-relative context: if I believe that there exists an absolute, unique, observer-independent reality out there, and that the task of science is to find out about this observer-independent reality, then in a sense I don't need to be too concerned about the subjective experiences of other observers. I do still need to rely on *reports* conveyed to me by other observers because I need to know the results of experiments performed in other places and times, but it's less crucial to speculate about the subjective experience of those other observers, for at some level what matters for empirical confirmation is simply that an experiment took place and had a unique outcome, and the other observer created a faithful record of that outcome and conveyed that information to me. For all I know, those other observers could actually be mindless automata with no conscious experiences at all – as long as they are nonetheless functioning as accurate machines for performing measurements and conveying information about their outcomes, I can still make use of their observations in much the same way as I make use of my own. Whereas we saw in chapter 9 that if the world is as described by the observer-relative interpretations then there is nothing for scientific knowledge to be *about* other than the content of people's subjective experiences, so scientific knowledge will be unacceptably diminished if we do not have grounds to think we can find out about the perspectives of other observers by interacting with them.

However, even in the non-observer-relative case, it is still the case that scientific knowledge, including empirical scientific knowledge, would be less robust if I were the only knower of it and all of the rest of you were just automata. For a start, I myself am only able to understand a tiny proportion of the totality of the empirical scientific knowledge that humankind has accrued; perhaps there was a point in history at which an individual could hope to fully learn and comprehend all scientific knowledge of the day, but that time is long past. Furthermore, even when it comes to scientific theories

Saving Science from Quantum Mechanics. Emily Adlam, Oxford University Press. © Oxford University Press (2025).
DOI: 10.1093/9780197808887.003.0010

that I do know reasonably well, I do not accept those theories on the basis of my own judgement alone: my confidence in the soundness and empirical adequacy of a theory like quantum mechanics does not come entirely from my own analysis of the mathematics and the experimental results, but also from the knowledge that many other members of my epistemic community have analysed the same results and come to the same conclusions. Indeed, it is not just individual judgements about specific scientific matters, but even the languages in which I think (both linguistic and mathematical) and the framework of thought that I use to make judgements in the first place which I have inherited from my epistemic community, so it is hard to imagine how I could possibly formulate any one of our current scientific theories, let alone accrue the kind of overwhelming evidence for it that we now have, were I nothing more than a lone human animal without any social context. So, in fact, in the process of doing science we are not just relying on other observers to report measurement outcomes to us like automata – we also rely on them to exercise judgement and to report conclusions that we take seriously because we regard other observers as our epistemic peers.

The social aspects of scientific knowledge are studied in the field of social epistemology. As Solomon (1994, p. 336) puts it, in traditional scientific epistemology it is often assumed that '*if rationality or normativity is found, it will be instantiated in the decision making process of each individual scientist (or at least, the vast majority) at consensus*,' but on this view of scientific activity it is somewhat mysterious where scientific knowledge gets its special epistemic authority. For although I argued in chapter 6 that it is possible and important to distinguish between pragmatic and epistemic rationality, nonetheless it is clear that the epistemically rational agent is a fiction. No real person is ever motivated *purely* by the goal of believing propositions which are well-supported by the evidence and disbelieving propositions are which are not well-supported by the evidence; no matter how disinterested we may try to be, we all have pragmatic needs that we must attend to, as well as aesthetic, emotional, and moral motivations. And it is clear that these other motivations often do impinge on our beliefs – the phrase 'wishful thinking' was coined to describe precisely the scenario in which emotion gets in the way of epistemic rationality. This difficulty has been emphasized in feminist critiques of the notion of 'objectivity' which have documented the many ways in which pragmatic, emotional, moral, and aesthetic considerations are liable to creep into judgements which are claimed to be – which indeed the originators of the judgements may sincerely believe to be – purely objective (Haraway, 1988; Tavris, 1992; Antony, 2022). For as Longino (1990) points out, the failure of positivism puts paid to the hope that there might exist completely unambiguous, purely logical relations of evidential support, and thus whatever epistemic rationality entails, it is more nuanced than simply logical deduction, and hence more liable to be influenced by external factors.

Thus within the field of social epistemology, it has been argued that the distinctive epistemic authority attached to science may stem in part from its features as a community-based practice. For example, Solomon (1994, p. 337) argues that the social processes by which scientists reach consensus often have the result that the consensus is 'normative,' by which she means that the theory on which consensus occurs is the most empirically successful theory: '*Not even one individual scientist need make unbiased judgments of empirical success for a consensus to be normative.*'

Similarly, Longino (1990) argues that different members of the scientific community will have different histories and preferences which shape the background beliefs relative to which they take scientific theories to be empirically confirmed, but through the process of subjecting theory and evidence to transformative criticism from the scientific community as a whole, some of the biases inherent in the specification of background beliefs will essentially be averaged out, leaving behind beliefs which are more well-founded and epistemically secure than the judgements of any one individual. Longino thus locates the 'objectivity' of science in its social dimensions: '*The objectivity of scientific enquiry is a consequence of this inquiry's being a social, and not an individual, enterprise*' (Longino, 1990, p. 67).

Now, we should be careful with the term 'objectivity' here because the proponents of observer-relative interpretations have sometimes expressed willingness to jettison the notion of 'objectivity' altogether, insofar as that word is understood as making some claim to the effect that the deliverances of science are true in an absolute, observer-independent way. However, as Longino (1990, p. 62) emphasizes, there are two importantly different conceptions of objectivity: '*In one sense objectivity is bound up with questions about the truth and referential character of scientific theories ... In the second sense objectivity has to do with modes of enquiry.*' Although proponents of observer-relative interpretations perhaps are less concerned with objectivity in the first sense, they presumably still want to maintain that there is something specially 'objective' about the scientific method in the second sense of the word, since they still appear to believe that science is in some way importantly different from opinion, or fiction, or dreams. And Longino's argument is precisely that the social aspects of scientific practice are an important contribution to the objectivity of science in this second sense – '*Only if the products of inquiry are understood to be formed by the kind of critical discussion that is possible among a plurality of individuals about a commonly accessible phenomenon, can we see how they count as knowledge rather than opinion*' (Longino, 1990, p. 74). Thus her arguments are still very applicable even in the context of the observer-relative interpretations, and therefore the role of the epistemic community cannot be disregarded in that context.

van Fraassen (2008) makes a similar point in his discussion of the distinction between public and private hallucinations. A rainbow, for example, is a public hallucination – when we see a rainbow we are not directly observing any particular object, and each observer sees the rainbow slightly differently, but nonetheless it is 'public' because the observations of different observers are related in systematic, predictable ways, and such observations can even be made indirectly using a camera. Moreover, it is not just phenomena like rainbows which are relevant here, for van Fraassen points out that many other kinds of observations can also be regarded as having the character of public hallucinations – for example, when we look into a microscope we are not just passively looking through a window into previously unobservable parts of reality, rather the microscope is creating new kinds of phenomena at observable scales, somewhat akin to public hallucinations. In some sense every observation we ever make has something of the character of a 'public hallucination' – as van Fraassen (p. 99) puts it, '*If appearances are what appear to us then, by definition, we never do see beyond the appearances ...!*' But the key point is that 'public hallucinations' are regarded as suitable objects of scientific study, whereas private hallucinations like dreams and visions are not – or at least, they belong to the realm of psychology rather than physics. And thus the possibility of distinguishing between public and private hallucinations is crucial to our idea that science is 'objective' in the sense of being importantly different from opinion or fiction. Yet the distinction between public and private hallucinations lies precisely in the fact that public hallucinations, while differing between different observers, are nonetheless intersubjectively accessible, and the experiences that different observers have of them are related in systematic ways. If we reduce science down to descriptions relative to individual observers, we run the risk of erasing the distinction between public and private hallucinations altogether, thus losing any meaningful notion of objectivity in science.

Of course, appreciating the social dimensions of scientific practice does not oblige us to say that conclusions reached by the scientific community as a whole are infallible, or by default epistemically rational. Individual scientists certainly can and should think independently and draw their own conclusions on scientific matters – indeed, the accounts of social objectivity and normative consensus offered by Longino and Solomon suggest that the social features of science are relevant to its epistemic authority precisely *because* combining critiques from many independent thinkers help to produce conclusions with epistemic virtues that individual judgements lack. But nonetheless, even when a visionary thinker criticises some element of the scientific consensus, this necessarily still takes place within a social context: the criticism is always expressed relative to a large body of shared knowledge, without which there is simply no way to articulate the criticism, argue for it, or imagine alternatives. No thinker, no matter how revolutionary, can function completely independently of their epistemic community.

Now, proponents of observer-relative interpretations may object that their approach does not prevent us from engaging in social practices of enquiry. For in the observer-relative picture I can still submit my observations and theories to other observers for critique – the responses I receive may have nothing to do with what is going on in their minds, but I still do receive a response, and we know from experience that those responses often seem to make sense! However, an important presupposition of the practice of appealing to an epistemic community to arrive at a higher level of objectivity or epistemic rationality is that we must see other observers as our epistemic equals – we must believe that they have minds which are just as capable as our own of producing good judgements about scientific evidence, and of course we must believe that we are actually able to make contact with those other minds and share our evidence with them via physical interactions. Longino emphasises the central importance of intersubjective sharing of evidence in her account of the socially grounded objectivity of science: '*The states of affairs to which theoretical explanations are pegged [in evidential and explanatory relationships] are public in the sense that they are intersubjectively ascertainable ... this does not require a commitment to a set of theory-free, eternally acceptable observation statements but merely a commitment that two or more persons can agree about the descriptions of objects, events, and states of affairs that enter into evidential relationships*' (Longino, 1990, p. 70). Whereas the observer-relative picture tells us that we are not actually communicating with or sharing evidence with other agents, we are merely interacting with some sort of simulacra of people which spit out superficially convincing but essentially random responses – and why should such interactions serve to give us any particular faith in the resulting judgements? We cannot regard the social dimensions of science as meaningfully adding to its epistemic authority unless we believe that we are genuinely part of an epistemic community whose members work together to arrive at shared conclusions, rather than a different conclusion relativized to each individual observer.

To give a more specific example, within an observer-relative picture it is clearly true that we can in some sense distinguish between public and private hallucinations. For in the observer-relative context I can ask you about your observation, and then I will have access to a fact, relative to me, about what I perceived you saying about your observations. So in this context we can simply say that public hallucinations are observations such that, if I ask another observer about them, I will perceive those observers as reporting having observed something similar to what I have observed, whereas private hallucinations are observations such that, when I ask other observers, I do not perceive them as reporting having observed something similar. But although this formal distinction can be made, it's entirely unclear why it should have any significance in the observer-relative picture. After all, the distinction ultimately comes down to saying that certain sorts of facts relative to me will be correlated with other sorts of facts relative to me, while other sorts of facts relative to me will not be correlated in this way – but still, in the observer-relative picture these are all legitimate relative facts, and since they are all in the end just relativised to a single observer, it seems hard to deny that they are all equally real, and thus there would seem to be no reason why one should be a suitable object of scientific study and the other not. Therefore the observer-relative picture flattens or erases many of the distinctions that we need to make sense of the notion that science has a distinctive subject matter.

As another example, note that in the practice of science we appeal to our epistemic community not only to make judgements but also to shape practical decisions about research in science. Our scientific conventions are set up in such a way as to ensure that decisions involving funding, collaboration, and prioritisation are not in general made by just one person but are distributed across the community – so for instance, grant proposals are typically sent out to a number of peer reviewers whose judgements are collated in order to make a decision. Thus scientific practice involves not merely intersubjectivity, in the sense of passively sharing information, but also what Roth calls 'practical intersubjectivity,' in which intentions and judgements belonging to one agents have the status of a rational constraint on the reasoning of other agent. And Roth emphasizes that an essential presupposition of practical intersubjectivity is '*the symmetry or equality of authoritative status: each individual is in a position to issue intentions that serve as rational constraints for the rest*' (Roth, 2003, p. 69). That is, the way in which we practice science makes sense only if we regard other agents as our intellectual equals, which requires that we believe in other minds and we believe that we are actually able to make contact with those other minds, in order that their judgements can serve as constraints on our own theoretical and practical reasoning.

10.1 Schematising the Observer

In chapter 4 I argued that a viable scientific theory must ultimately include a 'schematisation of the observer,' in the terminology of Stein (1994) and Curiel (2020) – we must put the observer into the theory, to '*embod(y) the possibility of the epistemic warrant we think we construct for our theories from such contact and connection*' (Curiel, 2020, p. 6).

But focusing on observer-relative interpretations has emphasized the fact that there are a number of different ways in which observers can be put 'into' the theory. One possibility is to give a 'first-person schematisation' in which we simply define scientific concepts in a functional or perspectival way, making their epistemic accessibility explicit by building the evidence for them directly into their definitions. This is the approach taken by perspectival and relational approaches to quantum mechanics, in which the whole theory is understood as being centered on some observer relative to whom all its concepts are defined; and it is also the approach we are taking when we write a theory in an operational form, specifying it purely in terms of operations that an observer can perform and observations that can be expected to follow those operations (Abramsky and Heunen, 2013; Hardy, 2016; Di Biagio et al., 2021). Whereas the alternative approach involves giving a 'third-person schematisation,' providing an account of reality in which observers play no special role, and then subsequently showing how to locate observers within this observer-independent picture of reality, thus implicitly making it clear how various features of reality are related to possible observations. In the third-person approach meaning is given to scientific concepts not by defining them directly in terms of experience, but rather in a holistic way, by means of successfully accommodating observers and their experiences within an account of reality which has empirical significance only in its totality.

A good example of a third-person schematisations is given by the 'IGUS' interpretation of the consistent histories formalism. In chapter 13 we will return to the SWRCH version of consistent histories described in section 3.3, but the IGUS formulation is an alternative suggested by Gell-Man and Hartle (Gell-Mann and Hartle, 1996) whose central notion of an 'Information Gathering and Utilizing System' – '*a type of creature which is coupled by some form of sensory organs to its environment, able to model the local environment by some form of logical processing, and able to act on the results of its computations*' (Dowker and Kent, 1996, p. 57, online version) – is clearly a schematisation of the observer, capable of accommodating both ourselves and also other physically possible observers. Gell-Man and Hartle suggest an interpretation of the formalism in which each IGUS is associated with its own consistent set of coarse-grained operators, defining the quasi-classical domain in which it lives. Thus the interpretation provides us with a *third-person* schematisation of the observer: IGUSes are defined in external, physical terms, and criteria are given to locate them within the third-person description of reality offered by the consistent histories formalism. And one important consequence of this schematisation is that it is immediately clear that meaningful communication between the agents will be impossible: as Dowker and Kent put it, '*There is, in this interpretation, no correlation between the experiences of these splendidly isolated IGUSes; each may well believe itself in communication with others, but the others may be experiencing a quite different history, or nothing at all*' (Dowker and Kent, 1996, p. 61, online version). So the third-person schematisation plays an important role here in helping us see that there is an intersubjectivity problem, and conversely, in other kinds of approaches a third-person schematisation can be used to affirm that there is *not* an intersubjectivity problem.

Now, the relation between first-person and third-person schematisations is complex, since neither is straightforwardly derivable from the other. Given a third-person description of reality which includes a schematisation of conscious observers, you will typically be able to arrive at some 'objective' description of what each agent will observe, e.g. the results they will obtain when they perform measurements, but as often pointed out in the philosophy of consciousness, there appears to be an element of 'subjectivity' which is not easily explained from the third-person point of view (Nagel, 1974). For example, Jackson famously argued that even if we can predict from our third-person description that some observer will have an experience which is describable as 'seeing red,' nonetheless there is something about the subjective experience of seeing red which is not captured by this third-person description (Jackson, 1986). But things are arguably even worse if we try to go in the other direction – for given a first-person description or a set of first-person descriptions, it is not always obvious how to arrive at any consistent third-person description encapsulating all the first-person experiences, and certainly there may not be any *unique* way to achieve this because there may be complexities of global structure which are not evident from the first-person point of view. And indeed, arguably this is exactly the problem we are encountering in Extended Wigner's friend scenarios like the ones described in section 9.7 – we have a set of first-person descriptions given by applying unitary quantum mechanics from the point of view of each of the individual observers in the scenario, but we do not know how to put all of these descriptions together into a joint third-person picture characterising the experiences of all of the observers in a way that satisfies all the properties we would intuitively expect such a description to have.

In light of these difficulties, it is perhaps not surprising that it has become popular to suggest that quantum mechanics is telling us we should simply stop at a first-person observer-relative approach in which we refrain from even attempting to give a third-person schematisation. For example, Cuffaro suggests that '*what Stein and Curiel have, on the grounds of practical and epistemic necessity, claimed to be required in classical theory should be understood, for a (neo-)Bohrian, to be elevated within quantum theory to the level of a postulate, in the sense that interpreting the outcome of a measurement interaction as providing us with information about the world requires, as a matter of principle, the specification of a schematic representation of an observer*' (Cuffaro, 2023b, p. 4). That is, Cuffaro thinks we should be content with and indeed build our understanding of reality around a purely first-person schematisation of the observer. These kinds of arguments raise an important question about how we should think about the epistemology of science: does an adequate schematisation of the observer require a third person description of observers, or does it require a first-person description, or will either do, or are both needed?

In fact, it has often been pointed out that the content of science cannot be exhausted by third-person descriptions; we need first-person descriptions in order to make practical use of scientific theories. Kant (1768, p. 39) emphasized this point, writing, '*No matter how well I may know the order of the compass points, I can only determine directions by reference to them, if I am aware of whether this order runs from right to left or from left to right, and the most precise map of the heavens ... would not enable me [without this orientation] to infer ... on which side of the horizon I ought to look for the sunrise.*' Similarly, van Fraassen (2008, p. 71) highlights the need for first-person descriptions in the context of special relativity: '*If someone is to use Einstein's theory to predict the behavior of electrically charged bodies in motion, bodies with which s/he is directly concerned, choice of a coordinate system correlated to a defined physical frame of reference is required. The user must leave the God-like reflections on the structure of spacetime behind in order to apply the implications of those reflections to his or her actual situation.*' We must make use of first-person, indexical descriptions of physical situations in order to be able to actually apply the tools of science to the problems that interest us, and thus first-person, indexical descriptions can never be fully eliminated from science, nor should we be seeking to eliminate them.

However, those who argue for the indispensability of first-person descriptions have not typically sought to deny the importance of third-person descriptions. In fact, both kinds of descriptions appear frequently in science – for example, in special relativity we often find it useful to adopt a particular reference frame and describe phenomena relative to that reference frame in an indexical way, but it is also often useful to describe phenomena in terms of the third-person, objective Minkowski space picture. Moreover, Shimony (1993a, p. 40) emphasizes that the ability to map back and forth between first and third person descriptions is essential to the epistemology of science – for the methodology of closing the circle '*envisages the identification of the knowing subject (or more generally, the experiencing subject) with a natural system that interacts with other natural systems. In other words, the program regards the first person and an appropriate third person as the same entity.*' That is, if our scientific theories are ultimately to affirm the reliability of the methods of enquiry that we use to arrive at them, we must be able to identify our first-person experiences with certain events describable in the third-person, at least in very broad outline, otherwise the methods of science will be incapable of telling us anything about the reliability or otherwise of these experiences.

Moreover, it should be emphasized that the indispensability of first-person descriptions in science does not in and of itself entail that third-person descriptions are in any way incomplete. Van Fraassen puts the point as follows: '*Maps normally do not have an arrow labeled "You are here." But even if the map does have that, the problem is really the same: I have to locate where I am with respect to that arrow ... Something more than what is contained in the printed map, physically constructed model, or computer monitor display is needed. But what is this "more"? Not a mysteriously different sort of fact which cannot be encoded on a map! The scientific story can be complete in the sense of describing all the facts, including that someone does or does not have the "extra" needed for him or her to draw on a particular bit of science. It is just that* ***describing the having of it*** *is no substitute for* ***the having***' (van Fraassen, 2008, p. 83). That is, the need for indexical first-person descriptions in science does not mean science must involve some kind of irreducibly indexical first-person facts; these descriptions are necessary simply because we want to be able to *use* the third-person descriptions, not because the first-person description has content that the third-person does not.

Indeed, the idea that all or part of the content of science is ultimately first-person is very much at odds with the actual nature of the data on which scientific theories are based. For as Suppes has emphasised, scientific theories are not typically confirmed directly by specific individual observations; before we compare experimental results to theories, we create simplified summaries of the experimental data, abstracting away details which are considered irrelevant. Suppes refers to these summaries as 'data models': '*The maddeningly diverse and complex experience which constitutes an experiment is not the entity which is directly compared with a model of a theory. Drastic assumptions of all sorts are made in reducing the experimental experience ... to a simple entity ready for comparison with a model of the theory*' (Suppes, 1961, p. 15). In particular, one common step in arriving at a data model is to get rid of specific information about the order in which various outcomes occurred and replace it with information about the relative frequencies of various outcomes – and certainly, this step is a crucial one in the creation of the kinds of data models used for the empirical confirmation of quantum mechanics, since quantum mechanics is a probabilistic theory which typically predicts only relative frequencies rather than specific outcomes or the order in which outcomes will occur.

The empirical confirmation attached to quantum mechanics thus comes from its ability to reproduce the statistics represented in a data model, not from its ability to model specific individual events: in van Fraassen's words, '*These models – the theoretical models – are provided in the first instance to fit observed and observable phenomena. Since the description of these phenomena is in practice already by means of models – the "data models" or "surface models", we can put the requirement as follows: the data or surface models must ideally be isomorphically embeddable in theoretical models*' (van Fraassen, 2008, p. 168). But as Suppes emphasizes, a data model is already several steps of abstraction away from our immediate sensations, so data models have built into them a number of nontrivial assumptions which go beyond what we can know for sure on the basis of our direct experience. In particular, in order to arrive at a data model giving summaries of relative frequencies, we must at the very least assume that our present memories and records of past events are largely accurate descriptions of the events that actually occurred, which means presupposing that observers are able to share information with past versions of themselves. Moreover, the data models used to empirically confirm quantum mechanics are typically understood as summaries of relative frequencies not just of experiments by one particular experimenter but in experiments by many different experimenters at different times and places, so in order to arrive at these data models we must also suppose that we are able to exchange information about relative frequencies with other observers. Thus the idea that quantum mechanics is a first-person theory is at odds with the nature of '*the* empirical evidence' for quantum mechanics; we confirm the theory not by comparing it to individual first-person experiences, but by comparing it to data models summarising experiences across an entire epistemic community, so the theory cannot really be 'first-person' or relativized to an observer unless the 'observer' we have in mind is actually the entire epistemic community of human scientists.

From this point of view, observer-relative approaches to quantum mechanics involve a severe misunderstanding of the nature of scientific knowledge. These approaches would have us believe that quantum mechanics – a scientific theory which has been arrived at, refined, and empirically confirmed by the efforts of an entire scientific community – could somehow turn out to be nothing more than a first-person description of the experiences of an individual observer. But as Longino puts it, '*What is called scientific knowledge, then, is produced by a community … and transcends the contributions of any individual or even of any subcommunity within the larger community*' (Longino, 1990, p. 69). Scientific knowledge is not and could not possibly be first-person, and therefore any plausible solution to the measurement problem must be capable of properly accounting for and doing justice to the relations between individual observers and the means by which the community as a whole arrived at the theory of quantum mechanics.

10.2 Measurements

Now, at this juncture the proponent of a purely first-person or perspectival view of science might object that although data models may appear to present scientific data in a third-person way, nonetheless measurements are, in fact, irreducibly perspectival and contextual – as van Fraassen puts it, '*The outcome of the measurement operation is a representation of the target, but it represents the object as it appears in that measurement set-up*' (van Fraassen, 2008, p. 176). One might argue, therefore, that ultimately all of our evidence is in a first-person form relativized to a certain context, and thus our evidence doesn't actually justify us in believing in the existence of an objective, third-person description of reality.

However, as van Fraassen points out, '*It does not follow [from the perspectival nature of measurement] that the appearance there is different from the appearance of the same thing in other such set-ups. For after all there are invariants too! … There are parameters, even ones definable from the results of local distance and time measurement outcomes, that have the same value in different frames … There are therefore measurement outcomes that have no relativity left*' (van Fraassen, 2008, p. 176). The existence of invariants and relations between different perspectives is precisely what allows us to generalise beyond our limited perspectival evidence to arrive at an 'objective,' third-person view. Moreover, invariants of various kinds are not just incidental facts which must be accommodated; identifying invariants and showing how the values of non-invariants transform across different contexts is often central to the insight that a theory gives us into physical reality. Defining a scientific concept frequently requires saying not only how it appears to one observer, but also how it transforms as we move to the perspective of different observers, thus allowing us to come to an understanding of the underlying invariant structures that all observers can agree on.

The obvious example is special relativity (Einstein, 1905), since the central content of that theory are the Lorentz transformations, which tell us how to transform spatial and temporal distances as we switch between reference frames. These transformations reveal that a particular combination of spatial and temporal distance known as the 'spacetime interval,' is invariant between different frames – and such facts about transformations and invariants are right at the heart of what special relativity has to say about the nature of spacetime. So there is often a great deal of scientifically relevant content in the relation between different perspectives, and indeed, it has sometimes been argued that only quantities which are invariant under the symmetries of a theory are physically real according to that theory – Dirac (1930), Weyl (1952) and Nozick (2001) all seem to espouse some version of this view. Thus for most scientific theories, it will not be enough to schematise just one observer; we have to schematize *the epistemic community*, and understand how the different perspectives present in the community are related to each other in ways that can then be reflected in the invariances and symmetries of the theory that we ultimately arrive at.

In this connection, note that proponents of observer-relative approaches have sometimes argued that their approach is supported by the history of physics, since 'relativizing' physical quantities has often led to progress (Rovelli, 1996; Oldofredi, 2021). Here again the most obvious example is special relativity, which can be understood as a 'relational' approach, since it tells us that facts about simultaneity are not absolute but relational. But as Pienaar (2021) argues, there's also a sense in which the example of relativity, in fact, supports exactly the opposite conclusion. It is true that special relativity tells us that 'simultaneity' is not absolute but relativized to reference frames, but this picture involves only a fairly moderate kind of relationality, since it also tells us that there are absolute, observer-independent facts about the locations and velocities of objects and observers in spacetime, and the relational facts about simultaneity are derived from these facts. Indeed, many people would argue that the theory's most central content is the part of it which is invariant under transformations between different reference frames, i.e. the spacetime interval and the underlying structure of Minkowski spacetime (Balashov and Janssen, 2003). So unlike the observer-relative views, special relativity is certainly not denying the existence of a third-person, observer-independent reality – its relationality simply consists in moving facts about simultaneity out of third-person observer-independent reality and into the perspective of observers, whilst still maintaining that those perspectives are themselves grounded on an absolute reality.

The proponents of observer-relative views might perhaps contend that arriving at 'everything is relativized to an observer' is the natural endpoint of the progressive relativisation of more and more physical quantities. However, it must be emphasized that the difference between the moderate approach taken by special relativity and the approach which denies the existence of a third-person reality altogether is not simply a matter of degree. As Berenstain (2020, p. 31, online version) observes, '*The history of modern science can be viewed as a series of discoveries that have continually dethroned humankind from a presumed special and unique place in the universe,*' and thus in most historical examples, the purpose of relativizing quantities in physics has not been to make the observer more central, but rather to arrive at a more objective, observer-*independent* description of reality, by removing from our fundamental physical descriptions quantities that we now understand to be observer-dependent, in order that we can better understand the absolute, invariant reality that grounds the relational quantities and the transformations between them. This is exactly what Minkowski space does for us in the context of special relativity. So an observer-relative approach which denies the existence of a third-person reality is not simply an extreme version of the well-established process of relativizing things – it is a completely different kind of project which must have completely different motivations. Indeed, an alternative interpretation of these historical episodes would see them as evidence that looking for a more objective, observer-independent description of reality is well-supported by previous scientific successes, in which case, rather than lending support to an observer-relative approach, these examples, in fact, suggest that the observer-relative picture may be precisely the wrong way to go.

10.3 Empiricisim

These reflections on the relation between first and third-person perspectives in science also have interesting consequences for the long-standing debate between scientific realism and empiricism. Here, I will take as an exemplar of empiricism the 'constructive empiricism' developed by van Fraassen, which suggests that acceptance of a scientific theory involves believing only that it is empirically adequate, i.e. that it is correct in what it says about observable things (van Fraassen, 1980). And I will understand scientific realism as simply the view that full acceptance of a scientific theory involves holding some beliefs about things which are not directly observable.

Now, a common objection to empiricism of various flavours is based on the idea that the line between the 'observable' and the 'unobservable' is not well-defined; this argument is in effect a direct descendant of Quine's objection to the verificationist's criterion of reality. But many empiricists are not particularly moved by such objections, noting that although there may be borderline cases, there are also obvious examples of 'observables' about which no reasonable person could have much doubt – as Teller (2001, p. 131) puts it, '*"Observable," like "fragile" and "portable" which are similarly relative to human capacities, is not a completely precise predicate. This fact does not impugn the useful application of such predicates in most normal contexts.*'

But a more nuanced version of this argument is made by Shimony, who notes that although van Fraassen seems to advocate a naturalistic approach to epistemology in which the limits of observability are demarcated by empirical science rather than philosophical analysis, this seems incompatible with his constructive empiricism. For in order to adopt such a naturalistic approach we must be able to map our first-person experiences as epistemic agents onto some third-person descriptions of the process of observation appearing in science, and as Shimony puts it, '*If the first and third-person aspects of the knowing subject are identified by van Fraassen, then his agnosticism about any claim for a scientific theory beyond empirical adequacy is breached, for an element in a theory would be identified with something that exists in an exemplary way*' (Shimony, 1993a, p. 45). That is, adopting a naturalistic approach to epistemology seems to demand of us some commitment to the unobservable things which appear in a scientific description of the process of observation.

And this point can be taken further in light of the discussion of this chapter, for the difficulties for observer-relative interpretations as discussed in chapter 9 spring in part from the fact that strictly speaking, 'the unobservable' also includes relations between the experiences of different observers, and even the relations between our current records and the experiences of our past selves. That is, although the observations made by any individual observer are, of course, observable, nobody can directly observe the *relation* between those observations. Shimony's point can therefore be extended: if we wish to adopt a naturalistic account of the relations between observers and different temporal parts of the same observer, then again we will need to be able to map between first and third person descriptions of those observers, and this will demand of us commitments to features of reality appearing in the third-person description which are not directly observable.

Now, considering the relation between the observations made by a pair of observers, Alice and Bob, the constructive empiricist might hope to avoid the above problem by introducing a third observer, Chidi, who asks Alice and Bob what they have observed and thus 'observes' the relation between Alice's observations and Bob's observations, thus admitting this relation into the realm of the observable. But as noted in section 3.2, we cannot interpret this observation as providing information about the relation between the observations of Alice and Bob without making assumptions about the relations between Alice's experiences and Chidi's experiences, and between Bob's experiences and Chidi's experiences, so it would seem that there is no possible way anybody can arrive at beliefs about the relation between observations made by two different observers without substantive theoretical assumptions. Likewise, although my past observations and my current records of my observations are, of course, both observable, nobody can directly observe the relation between them, so we can't learn anything about the relation between them without substantive theoretical assumptions.

Evidently this way of thinking about what is really 'observable' poses questions for the idea that acceptance of a scientific theory only involves believing what it says about observable things; for as noted in section 10.1, the 'data models' that we use to empirically confirm scientific theories encode a large number of assumptions about the relations between observations made by different observers, as well as different versions of the same observer over time. In particular, when we specify a relative frequency we are saying something about the relations holding within a large set of different observations, which usually cannot all be made by a single observer at a single point in time. So in the strictest sense, what a theory says about 'observable things' does not include the predictions that it makes about relative frequencies; and thus there is a sense in which a maximally pure version of empiricism would preclude accepting a large amount of the 'empirical' content of quantum mechanics and other probabilistic theories whose predictions are primarily verified via observations of relative frequencies.

Therefore it seems that any empiricist who wants to admit the possibility that we could meaningfully 'accept' a probabilistic theory like quantum mechanics must agree that accepting a theory involves believing not only what the theory says about observable things, but also what it says about the unobservable relations between the observations made by different observers, or different temporal parts of the same observer. In the context of probabilistic theories in classical physics, this addition perhaps seemed so trivial as to be unremarkable, but the interpretative difficulties of quantum mechanics, as highlighted by the observer-relative interpretations and the Extended Wigner's Friend theorems, draw our attention to the fact that believing what a theory says about relations between observations actually involves significant theoretical commitments to unobservable relations between observations, and thus a maximally pure empiricism may be unsustainable in the context of probabilistic theories.

Now, possibly empiricists would want to make a distinction between beliefs about unobservable *entities* like atoms and quarks, and beliefs about unobservable *relations* between observable things, in order to argue that it is reasonable to accept commitments to the latter and still reject commitments to the former. For example, one might argue (borrowing an argument from structural realists like Maxwell (1970)) that we have no way to get a grasp on the intrinsic nature of unobservable entities in and of themselves, and thus we are simply not capable of formulating propositions which successfully refer to such things. Whereas one might perhaps contend that relations between observations need not have any intrinsic nature over and above the intrinsic nature of the observations themselves, particularly if they are defined extensionally, and thus we should be able to formulation propositions which successfully refer to such things in virtue of our ability to refer to observable things. However, it's not clear that commitments to unobservable entities really require us to define such entities by specifying their intrinsic natures – many realists argue that we can refer to unobservable entities by giving a purely structural or functional characterisation of them, or they simply deny that scientific realism requires successful reference to theoretical entities at all (Cruse and Papineau, 2002). Relatedly, it's also unclear that such an argument would ultimately justify some form of empiricism rather than simply structural realism.

Alternatively, one might make a pragmatic point to the effect that the argument above shows that we cannot do without commitments to unobservable relations between observable things, but we *can* do without commitments to unobservable entities, so the epistemically cautious approach is to refrain from commitments to the latter – as van Fraassen (1980, p. 72) puts it, '*It is not an epistemological principle that one might as well hang for a sheep as for a lamb.*' However, it is not entirely clear that we can sensibly pull the two kinds of commitments apart in this way. For as emphasized by the bootstrapping account of scientific practice presented in chapter 4, our confidence that we have knowledge about the experiences of others and our past selves is ultimately underwritten by the progressive development of an account of reality which provides us with an understanding of the physical process by which we interact with other observers and exchange information; and since no one can directly observe their own process of observation, this account of reality will necessarily say something about unobservable entities and/or structures. So if we are obliged in any case to have epistemic commitments outside of what is strictly speaking observable, we may ultimately be better off taking the whole set of beliefs about unobservable reality as a package which provides a coherentist grounding for the beliefs we need about unobservable relations, rather than artificially divorcing the beliefs about relations from the other beliefs in a way which makes the overall picture significantly less coherent.

10.4 Everett Again

Let me close by noting that these issues related to the relation between first and third-person views of science also crop up in the context of the Everett interpretation. For we saw in chapter 6 that in order to sensibly understand mod-squared amplitudes as probabilities, Everettian observers must be able to see themselves as random samples from their set of post-measurement successors, and arguably the only way to achieve this involves postulating a radically new kind of irreducibly indexical information. What makes this information so radically new is precisely that it cannot be described from a third-person point of view: when we ask what outcome 'this' observer will observe, there are no physical facts we can point to in order to specify which observer we mean without effectively saying what outcome the observer obtains, so, if this question can be asked at all, it can *only* be asked from the first-person perspective. Whereas in more ordinary cases of indexical individuation, this is not the case – in all familiar examples there are at least facts about spacetime location or relations of relative distance which allow us to identify distinct observers from an external point of view, even if they are having qualitatively identical experiences. So there is a sense in which Everettian attempts to solve the probability problem end up denying the possibility of giving a third-person description of the content of quantum mechanics.

Now, the Everettian version of this problem is not as extreme as the observer-relative version, since Everettians do accept that there exist some observer-independent facts, such as facts about the wavefunction of the whole universe. So they are certainly not committed to the idea that nothing whatsoever can be said about reality from a third person standpoint. However, the Everettian view does apparently have the consequence that any third-person description is necessarily incomplete, since it cannot include the irreducibly indexical self-reference involved in making sense of the Everettian probabilities. Moreover, what is missing is not just some subjective sensation like the experience of seeing redness, which we usually imagine can be safely ignored in the formulation of a physical theory – in the Everettian picture the information which is missing from the third-person description is central to the formulation and the empirical confirmation of the theory, since most of the theory's content is in its probabilistic predictions, and we can apparently only make sense of those predictions in terms of indexical self-reference. Thus if the Everettians are really required to make use of irreducibly indexical information in this way, the empirical confirmation of Everettian quantum mechanics will be dependent on a form of information which cannot possibly be communicated between observers because it can never be described in a non-indexical way.

This suggests that the point I have been making in this chapter may also be a problem for the Everett interpretation – just like the proponents of observer-relative interpretation, the Everettians are ultimately forced to say that a theory which has been arrived at and confirmed by the efforts of the epistemic community as a whole can nonetheless only be properly understood from the first-person point of view.

While the Everettians do not need to deny all possibility of communication between different observers, nonetheless the centrality in their understanding of the theory of a form of information which cannot be communicated is certainly in tension with the reality of science as a community activity, and this is a further reason to be suspicious of the appeal to irreducibly indexical facts in making sense of Everettian probabilities.

11
Primitive Ontology, Fundamentality, and Scale

I have thus far argued that a number of unitary-only solutions to the measurement problem involve severe epistemic weirdness which may make them untenable. In this chapter, I will discuss some existing approaches which are *not* unitary-only. We will see that these approaches do not seem to encounter the same kinds of epistemic problems that we have seen so far, but unfortunately there appear to be significant technical problems standing in the way of their full acceptance. Thus this discussion will sharpen the dilemma before us: the unitary-only approaches have severe epistemic problems, while many of the non-unitary-only approaches have significant technical problems, and therefore it's unclear that we currently have *any* fully-developed solution to the measurement problem which avoids epistemic difficulties whilst still being adequate at a technical level. This will then lead us to a more general discussion of the epistemic challenges involved in dealing with a non-fundamental theory, including the question of what 'ontology' means in this context, and the role of autonomy of scales.

Because there exist a large number of proposed approaches to the measurement problem, this chapter cannot possibly cover them all. For example, I will not discuss the Diósi-Penrose model (Diósi, 1989; Penrose, 1996), the stochastic interpretation (Nelson, 2014), the transactional interpretation (Kastner, 2017; Kastner and Cramer, 2018), or modal interpretations (Vermaas, 1999). This is not intended as a dismissal of these other approaches; I have simply chosen to focus here on the research programs which are currently most well-known and popular. At this time it appears that no approach outside of the unitary-only category has been able to reproduce all of the predictions of quantum field theory (QFT), and I think it is likely that the concerns regarding QFT and emergence discussed in this chapter would also present difficulties for some of the approaches I have just listed, but the details are yet to be established – in particular, some of these approaches may not quite be 'primitive ontology' approaches in the usual sense, so they may have additional flexibility to deal with these challenges. Therefore it would certainly be a useful project to look further into the possibility that these other approaches can be expanded to reproduce all of the quantitative predictions of QFT.

11.1 Primitive Ontology

The 'primitive ontology' (PO) approaches to the measurement problem proceed by postulating some ontology which '*lives in three-dimensional space*' and '*constitutes the building blocks of everything else*' (Allori, 2015, p. 1).

Saving Science from Quantum Mechanics. Emily Adlam, Oxford University Press. © Oxford University Press (2025).
DOI: 10.1093/9780197808887.003.0011

One well-known example is the de Broglie-Bohm approach (Holland, 1995; Esfeld et al., 2014), whose ontology is given by a set of classical particles which move through spacetime according to the following guidance equation:

$$\frac{dq_k}{dt} = \frac{\hbar}{m_k} Im(\frac{\delta_k \psi}{\psi}(q_1, q_2, q_3))$$

Here, q_k is the position of particle k, m_k is the mass of particle k, $\hbar$ is a constant, and ψ is the quantum-mechanical wavefunction of the particles. The important thing about this equation is that it entails that the motion of the particles through space at a given time t is determined by the wavefunction at time t, so the particles exhibit certain quantum-mechanical behaviours, despite being essentially classical themselves. It can be shown that this formulation reproduces the predictions of standard quantum mechanics in non-relativistic scenarios, provided we assume that the initial positions of the particles at time t_0 are distributed in proportion to the mod-squared amplitude of the wavefunction at time t_0.

Another example is the spontaneous collapse approach, which postulates that the wavefunction randomly undergoes spontaneous collapse events in which it becomes localised around a point in spacetime (Ghirardi et al., 1986; Tumulka, 2006; Frigg, 2009). The most well-known version is known as the GRW model, in which collapse is modelled by an operator, which is defined as follows for a collapse at a position x:

$$\Lambda_i(x) = \frac{1}{(2\pi\sigma^2)^{3/2}} \exp[\frac{-1}{2\sigma^2}(Q_i - x)^2]$$

Here, Q_i is the position operator for particle i, and σ is a parameter which we can set to decide how tightly the wavefunction is localised about x after the collapse. One important consequence of this equation is if a set of quantum systems are entangled, then whenever one of those systems undergoes a collapse of this kind, all the other entangled systems will also undergo a collapse. And since macroscopic systems are typically made up of a large number of entangled particles, it follows that if we set the parameter governing the rate of collapses sufficiently high, then it is virtually guaranteed that any macroscopic systems we encounter will be found in a definite localised position, thus explaining why we always experience the objects around us as being in definite classical states. It can be shown that if the probability of a collapse at a point x is specified to be proportional to the expectation value $\langle\psi|\Lambda_i|\psi\rangle$, then the particle is most likely to collapse at a position x if the usual quantum description says it is likely to be at position x, and therefore this approach reproduces the statistical predictions of standard quantum mechanics in non-relativistic scenarios.

Now, to make spontaneous collapse models into primitive ontology approaches, they need to be supplemented with some specification of an ontology, which is usually either a mass-distribution over spacetime whose value at a point is given by the mod-squared amplitude of the wavefunction (Egg and Esfeld, 2015), or a set of pointlike events with one event at the spacetime location of every collapse and no physical ontology in between the collapses (Allori, 2013). Either way, much as in the de Broglie-Bohm case, we end up with an essentially classical ontology which exhibits quantum-like behaviours due to the role of the quantum wavefunction in defining its distribution over spacetime.

Now, clearly the standard models of unitary quantum mechanics do not include classical particles, classical mass distributions, or classical pointlike events, nor anything that could be described as primitive ontology. Therefore the de Broglie-Bohm approach, the spontaneous collapse approach, and other such PO approaches are not just interpretations of unitary quantum mechanics but rather package deals: each PO approach offers a new theory with its own set of models, together with an intuitively natural interpretation of these models. The existing PO approaches are for all practical purposes empirically equivalent to standard non-relativistic quantum mechanics, but in some cases they make different predictions for experiments that we are not yet able to perform – for example, some versions of the de Broglie-Bohm approach predict deviations from standard quantum behaviour in the early universe which could potentially have left traces that we might be able to detect via cosmological measurements (Valentini, 2001). Moreover, as we will shortly see, most existing PO approaches have not yet been fully extended to quantum field theory, and hence at the moment they are not empirically equivalent to standard quantum mechanics in relativistic domains where we must make predictions using QFT.

11.1.1 Classical Ontology

Although this is not explicitly stated in the definition of 'primitive ontology,' there seems to be an implicit assumption within the program that the basic building blocks which form the ontology should be explicable in classical terms, so these building blocks are typically things like classical particles, mass distributions, point-like events, etc. Such entities may sometimes behave in non-classical ways – there may be non-local influences (see section 4.0.1), or contextuality (see section 2.4.1), or retrocausality (see section 12.2) – but nonetheless they are essentially classical. Most crucially, they never enter into superpositions which combine two or more different classical positions, i.e. they always have a unique, well-defined spatial location.

Ontologies of this kind are naturally appealing to us because they are intuitively comprehensible, but this is not their only advantage: such ontologies also offer a straightforward solution to the epistemic version of the measurement problem. For we need only specify that our conscious experiences, and/or our macroscopic reality, supervene directly on this classical ontology – so in the de Broglie-Bohm approach our experiences supervene on the positions of the particles, and in the spontaneous collapse approach our experiences supervene on the mass distribution or the pointlike events. Then, since this classical supervenience basis cannot enter into superpositions which combine several different classical positions, it follows that our experiences will also never enter into superpositions combining different classical positions. That is, when we look at the needle on the dial of our measuring instrument, we will perceive it as having a unique determinate position, thus ensuring that we see a well-defined unique outcome for the measurement. Therefore it follows immediately from all such PO approaches that measurements have unique, definite, intersubjectively accessible outcomes, so such approaches avoid the epistemic problems we have discussed in this book by means of keeping the ontology entirely classical and banishing quantum weirdness to the domain of the laws which govern how that ontology behaves.

Now, of course, there are still some ways in which measurement in a PO approach differs from the usual conception of measurement in classical physics. In particular, PO approaches typically have the consequence that we are at least to some degree mistaken about *what it is* that we're measuring. For example, when we measure the spin of a particle, it's natural to suppose that we are just learning about the pre-existing value of spin for that particle; but in the de Broglie-Bohm interpretation, the only physically real systems are the de Broglie-Bohm particles, and the only property they have is their position in space. So when we measure the spin of a system in the de Broglie-Bohm approach, the result of the measurement does not actually reveal the spin of anything – ultimately the result depends on the initial position of the de Broglie-Bohm particles and potentially also on some details of the configuration of the measuring instrument (Tastevin and Laloë, 2021). Moreover, this is not just an incidental feature of these particular PO approaches: the phenomenon of quantum contextuality (see section 2.4.1) indicates that such behaviour will arise generically in any approach postulating some kind of quasi-classical reality underlying quantum mechanics, since quantum measurements cannot always be understood as revealing the pre-existing value of some variable defined within the underlying classical ontology.

But should contextuality of this kind be classified as benign weirdness or epistemic weirdness? Well, it is clear that contextuality may serve to undermine certain kinds of theoretical scientific knowledge: we may have thought that our measurements were giving us information about the spins of particles, but, in fact, we are getting some other kind of information altogether. So the phenomenon of contextuality should perhaps lead us to be more cautious about the kinds of inferences about unobservable features of reality that we make from measurement outcomes.

However, the existence of contextuality in the PO approaches does not prevent them from providing measurement models which make it clear what is going on during a quantum measurement. For example, in the de Broglie-Bohm picture, when we perform a spin measurement, although we are not actually learning the pre-existing value of a property like 'spin,' we *can* write down a measurement model which shows how the result of the measurement is determined by the initial configuration of the particles, the wavefunction, and the guidance equations, so we can see that the measurement is providing meaningful information about the theoretical structures which play a central role in the de Broglie-Bohm account of reality – that is, the initial positions of the particles and the structure of the wavefunction that guides them. That is, in the terminology introduced in section 4.2, the indications produced by the measurement device stand in a particular mathematical relation to the initial condition and wavefunction, and while the relation is not so straightforward as it is in typical classical cases – we cannot directly infer the whole wavefunction just from the measurement result, for example – nonetheless it seems clear that meaningful information about the theoretical structures of the de Broglie-Bohm models is conveyed by the indication.

Thus the PO approaches do seem to be capable of providing an account of the physical nature of measurement which affirms that our methods of enquiry reliably inform us about the theoretical structures that we are enquiring into, as long as we appreciate that the relevant theoretical structures are the ones encoded in the actual models of the PO approach in question, rather than whatever we might have naively expected them to be within a classical picture. Thus despite some surprises with regard to theoretical scientific knowledge, the PO approaches do not appear to pose any particular threat to our *empirical* scientific knowledge because they contain within them suitable models of measurement which can reassure us as to the meaningfulness of our measurements and the stability of the resulting empirical regularities exhibited by measurement results. And in light of the preceding discussion of the unitary-only approaches, it is notable that the PO approaches achieve this by postulating a unique, shared, stable reality in which all of the measurement outcomes live, ensuring that these outcomes are in most cases unique, shareable, and persistent, allowing us to straightforwardly access veridical information about general empirical regularities.

11.1.2 QFT

But although the primitive ontology approaches are attractive in many ways, there are sticking points. One common criticism is that a number of primitive ontology accounts look to be in conflict with special relativity, or at least somewhat in tension with it. For as discussed in chapter 5, perhaps the most foundational principle of special relativity is the idea that there is no absolute simultaneity, and thus there are no preferred reference frames; whereas a number of PO approaches appear to require preferred reference frames. For example, we saw above that the standard version of the de Broglie-Bohm formalism stipulates that the motion of a de Broglie-Bohm particle at some spacetime point x is determined by the global state of the wavefunction *at that time*, but in order to determine a global state of the wavefunction at a single time, we must first select a preferred reference frame which specifies whether or not a given distant event occurs at the same time as x. However, considerable progress has been made on reformulating PO approaches to be more compatible with relativity, including Bohmian approaches in which the preferred reference frame is determined in a relativistically covariant way by something like the global distribution of matter (Dürr et al., 2014), and the relativistic GRW flash approach formulated by Tumulka (2006, 2021), and thus relativistic covariance may not be an insuperable obstacle for the PO approach.

But nonetheless, there are other problems in the vicinity of relativity. Issues particularly arise when we consider the relativistic generalisation of quantum mechanics known as quantum field theory, or QFT, which models fields as collections of quantum systems at every point of spacetime, with dynamics encoded in a mathematical object known as a 'Lagrangian' (Peskin and Schroeder, 1995). We have to use QFT any time we wish to use quantum mechanics to describe the behaviour of fields – in particular, the theory of quantum electrodynamics, describing the interaction of light and matter, is a QFT (Berestetskii et al., 2012), and the standard model of particle physics is a QFT (Burgess and Moore, 2006), and both of these are extremely successful theories whose empirical predictions have been verified to an extremely high level of accuracy.

The difficulty is that at the moment there does not appear to be any PO approach which is able to reproduce all of the empirical predictions of QFT. There are approaches which work for certain special cases – for example, the de Broglie-Bohm interpretation has some relativistic generalisations in which the primitive ontology of particle positions is replaced with an ontology more suited to QFT, such as a variable encoding both the number of particles and the positions of those particles (Dürr et al., 2004), the electromagnetic field (Struyve and Westman, 2007), or Grassman fields (Valentini, 1992). But as Wallace (2022) points out, these approaches will only work in regimes in which there is a sensible description of the QFT in terms of particles, classical electromagnetic field states, or Grassman field states, respectively. And the problem is that none of these things exist at a fundamental level in QFT – for example, particles in QFT are not fundamental but rather are simply excitations in fields which emerge at certain scales and disappear at others. Similarly, classical field states in QFT can likewise only be defined in a certain limit known as the 'free field limit,' and thus they also appear at certain scales and disappear at others. Thus it appears that many existing generalisations of PO approaches to QFT rely on special features appearing in certain kinds of scenarios at certain scales, meaning that they can't be generalised beyond those scenarios and thus can't reproduce all of the results of QFT.

Indeed, Wallace (2018, p. 21) argues that there are structural reasons to think it may be impossible for a primitive ontology approach to adequately reproduce all of QFT. The fundamental problem is that, as Wallace puts it, the primitive ontology approach '*is committed to a description of [the] theory's empirical content directly in its microphysical vocabulary*.' That is, the PO approach presumes that we can find some ontology specifiable at the most fundamental level of the theory such that our experience can supervene directly on something which arises from that ontology when we 'zoom out' to macroscopic scales, in a process known as 'coarse-graining.' For example, in a QFT version of the de Broglie-Bohm approach, we must find some precisely-defined microscopic variable, analogous to the positions of the de Broglie-Bohm particles in the non-relativistic case, on which our macroscopic reality ultimately supervenes; and if this is to work out, then it must be the case that as we zoom out, these variables start to look suitable to characterise our experience of the world, which probably means that they must turn into something like coarse-grained position variables in an appropriate limit. Similarly, in a spontaneous collapse approach, the dynamical mechanism producing collapse will presumably have to be defined at the smallest possible scales, but we also want it to be the case that in the appropriate coarse-graining limit we get effective collapses which appear to take place in the position basis, as described in the equation in section 11.1 – that is, in the regimes relevant to ordinary quantum-mechanical experiments collapses should always result in wavefunctions becoming localised around specific positions, rather than some other kind of localisation.

But the problem is that in quantum field theory, the relation between the most fundamental level of the theory and the macroscopic world is, as Wallace (2018, p. 21) puts it, '*complicated, indirect, dynamically mediated and cutoff-dependent*.' One example of this problem is the fact that the parameters of the Lagrangian are scale-dependent, which means that the solutions to the dynamical equations change drastically as we move through scales, as terms in the Lagrangian become more or less important, or even vanish completely. So it is prima facie very unlikely that any variable which can be defined in relatively simple terms at the most fundamental level of the theory would, under the various complicated scale transformations needed to get from the smallest possible scales to macroscopic scales, *always* give rise to a relatively simple macroscopic variable which would be suitable to characterise our empirical experience. It may be possible to do this in some special cases, but it is hard to imagine anything that would work universally. For example, suppose we agree that our macroscopic experience should be described in terms of coarse-grained particle position; then if our experience is to supervene on some fundamental primitive ontology, that ontology must be transformed into something like a coarse-grained position variable by the action of the relevant scale transformations in the macroscopic limit. But at present it seems quite unlikely that there is some relatively simple variable or class of variables which can be defined at the most fundamental level of QFT which always turns into coarse-grained particle position in the relevant macroscopic limit, regardless of the specific details of implementation and context. So the whole strategy of the PO approaches seems to be in trouble when we move to QFT.

Now, Wallace's comments on difficulties for PO approaches in reproducing QFT are focused on the de Broglie-Bohm and GRW approaches, so it is, of course, still possible that some other PO approach that I have not discussed here will turn out to work for QFT. However, the structural argument given above seems likely to apply to other PO approaches as well, and the proponents of such approaches do not often discuss explicitly the challenges posed by scale transformations in QFT. So at the very least we have some reason to take seriously the possibility that primitive ontology approaches may be unable to overcome this obstacle.

Of course, one might argue that the existing primitive ontology approaches should be regarded as simply proofs of principle, demonstrating that something like non-relativistic quantum mechanics can emerge out of underlying quasi-classical hidden variables. Thus one might regard them as evidence for the possible existence of some as-yet unknown primitive ontology approach which would, in fact, reproduce all of quantum mechanics and QFT. But a proof-of-principle for a special case is not worth much if it relies crucially on special features of that special case which are unlikely to apply in the general case; and many existing primitive ontology approaches do indeed seem to rely on special features of non-relativistic quantum mechanics which are unlikely to apply in an extension to QFT. Thus, based on current understanding, it is reasonable to have some doubts about whether these approaches will ever be able to reproduce all of QFT.

Now, some proponents of primitive ontology approaches have argued that it is not very important to be able to reproduce QFT. This is because they consider the status of QFT to be somewhat murky – it is after all only an effective field theory, which means that it is not precise or fundamental; rather it is an approximation which works by treating quantum systems as continuous at some scale by simply ignoring effects below that scale (Rivat and Grinbaum, 2020). However, this doesn't mean we don't need to take QFT into account in solving the measurement problem – for even if QFT is not fundamental, nonetheless its predictions have been confirmed to a very high level of accuracy, and therefore any account of reality or package deal which can be considered a viable candidate to describe our actual reality must ultimately have some way to explain why these predictions work. Moreover, insofar as QFT is not fundamental, non-relativistic quantum mechanics itself is also not fundamental – indeed, it emerges *from QFT* in the non-relativistic limit, so in a sense it too is just an effective theory. Thus it doesn't make much sense to work on solutions to the measurement problem for non-relativistic quantum mechanics but then say that it is premature to worry about QFT. Perhaps there is a justification for this if one thinks of the measurement problem as fundamentally a problem about how to tell a story which will give us a subjective feeling of understanding, in which case one might think that having the feeling of understanding non-relativistic quantum mechanics is better than nothing, but if the measurement problem is understood as more than merely pedagogical, there is no good reason to distinguish QM and QFT in this way.

Indeed, with respect to the *epistemic* versions of the measurement problem, we don't really have the option of simply disregarding QFT. For we know that quantum mechanics and presumably the quantum measurement process must arise out of QFT, or at least the two must transition smoothly into each other. So if I come up with some account of the measurement process which has the consequence that measurements yield reliable evidence about reality, and then you counter with a demonstration that this account is incompatible with what we know about QFT, it would not be reasonable for me to then say 'Well, nonetheless, I am reassured that our measurements are providing reliable information about reality.' In this case it would seem that the empirical evidence is clearly telling me that my proposed account of the physical nature of measurement is wrong, and therefore that account can't provide any meaningful reassurance as to the reliability of the measurement process. Thus, if we cannot at least dimly envision how the PO approaches could be made consistent with quantum field theory, they are not viable solutions to the measurement problem.

11.2 Ontology for Non-Fundamental Theories

The preceding discussion of PO approaches reminds us of an important fact that should be kept in mind in attempting to solve the measurement problem: quantum mechanics is not fundamental. Certainly, non-relativistic quantum mechanics is not fundamental, since it emerges from QFT, but QFT is most likely not fundamental either, since it is commonly believed that it emerges out of an underlying theory which takes over at or below some small-distance 'cutoff' (Wallace, 2001).

This means that any ontology that we might choose to associate with quantum mechanics – including any story we might tell about the physical nature of quantum measurements – probably cannot be regarded as 'fundamental.' It will instead involve approximately-defined entities or structures which emerge from some underlying theory at scales relevant to quantum mechanics but which are not present at smaller scales. After all, if QFT emerges from some other physics below the cutoff, then even if there exists a unique fundamental ontology which can be associated with the theory below the cutoff, it seems very unlikely that we will be able to infer what it is from the structure of quantum mechanics or QFT alone – for if we could, then the problem of finding a successor theory to quantum mechanics would probably be much easier than it has turned out to be!

Moreover, this is not just a temporary situation. Not only are we not currently in possession of a candidate for a complete, final, fundamental theory of physics, it is very likely that we will *never* be in possession of such a thing. So it would not be good strategy to simply defer the measurement problem and other questions about epistemology until the fundamental theory is arrived at; we need to be able to reassure ourselves as to the meaningfulness of our scientific knowledge in a way which does not depend on the claim that we have identified the fundamental building blocks of nature.

In particular, the fact that quantum mechanics is not fundamental has consequences for how we should think about the measurement problem because it means we have to be careful about what we mean when we ask questions like 'What is the ontology of quantum mechanics?' For if the regimes in which quantum mechanics applies are not, in fact, where the ultimate building-blocks of nature are to be found, this question probably cannot be answered by giving an accounting of the basic building-blocks of nature. Instead it must be understood as asking about patterns or structures that emerge at scales relevant to quantum mechanics – and of course, at a purely formal level there may well be many different ways of identifying 'patterns' at a certain scale, so in order for this question to make any sense we need some way of saying what it is for a particular pattern or structure to represent 'the' ontology of the theory.

Now, quantum mechanics is obviously not unique in this regard. No other extant theory of physics is 'fundamental' either, so the ontologies that we associate with those theories are necessarily composed of non-fundamental entities. There is nothing particularly puzzling about that – the everyday world of tables, cats, and strawberries is also not fundamental and yet we have no difficulty associating an ontology with it. Many philosophers have emphasized that the fact that entities are not fundamental does not mean they are not 'real' – as Hofstadter and Dennett (1982, p. 6) put it, '*Our world is filled with things that are neither mysterious and ghostly nor simply constructed out of the building blocks of physics. Do you believe in voices? How about haircuts? ... These things are not physical objects with mass, or a chemical composition, but they are not purely abstract objects either ... We must not suppose that science teaches us that every thing anyone would want to take seriously is identifiable as a collection of particles moving about in space and time.*' Building on Dennett's work, Wallace (2011, p. 13) has proposed a criterion to help identify what counts as 'real' or 'ontological' in a non-fundamental description of reality: '*A macro-object is a pattern, and the existence of a pattern as a real thing depends on the usefulness – in particular, the explanatory power and predictive reliability – of theories which admit that pattern in their ontology.*'

But it seems harder to identify patterns in this way in the context of quantum mechanics than it is for the everyday world of tables, cats, and strawberries. For as Dennett (1991, p. 36) notes, our identification of 'real patterns' in the macroscopic world is often driven by pragmatic considerations: it may depend '*on how swiftly and reliably we can discern the simple pattern, how dangerous errors are, [and] how much of our resources we can afford to allocate to detection and calculation. These "design decisions" are typically not left to us to make by individual and deliberate choices; they are incorporated into the design of our sense organs by genetic evolution, and into our culture by cultural evolution ... The ontology generated by the manifest image has thus a deeply pragmatic source.*' That is, although there are many different ways in which we could group parts of the macroscopic world into 'patterns,' in practice there are usually obvious and natural ways of identifying pragmatically useful 'real patterns' in the macroscopic worlds, and indeed most of the time preprocessing in our brains makes these identifications for us without conscious effort.

By contrast, we don't directly perceive anything in the regimes relevant to quantum mechanics, so we can't rely on preprocessing in our brains, or indeed pragmatic utility, to identify an 'obvious and natural' ontology for the theory. True, we may be able to identify certain structures in functional terms by means of our interactions with them – for example, in certain experimental situations we can identify some structures which behave a lot like particles, so it is natural to propose an ontology for quantum mechanics which includes 'particles' at least some of the time. But part of the difficulty of interpreting quantum mechanics comes from the fact that quantum-mechanical systems seem to behave differently when we are interacting with them compared to when we are not – for example, the famous phrase 'wave-particle duality' refers precisely to the observation that some quantum systems seem to behave more like particles when we measure them and more like waves when we do not measure them (Anastopoulos, 2008). So it's unclear that we can simply extrapolate entities identified functionally by means of interactions to parts of the quantum world that we are not currently interacting with; and yet unless we are willing to adopt some kind of strong operationalism or idealism, we will still want to say that there is *something* around when we are not interacting with quantum systems, so if we are to have a complete picture of quantum mechanics we need some way of identifying 'real patterns' which are present *outside* of measurements.

Thus quantum mechanics appears to lie in an awkward intermediate zone – too small for pragmatic considerations to deliver an obvious and natural ontology, but too big to have a unique ontology composed of the fundamental constituents of nature. Moreover, since it's likely that we will never reach a final, fundamental description of nature, we may well be stuck in that intermediate zone forevermore. So how should we proceed?

Of course, one option is to simply give up on identifying a unique ontology for the theory – and indeed, there is surely something right about that. Having recognised that we are not dealing with a truly fundamental theory, it is tempting to adopt a pluralist view of ontology, allowing that a number of different ways of thinking about the ontology of quantum mechanics may all identify important and relevant structures, and there may be no fact of the matter about which one is correct.

However, this doesn't mean that all proposed ontologies are equally valid. After all, as discussed in chapter 2, the epistemic version of the measurement problem remains a problem even if we are only concerned about empirical scientific knowledge and not about theoretical scientific knowledge pertaining to unobservable parts of the ontology; and yet it also seems likely that a solution to the epistemic version of the problem will involve saying something at least incidentally about ontology, because the epistemic problem is about how to give a physical account of what is going on in the process of measurement, and even if we do this in a purely mathematical or structural way it will surely have some ontological implications. So we likely still need some way of narrowing down what kinds of patterns in the quantum-mechanical regimes should be elevated to the status of 'real patterns' or 'ontology.' But fortunately, focusing on an epistemic conception of the measurement problem also helps clarify the kinds of patterns we ought to be looking for: ones which solve the (epistemic) measurement problem! That is, we are interested in defining structures in which we can embed a description of the measurement process in a natural and coherent way – structures which are related in a systematic way to measurements and measurement outcomes, and which mediate the relation between these measurements and the rest of the theoretical entities or structures that we have supposedly found out about by means of measurement.

11.2.1 Primitive Ontology and Fundamentality

To better understand what this criterion means for the kinds of structures that can be regarded as solving the measurement problem, let us reconsider Wallace's objection to the PO approaches in light of the preceding discussion. Wallace appears to suggest that if an ontology for quantum mechanics is to be taken seriously, it must be related in a relatively direct way to the fundamental building blocks of nature. That is, the de Broglie-Bohm particles cannot be the correct ontology for quantum mechanics unless they have a relatively direct analogue at more fundamental scales – something like '*a microphysically-stateable, precisely-defined dynamical variable which, on coarse-graining and restriction to the non-relativistic particle-mechanics regime, nonetheless delivers coarse-grained particle position or some appropriate surrogate*' (Wallace, 2022, p. 10).

But why require this? After all, there is no doubt that structures isomorphic to the de Broglie-Bohm particles do emerge in some sense within quantum mechanics, for we know that in non-relativistic regimes the de Broglie-Bohm formalism can reproduce all of the empirical predictions of quantum mechanics. And the 'real patterns' approach to ontology would seem to suggest that such structures can be regarded as 'real' to some extent, since a theory admitting them as ontology will have considerable predictive and explanatory power at least in the non-relativistic regime. Perhaps these structures emerge in a somewhat indirect way from the fundamental building blocks, but what of it? Tables and chairs also emerge in a somewhat indirect way from the fundamental building blocks of nature, but that doesn't stop us from thinking of them as real. If any ontology for quantum mechanics is necessarily going to be non-fundamental, what grounds do we have to insist that a viable ontology must emerge in one particular way rather than another?

It is helpful here to consider exactly what role the de Broglie-Bohm particles must play if they are to solve the (epistemically construed) measurement problem. This issue can be framed in terms of a prior debate about whether de Broglie-Bohm is just 'Everett in denial.' For example, Brown and Wallace (2005) make an argument based on considering what we should say in the de Broglie-Bohm approach about the case of performing a quantum measurement on a system which happens to be in an 'eigenstate' of the measured variable, meaning that only one outcome of the measurement is assigned a non-zero probability by the standard formalism. They argue that in this case, the single branch of the wavefunction (or 'wavepacket') associated with that outcome seems fully capable of representing a scenario in which a definite measurement outcome actually occurs, with or without the de Broglie-Bohm particles. And thus they contend that in the more common case where there is more than one possible outcome, and hence more than one branch, nonetheless '*each of the non-overlapping packets in the final joint-system configuration space wavefunction has the same credentials for representing a definite measurement outcome as the single packet does in the predictable case.*' Brown and Wallace (2005, p. 6) They argue, therefore, that since the de Broglie-Bohm approach includes multiple 'wavepackets,' all of which are capable of representing a scenario in which a definite outcome occurs, then in the de Broglie-Bohm picture all of those outcomes actually do occur, and hence the de Broglie-Bohm model collapses into the Everettian approach.

However, Valentini (2010, p. 8) objects that this criticism fails to interpret the de Brogle-Bohm picture on its own terms because the approach does, in fact, offer an explicit specification of the relation between its mathematical structure and observed reality: '*Physical systems, apparatus, people, and so on, are "built from" the configuration q [of the de Broglie Bohm particles] ... macroscopic systems – such as experimenters – will usually supervene on q under some coarse-graining.*' That is, in order to resist the claim that the other branches of the wavefunction are equally real, the de Broglie-Bohm theorists must insist that the de Broglie-Bohm particles have a special status as the foundation of our unique macroscopic experience, and that the wavefunction without the particles cannot play this role.

Now, if we think of the de Broglie-Bohm particles as 'fundamental building blocks of nature,' then this seems like a reasonable move: we need only say that reality is, as a matter of fact, made up out of these particles, and the wavefunction is simply something like a rule telling these particles how to move about, so consciousness and the macroscopic world of our experience do not and could not possibly supervene on it. But if the de Broglie-Bohm particles are not fundamental building blocks of nature, this seems more difficult. Of course, it is entirely possible that consciousness supervenes on some non-fundamental pattern or structure – after all, consciousness appears to be associated primarily with brains, which are not typically considered to be fundamental – so there is nothing in principle stopping us from postulating structures which look like the trajectories of de Broglie-Bohm particles as higher-level structures and then hypothesizing that consciousness supervenes on these kinds of structures. But the problem then is that, if we look at the whole wavefunction, we will be able to find structures which look like the trajectories of de Broglie-Bohm particles in *all* branches of the wavefunction. Whereas for the de Broglie-Bohm approach to work, we need it to provide us with a structure which singles out just one branch, otherwise what we get really *is* just Everett in denial.

This is an instance of a more general difficulty. The unitary-only wavefunction description is relatively symmetric with respect to all branches of the wavefunction, so any kind of structure which could emerge out of the unitary-only wavefunction description on its own would inevitably end up emerging in all, or at least many, of the branches of the wavefunction. And therefore such a structure can't possibly prevent us from ending up in an Everett-style picture, with all the attendant epistemic problems. Thus if our aim is to somehow single out just *one* branch of the wavefunction, it seems unavoidable that we must add something to the wavefunction description. Moreover the standard way of thinking about relations between scales in physics – 'petty reductionism' as Weinberg calls it – assumes that all large-scale phenomena are ultimately reducible to smaller-scale phenomena, and if we accept this idea, it would seem to follow that anything we add to the wavefunction description must ultimately have its origins in some structure that exists at the smallest possible scales (see section 13.2 for further comments on petty reductionism). Hence Wallace's challenge: if anything like the de Broglie-Bohm approach is to work, we must be able to add something at very small scales which will, under appropriate coarse-graining, give rise to a higher-level structure which singles out *just one quasi-classical branch of the wavefunction*. That is, the concern here is not that the ontology required by the PO approaches could not emerge out of underlying physics – clearly in some sense it can and does – but that it could not emerge in such a way that it exists in only one branch of the wavefunction, which is what we need to avoid the epistemic problems of the Everettian approach.

Now, at this point one might object that it is unreasonable to expect the proponents of the de Broglie-Bohm approach to demonstrate how de Broglie-Bohm particle-like structures in just one branch can emerge out of some underlying physics which has not even been invented yet. It is, and indeed always has been, perfectly reasonable to work with an emergent ontology without knowing exactly how it emerges or what it emerges from. Consider thermodynamics, for example – for a long time physicists were unsure about or mistaken about the underlying physics of heat (Psillos, 1994), but that doesn't mean they were wrong or unjustified in thinking of heat as a real phenomenon.

However, the situation in quantum mechanics is disanalogous to earlier examples of emergent ontologies. In the thermodynamical case, for example, although physicists did not always know what kind of underlying physics the higher-level structures emerged out of, they didn't have any reason to think the higher-level structures could *not* emerge out of the underlying physics. Whereas with de Broglie-Bohm particles, spontaneous collapses, and so on, the problem is not just that we don't know how they emerge out of something more fundamental, but that our current understanding makes it look quite unlikely that they *could* emerge out of something more fundamental. For we have good mathematical evidence that non-relativistic quantum mechanics emerges out of QFT – and so if the ontology of non-relativistic quantum mechanics emerges from some as yet unknown underlying physics, it stands to reason that this emergence should be mediated via QFT. That is, we would expect that the unknown physics gives rise to some kind of intermediate structure which can be regarded as the ontology of QFT, and those structures then give rise to de Broglie-Bohm particles or spontaneous collapses or something of that kind in the regimes relevant to non-relativistic quantum mechanics. But we have seen that it is very challenging to define any relatively simple structures at small scales in QFT which would reduce to something like de Broglie-Bohm particles or spontaneous collapses in the position basis *in just one branch of the wavefunction* in the regimes relevant to non-relativistic quantum mechanics, so on the strength of our current evidence it seems quite unlikely that the emergence of de Broglie-Bohm particles or spontaneous collapses could actually be mediated that way. Therefore if these things do, in fact, emerge out of some underlying physics it would seem that this emergence would have to somehow bypass QFT altogether, and it's hard to make sense out of that idea – if non-relativistic quantum mechanics ultimately arises out of something other than QFT, is it just a coincidence that we can also derive it from QFT in the non-relativistic limit? Why is QFT so successful at making empirical predictions if our macroscopic experiences arise out of something else entirely?

Additionally, I have emphasized that the ontology posited in the primitive ontology approaches is essentially classical in nature, and that this is a central element of the solution they offer to the measurement problem. But the reason why we have a measurement problem in the first place is that it is difficult to get something classical to emerge out of something non-classical in a consistent and natural way. So if the ontology of the PO approaches emerges out of some as yet unknown underlying physics, it's likely that underlying physics must also be somewhat classical in nature – after all, if we *could* get the de Broglie-Bohm particles to emerge out of some non-classical or quantum substratum, then we might as well cut out the middleman and have classical physics emerge directly from a non-classical or quantum substratum! What exactly it means for the underlying physics to be sufficiently 'classical' that the ontology of the PO approaches could emerge from it is not entirely clear, but, for example, one might expect that it would have an ontology composed of localised entities which don't enter into superpositions. And is it really plausible that what lies beneath all the weird and wonderful non-classical structure that we have uncovered in quantum mechanics and QFT is something that looks classical again? We can't know for sure, but all current indications point to the 'unknown underlying physics' that gives rise to QFT and gravity being even *more* non-classical than quantum mechanics. So again, there are strong reasons to doubt that the ontology suggested by the PO approaches could possibly emerge out of something more fundamental.

So the problems for the PO approaches posed by QFT cannot be defused by simply deferring the problem to some unknown future physics – of course, it's still possible that future developments will work out in a way that is favourable to the PO approaches, but the evidence we have available to us right now makes it seem quite unlikely that the kinds of structures that the PO approaches require if they are to solve the measurement problem can actually emerge QFT or from whatever physics lies below QFT.

11.3 Autonomy of Scales

Another important issue that arises in the context of non-fundamental theories concerns the relationship between physics at different scales. For when the physics we are studying is non-fundamental, there is always the possibility that new physics discovered at smaller scales could subsequently alter or even undermine our understanding of this non-fundamental physics. We have seen exactly this in the case of the PO approaches: these approaches appear to be perfectly adequate descriptions of physics in the regimes relevant to non-relativistic quantum mechanics, but physics at smaller scales, in the form of QFT, may be incompatible with their proposed account of reality and in particular their account of measurement. This raises questions about how we can ever take ourselves to have stable scientific knowledge of physics at a given scale without knowing the smaller-scale physics from which it is assumed to arise.

In fact, in the practice of physics we commonly assume that the physics of smaller scales can for all practical purposes be ignored when we are studying physics at a larger scale. This idea is sometimes known as the 'autonomy of scales,' referring to the fact that the physics of our world appears to consist of a hierarchy of quasi-autonomous physical domains defined at different distance scales (Cao and Schweber, 1993). Here, following Franklin (2020), it is helpful to distinguish two different forms that the autonomy of scales might take. First, what Franklin calls 'autonomy from microstates' or autonomy$_{ms}$, which requires that happenings at larger scales do not depend too much on specific details of the state at smaller scales. That is, suppose we have a system prepared in some macroscopic state S at a scale x_1: typically there are a number of different microscopic states which give rise to the same macroscopic state, so we can identify a space of states $\{s\}$ at some smaller distance scale x_2 which are all compatible with S. Autonomy$_{ms}$ then requires that we can usually predict the evolution at scale x_1 of the state S without knowing which specific state s the system is in at scale x_2 – so, for example, perhaps we can predict that S always or with high probability evolves to some other macroscopic state S'. At the scale x_2, this means that that a 'typical' small-scale state s in the set $\{s\}$ compatible with S will evolve in a way that corresponds to an evolution of S to S' at the scale x_1. Autonomy$_{ms}$ is therefore a function of the laws of evolution and the typical distribution of microscopic states s, which means that it can be understood as a property of an individual theory, possibly in combination with a specification of a probability distribution over initial conditions, if that is not included in the theory itself.

Second, we have what Franklin calls 'autonomy from microlaws' or autonomy$_{ml}$, which requires that happenings at larger scales do not depend too much on details of physics at smaller scales – that is, any other possible world with microscopic physics which is relevantly similar to the physics of our own world will have effectively the same larger-scale physics. Now, if we allow any possible microscopic physics whatsoever it will, of course, always be possible to come up with some alternative microscopic physics which produces larger-scale physics different from our own world, and therefore autonomy$_{ml}$ is a property not of an individual theory but of a theory with respect to some class of admissible theories. Thus it must be studied within a framework which allows admissible theories to be formulated and classified – for example, a specific version of autonomy$_{ml}$ known as 'naturalness' is commonly studied in the framework of quantum field theory (Giudice, 2013; Williams, 2015). Theories in this framework are characterised by Lagrangians encoding their dynamics, and an individual theory is said to be 'natural' if it exhibits autonomy$_{ml}$ with respect to a class of admissible theories which includes the original theory and all other theories with Lagrangians of the same form as the original theory, but slightly different values of the Lagrangian parameters – for example, if the original Lagrangian for a field ϕ has the form $\frac{1}{2}(\partial_t\phi)^2 - \frac{1}{2}m^2\phi^2$ we might consider a class of admissible theories with Lagrangians of the form $\frac{1}{2}(\partial_t\phi)^2 - \frac{1}{2}k^2\phi^2$, where k is required to be relatively close to m. In other words, a theory is said to be natural if '*the observable properties of a theory (are) stable against minute variations of the fundamental parameters* (2019, p. 10, online version)'. Williams (2015, 2019) demonstrates that several apparently distinct concepts of 'naturalness' employed in theoretical physics can also be understood as mandating this version of autonomy$_{ml}$[1].

'Naturalness' has historically been invoked as a criterion of theory selection in fundamental physics, but there is now significant debate over the validity of this criterion (Grinbaum, 2012; Hossenfelder, 2019; Rosaler et al., 2019). The problem has some urgency because although the standard model of fundamental physics mostly satisfies naturalness, there are a few anomalies where parameters appear to be 'fine-tuned,' meaning that small changes to these parameters *would* significantly alter large-scale physics. In particular, our best current evidence suggests that the value of the cosmological constant Λ_0 at the smallest possible scale, and another constant v depending on vacuum fluctuations, almost cancel out such that the cosmological constant Λ_M at macroscopic scales is much smaller than it would otherwise have been expected to be. Thus a small change to v would lead to very significant changes in Λ_M because v and Λ_0 would no longer cancel out, so the observable properties of this theory do not seem to be stable against minute variations of the fundamental parameter v. Similarly, our best current evidence indicates that the value of the bare Higgs mass $m_{H,0}$ and quantum corrections c coming from all the other standard model particles must almost cancel out in such a way as to result in the observed value of the Higgs mass $m_{H,M}$ at the scales we are able to access, and thus again, a small change to $m_{H,0}$ or c would, in fact, lead to very significant changes to observable physics by changing the observed value of the Higgs mass. So the normative status of naturalness – and more generally, of autonomy$_{ml}$ – needs clarification, in order that we can decide whether or not these apparent failures of naturalness and of physics-based autonomy of scales should be regarded as a problem.

Now, one reason to expect that current and future theories should obey autonomy$_{ml}$ in general, and naturalness in particular, is simply induction on past successes – for example, the fact that apart from the mass of the Higgs boson, all of the couplings in the standard model are natural (Williams, 2019). However, this particular application of induction is not entirely analogous to the more common use of induction to make inferences about the future from the past because we have no particular reason to think the future will *not* be like the past, whereas we know for sure that physics looks very different at different scales, and one might think this fact somewhat undercuts the motivation for an inductive inference to the conclusion that physics at a given scale must be similar to physics at other scales in one specific way. Alternatively, another way to motivate a naturalness requirement is to show how it emerges from the structure of the effective field theory framework within which our QFTs are formulated (Williams, 2015). However, as Williams (2019) emphasizes, strictly speaking what emerges directly from the effective field theory framework is only quite a limited version of autonomy$_{ml}$, which probably does not entail that

[1] Williams also notes that some modern applications of 'naturalness' in fundamental physics have drifted away from the original motivation in terms of autonomy of scales – I will discuss some other notions of naturalness in section 12.3, but for now let us focus on the question of autonomy.

'fine-tunings' of fundamental constants are a problem. So doubts around the use of 'naturalness' or autonomy of scales as a criterion of theory selection do seem quite reasonable.

However, there are also some a priori *epistemic* reasons to insist on some version of autonomy$_{ml}$, so the argument for it need not rely entirely on induction or on the structure of existing theories. Here we must distinguish carefully between autonomy$_{ms}$ and autonomy$_{ml}$. It is clear that the possibility of stable empirical knowledge depends on the physics of our actual world exhibiting autonomy from microstates, autonomy$_{ms}$, because without such autonomy there could not exist any stable macroscopic regularities at all – in the absence of autonomy$_{ms}$, systems prepared in a given macrostate would exhibit wildly different macroscopic evolution at every individual preparation, so the macroscopic world would simply not exhibit any meaningful organisation. But as Hossenfelder (2019, p. 4, online version) points out, this argument does nothing to show that we also require autonomy from microlaws, autonomy$_{ml}$, because this second kind of autonomy '*quantifies a sensitivity to a virtual ("mathematical") change of parameters, not a sensitivity to a change that can actually happen.*' That is, it is entirely conceivable that there should exist stable macroscopic regularities in our actual world even though in a possible world with slightly different values of the fundamental parameters there would not exist stable macroscopic regularities, or the regularities would be very different. And thus since autonomy$_{ml}$ pertains only to counterfactual changes rather than actual ones, one might initially think that it cannot have any particular significance for the epistemology of science.

However, in fact, there are limitations on the extent to which we can tolerate failures of autonomy$_{ml}$ because measurements, measurement outcomes, and records of measurement outcomes all belong to the macroscopic world, and thus our analysis of them draws heavily on our pre-existing knowledge of macroscopic physics. Indeed, Curiel (2020, p. 29) argues on these grounds that '*one can never impose a clean, sharp demarcation, once and for all, between a given theory and auxiliary theories needed by experimentalists in the study and use of it.*' Consequently, we need to be confident that our effective description of the macroscopic world is largely reliable if we are to be able to interpret measurement outcomes and records of measurement outcomes as providing us with any meaningful information about fundamental physics, and therefore we depend on autonomy from microlaws, autonomy$_{ml}$, at least to the extent that we would like to be confident that pre-existing scientific knowledge about the macroscopic world will not be undermined by later discoveries about the microscopic world. If the dependence between microscopic physics and macroscopic physics were too strong, we would not be able to use macroscopic physics as an independent tool to probe microscopic physics because small differences in the microscopic physics would dramatically change the very macroscopic physics that we are presupposing in our enquiry about the microscopic physics – the system would in effect be too strongly coupled for any meaningful scientific enquiry to be possible. So from this point of view, the reason we ought to care about naturalness is not just because of counterfactual changes which will never actually occur, but because autonomy$_{ml}$ is important for the robustness of our system of scientific knowledge: significant failures of this kind of autonomy are problematic because they pose the threat of empirical incoherence, in a similar way to the other examples of empirical incoherence that we have been exploring throughout this book.

Now, at this juncture one might object that regardless of what is going on at microscopic scales, we already know the correct macroscopic physics from direct empirical enquiries at large scales, so no matter how strongly the microscopic physics is coupled to the macroscopic physics, it cannot undermine the macroscopic physics that we have already established on independent grounds. However, the examples discussed throughout this book make it clear that this kind of argument can only get us so far, since our belief that we know the correct macroscopic physics depends on assumptions about the accuracy of our macroscopic measuring devices, memories, records, and data obtained by other observers, and yet these things themselves could potentially be undermined by new developments in microscopic physics – indeed, in section 12.3 I will discuss in more detail the ways in which fine-tunings of the fundamental constants could plausibly undermine our knowledge of macroscopic physics.

One might also argue that even if widespread and generic failures of autonomy$_{ml}$ would, in fact, undermine the possibility of scientific enquiry, nonetheless the small number of fine-tunings that have been discovered thus far evidently do not, since after all we have been able to discover them empirically. However, these fine-tunings still pose a potential epistemic difficulty because there is arguably a problem with the overall coherence of a system of beliefs which accepts that there exist a few empirically observable fine-tunings but denies the existence of more dramatic fine-tunings which would undermine our means of enquiry in significant ways. I will return to this topic in section 12.3, but for now I will focus on placing the mandate for autonomy$_{ml}$ in a more general epistemic context.

Specifically, we can think of autonomy$_{ml}$ in the context of a more general *historically*-grounded notion of autonomy. The need for autonomy of this kind is particularly emphasised by Chang (2004, p. 224) in the context of his progressive coherentism, for as he puts it: '*In the framework of coherentism, inquiry must proceed on the basis of an affirmation of some existing system of knowledge.*' In order to make scientific progress we need to be able to build upon prior knowledge in a cumulative way, and thus in some sense the *epistemic* mandate for autonomy$_{ml}$ is only incidentally about scales: it is really a special case of a more general notion that we might call 'progressive autonomy,' which requires that as we expand our scientific enquiries to new regimes, our discoveries should not too dramatically undermine the pre-existing scientific knowledge that provided the starting point of those enquiries. As a consequence of the way that we ourselves happen to be physically embodied, we are obliged to start our investigations at macroscopic scales and then move to smaller scales, which is why autonomy of *scales* is important to us, but in some other possible world in which observers were differently embodied, progressive autonomy might take a different form, reflecting the realities of scientific enquiry in that world.

This way of thinking about autonomy of scales also gives a sense of why it is that we are interested in autonomy$_{ml}$ with respect to certain classes of theories and not others. Intuitively, if our enquiries lead us to conclude that the actual world is described by a certain physical theory, and then we notice that another *very similar* physical theory would lead to different macroscopic physics, that makes our scientific knowledge look quite unstable in a way that is close to self-undermining; whereas it seems less worrying that some physical theory which is *very different* from the one we have arrived at would lead to very different macroscopic physics. From a foundationalist point of view, it may seem hard to understand this distinction – for if these two possible theories both undermine the process of empirical enquiry in significant ways, it may be impossible to demonstrate empirically that either of them is false, and thus both would appear to pose equally severe threats to our empirical scientific knowledge. But from a progressive coherentist point of view we can understand the distinction: the key point is that we are allowing the process of empirical enquiry to tell us what kinds of possibilities we ought to take seriously, and hopefully what the process of empirical enquiry will tell us is that we do not need to take seriously the kinds of possibilities which would severely undermine the practice of science! So, for example, we can understand why 'naturalness' is defined in such a way as to encode autonomy$_{ml}$ with respect to a class of theories very similar to the one thought to describe the actual world – roughly speaking, the class of admissible theories should circumscribe the set of possible theories which are made relevant by the results of our empirical enquiries, because the hope is that these enquiries will ultimately reassure us as to their own reliability, in virtue of leading us to a class of admissible theories which all lead to the same conclusions about the reliability of those means of enquiry.

Evidently this historically-grounded notion of progressive autonomy has consequences for the practice of science – for example, Chang (2004, p. 225) suggests that scientists typically employ a 'principle of respect' which favours maintaining existing systems of knowledge, so even if they have reasons to reject an existing system, they may '*continue to work with it because they recognise that it embodies considerable achievement that may be very difficult to match if one starts from another basis.*' But as Sharlin and Sharlin (1979, p. 1) emphasise, there is a trade-off here, because although we do need to build on existing knowledge, we also don't want to allow existing knowledge to become unchallengeable dogma, so it is '*those who seek to escape the past without doing violence to the historical relationship between the present and the past [who] are able to maintain their independence and make original contributions.*' And understanding the 'principle of respect' in terms of the role of progressive autonomy in the epistemology of science is a helpful way to navigate this trade-off. For the epistemic framing emphasises that we don't necessarily have to uphold existing beliefs just for the sake of it, but if we *are* going to challenge existing beliefs, we must be cognisant of the extent to which those beliefs are epistemically foundational with respect to other scientific beliefs, and particularly with respect to the new results which are leading us to challenge them in the first place! So revisions to existing beliefs which lead to *epistemic* weirdness should be embarked on only with great caution, whereas we may be more cavalier about revisions producing only benign weirdness, since such revisions may well involve changing some of our very deeply held beliefs, but they don't lead to significant upheavals across our entire system of scientific knowledge.

For example, it can be argued that the Everett and observer-relative interpretations demonstrate violations of autonomy of scales in the sense of *progressive* autonomy. In the Everettian case, we started out with knowledge of macroscopic regularities within a single classical macroscopic reality, then used that knowledge to probe microscopic physics, and arrived at a theory which, if the Everettians are to be believed, tells us that our original picture of macroscopic reality was wildly mistaken, since, in fact, macroscopic reality is made up of many branching histories rather than a single classical reality. Thus the threat that failures of autonomy$_{ml}$ could undermine what we take to be established knowledge about macroscopic regularities is not an idle one: new developments in microscopic physics *can* significantly undermine our previous picture of the macroscopic world. Of course, it could well be that, if the probability problem could be solved, the experiences of 'the average' observer in an Everettian universe would be indistinguishable from the experiences of a single macroscopic observer in some comparable single-world theory, but from a third-person point person of view the two macroscopic realities are extremely unlike each other. And as argued in chapter 10, if our aim is to arrive at a satisfactory epistemology of science, it is not adequate to consider only the first-person point of view, so it is important to keep in mind the ways in which failures of autonomy$_{ml}$ can slip between the cracks, as it were, when we consider classes of admissible theories which maintain autonomy in a first-person sense but not a third-person sense. In section 12.3 we will take a more detailed look at what this notion of progressive autonomy means for the contentious issue of naturalness.

12
Superdeterminism and Naturalness

As noted in chapter 4, non-locality is a feature of quantum mechanics which some physicists and philosophers find unappealing, and thus many attempts have been made to come up with an approach to quantum mechanics which gets rid of non-locality altogether. In this chapter we will explore one such attempt, which makes use of a mechanism known as 'superdeterminism.'

Now, superdeterminism is not in and of itself a solution to the measurement problem, since it says nothing in particular about the physical nature of the measurement process. However, it does offer a unique way of thinking about the reality underlying quantum mechanics, and if the world were really superdeterministic, that would certainly have consequences for the correct solution to the measurement problem. So it makes sense for us to discuss superdeterminism here, and moreover the topic will also lead us onto some interesting related issues concerning the epistemology of retrocausality, fine-tuning, and naturalness.

12.1 Superdeterminism

As discussed in section 4.0.1, Bell's theorem sets out to prove that any correlations which can be explained by a local model, i.e. a model in which the correlations are fully accounted for by a common cause in the past, must obey a certain inequality (Bell, 1990). It is known that this inequality may be violated in quantum-mechanical experiments in which Alice and Bob share a pair of entangled particles and then perform various measurements on those particles at a distance. However, this does not immediately entail that we are forced to accept non-locality; the proof of Bell's theorem makes a number of assumptions about the specific way in which we must arrive at a local model of the correlations, so if we wish to avoid non-locality, we still have the option of denying one of these assumptions instead.

In particular, one key assumption of Bell's proof is 'statistical independence.' To formulate this assumption, let λ be the 'ontic state' of the entangled particles at the time of their preparation, before they are separated and sent to Alice and Bob – that is, λ represents the 'common cause' in the past that would be expected to fully explain the correlations, if the scenario could be modelled locally. This ontic state may simply be the quantum state in which the entangled particles have been prepared, but if we are willing to deny the completeness of unitary quantum mechanics, then it could also include additional information not encoded in the quantum state. Statistical independence then simply requires that the ontic state λ is uncorrelated with the choices that Alice and Bob later make about which measurements to perform. This assumption seems very natural: after all, we usually imagine that observers can make free choices

Saving Science from Quantum Mechanics. Emily Adlam, Oxford University Press. © Oxford University Press (2025).
DOI: 10.1093/9780197808887.003.0012

about their measurements, and in this case these free choices are made far in the future of the preparation of the particles, so surely these future free choices cannot influence the ontic state λ produced by the preparation, since the state λ has already been determined by the time the choices are made! Therefore denying statistical independence will certainly lead to some kind of weirdness – but is it benign weirdness or epistemic weirdness?

In fact, the answer likely depends on the specific way in which the failure of statistical independence is implemented. Since standard unitary quantum mechanics obeys statistical independence, any model which posits failures of statistical independence is necessarily different from the models used within unitary quantum mechanics, and thus approaches to Bell's theorem which deny statistical independence are package deals: they postulate a theory mathematically distinct from unitary quantum mechanics, as well as an associated account of reality within which locality is explicitly maintained. One way to achieve this is to introduce retrocausality, allowing us to say that the free choices made by Alice and Bob about their measurements may have a backwards-in-time influence on the ontic state λ at the time of the preparation, even though this preparation occurs earlier in time. Another possibility is what is known as superdeterminism. The most well-known version of superdeterminism, which I will refer to as 'superdeterminism 1.0,' stipulates that the correlations between the measurement choices and the ontic state λ are built right into the initial state of the universe(Bell, 2004; 't Hooft, 2015; Chen, 2020; Hossenfelder and Palmer, 2020). That is, superdeterminism 1.0 suggests that all the microscopic details of the initial state of the universe are arranged such that, however Alice and Bob go about deciding which measurements to perform, it always or nearly always happens that they choose to make measurements which are correlated with λ in just the right way to produce apparent violations of the Bell inequalities, even though no non-local effect actually occurs. For now I will focus on superdeterminism 1.0, though I will return to retrocausality in section 12.2.

Before moving on, I should also note that there exist more recent formulations of superdeterminism, such as Palmer's Invariant Set Theory (Palmer, 2016), which adopt an approach more sophisticated than simply writing all the correlations into the initial state of the universe. The epistemic concerns I raise in this chapter do not necessarily apply to approaches like this, although it is possible that related epistemic difficulties might arise, so this would have to be evaluated on a case-by-case basis.

12.1.1 Free Interventions

One reason to worry that superdeterminism 1.0 *may* be a form of epistemic weirdness comes from the fact that it prevents us from treating our interventions as independent from the systems on which we are intervening, since we are now accepting that our choices may be correlated with those systems in subtle, unavoidable ways. The concern, first raised by Shimony et al. (1976), is that the ability to make independent interventions is a crucial part of the scientific method. Maudlin (2019, p. 12) writes, '*If we fail to make this sort of statistical independence assumption, empirical science can no longer be done at all. For example, the observed strong robust correlation between mice being exposed to cigarette smoke and developing cancer in controlled experiments means nothing if the mice who are already predisposed to get cancer somehow always end up in the experimental rather than control group. But we would regard that hypothesis as crazy.*'

However, is this assumption about independent interventions really so crucial? After all, such independence assumptions are particularly associated with the first-person approach to schematizing the observer as discussed in chapter 10, where observers are seen as external to the theory, freely intervening on objects described by that theory and receiving responses which can then be analysed using some kind of statistical model. Whereas in a *third-person* schematization of the observer, in which the observer is modelled within the framework of the theory, there is no particular reason why we should be obliged to model interventions as independent from the systems that are being intervened on, since in that case intervention is treated as just another physical event which may or may not be correlated with various other physical facts. Of course, we do want the third-person description to be able to explain the qualitative features of our first-person experience of the world, but a successful accommodation need not necessarily reproduce all of our common-sense intuitions about the causal structure of our agency. Indeed, even without taking superdeterminism into account, one would naturally expect that in a fully realistic picture our actions will only be independent of the systems we are acting on at some sufficiently coarse-grained effective level which results from averaging over various microscopic correlations, so correlations of the kind required for superdeterminism might not necessarily be too far-fetched.

Now, it is probably true that if our interventions were often very strongly and also unpredictably correlated with the systems on which we were intervening, it could be too hard for us to gather enough meaningful information about the underlying physical structures to ever arrive at a third-person picture which would satisfactorily accommodate both the interventions and the systems, but that doesn't appear to be the case here, since superdeterminism only requires fairly mild and subtle correlations. Indeed, the fact that people have been able to write down well-formulated superdeterministic models for quantum mechanics shows that it's possible to get from a first-person picture based on free interventions to a third-person picture where those interventions are not entirely free, so regardless of whether or not these models are correct, they do seem to show that the mere *existence* of superdeterminism 1.0 wouldn't necessarily derail scientific practice. We can understand this in terms of a progressive, coherentist approach – we start out with the assumption of free interventions, but subsequent scientific developments allow us to refine that assumption and reveal the limits of its applicability, and such refinements to our initial assumptions need not be considered problematic as long as they leave the reliability of our methods of enquiry substantially intact.

That said, of course, there is one kind of scientific knowledge which would, in fact, be threatened by the failure of statistical independence: knowledge of causal structure. Learning about causal relations has always been an important feature of scientific practice, since one major pragmatic motivation for scientific research is the desire to know what kinds of interventions we would need to make in order to achieve various ends. And recently this practice has been formalised in the framework of causal modelling (Spirtes et al., 2000; Pearl, 2009), which is a methodology for making inferences about causal relationships from statistical data. A causal model is composed of a set of 'exogenous' variables $\{E_i\}$, whose values are determined by factors outside of the model; a set of 'endogenous' variables $\{X_i\}$, whose values are determined by factors within the model; and a set of structural equations $\{p(X_i|X, E)\}$, which express the value of each endogenous variable conditional on the values of the exogenous and/or endogenous variables. Often such models are represented as a directed graph G whose edges represent the causal relations between the variables. It is crucial to the methodology of causal modelling that we are able to identify variables which can sensibly be regarded as exogenous, i.e. such that it is reasonable to think that they have no causes within the model. And typically it is the variables on which conscious agents can intervene, such as choices about which measurement to perform, which are assumed to be exogenous. So if we can no longer treat interventions as exogenous, or at least we can't do this in quantum-mechanical contexts, it may be the case that we won't be able to make inferences about causal structure in these contexts, thus potentially undermining scientific knowledge of causal structure. And thus if one believes that knowledge of causal structure is a central pillar of our scientific knowledge, one may take the view that superdeterminism does indeed significantly undermine scientific knowledge.

However, in this book we have primarily been concerned with epistemic weirdness that threatens *empirical* scientific knowledge, and it isn't clear that knowledge of causal structure is really empirical knowledge. Agreed, there is an empirical element to the knowledge that intervening on X is a good way to achieve some (observable) effect Y, but it doesn't seem that any part of that piece of knowledge depends on the claim that X is 'the cause' of Y, since ultimately the relevant empirical knowledge pertains to the fact that doing X is reliably followed by Y. Perhaps it's actually the case that the universe is so arranged that whenever a person decides to do X, that action is the result of some event E in the past which reliably also causes some event Y – but from your point of view as a person who wishes to bring about Y, that is still good enough to conclude that you can make sure Y occurs by choosing to do X, provided you don't also do anything to screen off the effect of the event E. So although statements about causal structure do certainly include some empirical knowledge about regularities, they also include additional theoretical baggage which probably should not be considered a part of empirical scientific knowledge, and therefore we can potentially sacrifice the causal part of this knowledge without any significant loss of empirical scientific knowledge.

Thus if superdeterminism 1.0 does entail that we can't have scientific knowledge about causal structure in contexts where quantum mechanics is relevant, that would really be just another example of scientific advances rendering obsolete some kind of theoretical scientific knowledge which we previously imagined ourselves to have – in much the same way, we no longer believe that we have scientific knowledge about caloric (Chang, 2004), or the ether (Walter, 2018). Indeed, as noted by Adlam (2023b), there are already independent reasons to think that causal structure is not intrinsic to quantum mechanics, so we should perhaps be prepared for the eventuality that causal knowledge in the ordinary sense will have to be given up in this regime in any case.

12.1.2 Initial States

However, there is another way in which superdeterminism' 1.0 might give rise to epistemic weirdness. Specifically, if we allow that the initial state of the universe can be very carefully adjusted in order that every time a Bell experiment is performed, the brains of the experimenters end up in certain very specific states, which result in them making certain very specific measurement choices; then it would arguably be simpler to just have the initial state of the universe very carefully adjusted to directly give rise to specific brain states which have memories of sequences of Bell experiments violating a Bell inequality, even though no violation, in fact, occurred. So allowing possibilities of this kind leads to an issue similar to the Boltzmann brain scenario discussed in section 7.2: for any event at all that we might want to explain, we can just as easily explain it by saying that the event in question did not actually occur, it simply happens to be the case that the initial state of the universe was fine-tuned in such a way as to eventually give rise to a memory or record which makes it look as if the event occurred.

To say more about this concern, let us pause to reflect on the connection between scientific epistemology and the selection of an initial state. Superdeterminism 1.0 requires us to make a highly specific specification for the initial state of the universe – it must belong to a special set of states, *S*, such that all the states in this set include a complex network of correlations which will, under time evolution according to the appropriate laws of nature, consistently produce failures of statistical independence in many different times and places. But the proponents of superdeterminism 1.0 have argued that there is nothing wrong with making this highly specific specification for the initial state of the universe, for the set of states *S* can only be regarded as 'special' or 'unlikely,' if we have in mind some 'measure' over initial states for the universe relative to which the set *S* is small (Hossenfelder, 2020). This is because the set of possible initial states is presumably infinite, so if we simply try to calculate the proportion of possible initial states belonging to the set *S*, we will get nonsensical results; we must specify a 'measure,' which stipulates a way of assigning numbers to regions of the state space so that we can sensibly ask how much of the state space is taken up by states with a certain property.

In many cases the physical meaning of a 'measure' is similar to a probability distribution, i.e. measures are a generalisation of probability distributions which can be assigned over a continuous space rather than a discrete set of outcomes. Thus, for example, if we imagine a certain preparation *P* which can be applied to systems with a continuous state space, there might be some physically preferred 'measure' over the state space which can be interpreted as giving the probability that a system prepared according to preparation *P* ends up in various different regions of the state space, and we might be able to determine empirically what that measure looks like by sampling many systems prepared according to preparation *P* and looking at how the results are distributed. But it is more complex to adopt this kind of interpretation for a measure over the initial state of the universe, for even if we are willing to think of the initial state as being sampled from the state of possible states according to some measure, nonetheless we have only one sample and thus we certainly do not have enough data to learn anything substantive about the 'correct' measure, insofar as there is such a thing. So the proponents of superdeterminism have argued that there can be no scientific reason to choose any particular measure over initial states of the universe, and therefore they are free to choose a measure which assigns a high measure to the set of states *S* with the properties needed for superdeterminism, such that according to their chosen measure there is nothing special or unlikely about an initial state which happens to belong to *S*.

However, although there cannot be empirical reasons to select a particular choice of measure over initial states of the whole universe, there may be extra-empirical reasons – and in particular, there are certainly *epistemic* reasons for us to make certain choices about the initial state. For it follows from the mandate that our account of reality should affirm the reliability of our methods of enquiry that it must also affirm the reliability of the kinds of inferences we typically make about the past and/or future from the information available to us in memories and records, since inferences of this kind are a central feature of any scientific enquiry. Roughly speaking, we need it to be the case that 'deceptive' possibilities are very unlikely – for example, we want to be the case that when we encounter a record in a book, it is very likely that somebody really wrote that record with the intent of recording events that really happened, and thus we would like it to be very unlikely that the particles of a book should randomly rearrange themselves to form a record of events that did not actually happen!

But there are certainly possible ways to choose an initial microstate for the universe which leads to a course of history in which the particles of books *do* frequently rearrange themselves to form records of events that did not actually happen; and thus if we are to avoid this sort of 'deceptive' possibility, essentially our options are either to use a measure over initial states which has the consequence that this kind of possibility can be understood to be unlikely, or to completely leave behind the 'time evolution' paradigm in which the course of history is produced by evolution from a fixed initial state. And the standard formulation of superdeterminism 1.0 assumes that we are working in the time evolution paradigm, so it must be understood relative to a context in which we habitually use a measure over initial states to rule out these deceptive possibilities. Thus although empirical evidence by itself can't deliver a measure over initial states, the epistemology of science certainly gives some indications of the kind of measure we ought to choose.

What might that measure look like? Well, notice that in order for the particles of a book to randomly rearrange themselves to form a false record, the book must start out with a significant number of very specific fine correlations between the positions of various microscopic particles. And thus an initial state of the universe which will later lead to this deceptive rearranging must contain many highly specific fine correlations between the positions of the particles which will one day make up the book in question – correlations which will remain largely invisible until some much later time, when they will suddenly result in the particles moving around in a coordinated way to create the new falsified record. So, in order to rule out such deceptive possibilities, we must assume that the initial state does not contain or is unlikely to contain many fine correlations of this kind.

The standard way of formalising this assumption is to use a version of the Past Hypothesis (see section 7.2), which assigns very low probability to initial states containing a large number of fine correlations. For example, Wallace (2019) suggests a probability measure on phase space that takes the form $P(X) = \int_X f(x)d\mu(x)$, where f is any function on phase space specifiable '*in a reasonably simple way from the microscopic variables of the system.* (p. 6, online version).' Following Wallace, I will refer to such measures as Natural. It can be shown that combining a Natural measure over initial states with the standard microdynamics posited by statistical mechanics leads to the emergence of roughly the standard classical macrodynamics that we are familiar with, meaning that false records will not typically form spontaneously, and more generally anti-thermodynamic effects will not occur, or at least will occur very rarely.

Now, one might worry at this point that it is not particularly helpful to be able to derive the standard macrodynamics from some measure over initial states because there is only one actual initial state, so a derivation of this kind won't do anything to reassure us that the *actual* macrodynamics are the standard classical ones. However, note that a Natural measure could in principle simply assign probability 1 to a single initial state, provided that this initial state itself could be singled out '*in a reasonably simple way from the microscopic variables of the system*' (Wallace, 2019, p. 6). So we can also describe individual states as being Natural; and the standard macrodynamics that we can derive from a Natural distribution over initial states plus the microdynamics will also arise from any sufficiently Natural individual state plus the microdynamics. Thus the assumption we really need to make is that the *actual* initial state is Natural in this sense, but we typically encode that assumption in a measure over phase space, since we will never know the actual initial state, and in any case the whole point of the Naturalness assumption is that specifying a particular choice of Natural initial state would not change the macrodynamics.

But what justifies our assumption that the actual initial state is Natural, or equivalently drawn from a Natural distribution? Well, this is essentially the same case as discussed with regard to Boltzmann brains in section 7.2: the assumption of a Natural initial state is arrived at as a part of our theory-building process as we seek to unify thermodynamics with statistical mechanics, and then the Naturalness assumption subsequently plays an important part in the story that the resulting theory has to tell about the reliability of the methods of enquiry that we used in that theory-building process, particularly with regard to the process of consulting memories and records. So the Natural initial distribution is not an unjustified foundational assumption; it earns its credentials by playing a part in a more general account of reality which reproduces our empirical evidence whilst also affirming the reliability of the methods of enquiry we used to obtain that empirical evidence. The Naturalness of the initial state or measure over initial states is thus inextricably tied up with all of our other scientific beliefs, playing a crucial role in justifying our use of memories and records in a vast range of domains.

Superdeterminism 1.0, on the other hand, threatens to undermine this harmonious picture, since it requires us to postulate an initial state, or measure over initial states, which is *not* Natural. For in order to make sure that in any Bell experiment the measurement choices and the ontic state will always or usually be correlated in the way needed to violate statistical independence, regardless of what complex and abstruse methods the experimenters might employ to select their measurements, we are going to need a very large number of individual microscopic particles to be correlated in extremely complex ways in order to make sure that they all converge on exactly the desired measurement direction at the right moment. So adopting superdeterminism 1.0 involves explicitly rejecting Naturalness, which is worrying given that without a Naturalness assumption it seems very hard to justify the role of records and memories in scientific practice.

To be clear, the issue here is not that the specific correlations required to violate statistical independence will result in our records and memories being inaccurate. For if the initial state of our universe is drawn from a measure which mandates the existence of the correlations required by superdeterminism 1.0 but which is otherwise Natural in every way, then records and memories will indeed work just as they usually do – the particles of books will not in general spontaneously rearrange themselves to create false records. So the concern is not with these correlations specifically, but rather with the overall consistency of a view which postulates the existence of one particular set of fine correlations in the initial state, but nonetheless continues to deny the existence of other kinds of fine correlations in the initial state.

In particular, the problem is that there is no direct empirical test we can do to establish whether or not the initial state contains fine correlations which lead to the existence of false records of the past, for if the false records were sufficiently widespread and comprehensive, we would have no way to tell that they were false. So in order to affirm the reliability of our memories and records we must appeal to more general kinds of evidence, and one useful kind of evidence is given by the fact that in general, all existing evidence seems consistent with the hypothesis that the initial state does not generally contain highly specific fine correlations, i.e. that it is Natural or drawn according to a Natural measure. And this line of argument will be significantly undermined if we postulate that the initial state does, in fact, contain one particular kind of fine correlations; we will then have less reason to be confident that it does not also contain other fine correlations which would, in fact, interfere with the reliability of our memories and records. As discussed in section 11.3, part of the cumulative growth of scientific knowledge involves allowing the process of empirical enquiry to tell us what kinds of possibilities we ought to take seriously, and hopefully what it will tell us is that we do not need to take seriously the kinds of possibilities which would severely undermine the practice of science – whereas accepting the superdeterministic hypothesis pushes us towards taking seriously the kinds of possibilities which would undermine the empirical evidence which has led us to that hypothesis in the first place, which looks like a presumptive argument against the hypothesis.

12.2 Retrocausality

The epistemic issues discussed in section 12.1 may provide a useful way of deciding between retrocausality and superdeterminism 1.0. Sometimes retrocausality and superdeterminism 1.0 are regarded as being effectively the same thing – for example, in the theorem of Bong et al. described in section 9.7, the assumption 'No-Superdeterminism' is written in such a way as to rule out any violation of statistical independence, regardless of whether it is brought about by retrocausality, or superdeterminism 1.0, or any other kind of model. And indeed, if one is just writing down a mathematical model for the Bell experiments in isolation, all such approaches will end up looking very similar. But retrocausality and superdeterminism 1.0 are not equivalent from the *epistemic* point of view – in particular, the issues we have just discussed concerning the choice of measure over initial state do not arise in the retrocausal context, and indeed, in a retrocausal approach the choice of initial state may be less significant in any case because the initial state does not necessarily have any privileged role to play in a retrocausal picture.

Of course, this is not to say that retrocausality does not have epistemic problems of its own. In fact, whether or not retrocausality counts as a form of epistemic weirdness will probably depend sensitively on the way in which one implements it. In particular, Adlam (2022b) distinguishes between 'dynamical retrocausality,' where we have two distinct and independent arrows of causality operating in opposite directions, and 'all-at-once retrocausality,' which arises naturally in models where the laws of nature select the whole of history all together rather than producing it in a temporally ordered process of evolution; in such models the past and the future will depend on one another in a symmetric, reciprocal way, and thus the future does in a sense influence the past, even though there is no distinct arrow of causation pointing backwards in time.

In the case of dynamical retrocausality, Adlam (2022b) makes a further distinction, inspired by the philosophy of time travel (Inwagen, 2010). Many time travel stories, like the *Back to the Future* films, postulate *mutable timelines*, allowing that time travellers are able to change the past and hence also the present. But there are also some time travel stories, like Chiang's *Story of Your Life* (Chiang (2010)), which postulate *fixed timelines*, meaning that time travellers *can't* change the past because whatever they do has already happened and is already part of the single fixed timeline. And it is straightforward to imagine retrocausal analogues: we may choose between mutable timeline retrocausality, in which retrocausal influences can change parts of the past that have already been determined, and fixed timeline retrocausality, meaning that retrocausal influences cannot change parts of the past that have already been determined. Note that this distinction is relevant only to dynamical retrocausality – all-at-once models automatically have a fixed timeline, because if the course of history is determined all at once with all retrocausal influences already included in the selection process, then evidently no retrocausal influence can ever change any part of this timeline.

With this classification in hand, we can see that dynamical retrocausality *with a mutable timeline* does lead to various kinds of epistemic weirdness. For example, in a model allowing mutable timeline dynamical retrocausality, it could be the case that you initially witness your experiment having an outcome X, but then a retrocausal influence reaches into the past and changes your outcome to Y, and then those changes are propagated forwards such that your memories and records change to show outcome Y. So mutable timeline retrocausality potentially gives rise to problems similar to those exhibited in the consistent histories picture: the past is never fully stable, and memories and records may be arbitrarily altered such that we can never really be confident that our observations provide accurate information about the past. As discussed in section 3.3, this would potentially give rise to a number of serious epistemic problems.

What about dynamical retrocausality with a *fixed* timeline? Well, it's hard to comment on this case because it is actually quite difficult to come up with a consistent model implementing dynamical retrocausality with a fixed timeline. This is because dynamical retrocausality with a fixed timeline can easily give rise to causal paradoxes involving 'bilking,' which refers to cases where a person knows for sure that something is going to happen in the future and then takes action to prevent this thing from happening (Ismael, 2003). For example, the famous Grandfather Paradoximagines a time traveller going back in time and killing her own grandfather before he can father any children. But then the time traveller will not be born and thus can't kill her grandfather, so if we assume that all of the processes involved are deterministic, there appears to be no logically consistent way to resolve this course of events. And, of course, one can easily imagine a retrocausal analogue where this enterprising person uses retrocausality rather than time travel to commit the paradoxical murder, so the same kind of difficulty can arise in a retrocausal scenario. In order to rule out these kinds of logical paradoxes it would be necessary to impose various kinds of global consistency constraints on one's retrocausal model; yet the presence of global constraints entails that the two directions of causality are not, in fact, independent after all, and thus this approach is likely to collapse into all-at-once retrocausality rather than dynamical retrocausality. So because it's not clear how to write down a logically consistent account of reality implementing fixed timeline dynamical retrocausality without global consistency constraints, there's no way to say with confidence whether or not such a thing would exhibit epistemic weirdness; but in any case, because of the consistency issues with dynamical retrocausality, it seems likely that any viable implementation of retrocausality will ultimately take the 'all-at-once' form.

And fortunately, all-at-once retrocausality does not seem to lead to any obvious epistemic weirdness. Indeed, all-at-once retrocausality looks like a simple extension of the standard scientific picture of a single world containing a single intersubjectively accessible distribution of events governed by universal regularities. A theory of this kind can be empirically confirmed in a simple way – we need only look back at the distribution of events observed so far and then try to write down laws describing that distribution. Sometimes we may find that the global distribution of events can be decomposed into a step-by-step process in which the occurrence of an event at some spacetime point depends only on events occurring in the past lightcone of that spacetime point, in which case there is no need to postulate retrocausality; but sometimes we may find that the global distribution cannot be decomposed in that way, which suggests that we are seeing something like retrocausality, in the generalised, all-at-once sense. Thus this kind of retrocausality doesn't require any significant revisions in the epistemology of science: we can just carry on the standard scientific practice of finding systematisations of macroscopic events, understood as existing in a unique, stable, intersubjectively accessible macroscopic universe.

Thus, if we are going to violate statistical independence, all-at-once retrocausality looks like the most epistemically safe route. That said, it has been argued that all-at-once retrocausality is in some sense intrinsically non-local (Adlam, 2022b), so depending on one's specific conception of locality this move might not really count as restoring locality to physics, and therefore one might alternatively see this line of argument as demonstrating that there is not any epistemically stable route to rescuing locality by means of denying statistical independence. But at any rate, all-at-once retrocausality certainly offers us options for explaining violations of Bell inequalities in ways that are more easily made consistent with relativity than simply the naïve approach in which there is a collapse or change of the wavefunction occurring instantaneously everywhere, so regardless of whether or not this approach restores locality in a formal sense, it may nonetheless be a promising way of resolving the puzzles of Bell's theorem.

12.3 Naturalness and Fine-Tuning

We saw in section 12.1 that the assumption of a 'Natural' initial state, or measure over initial states, is essential if we are to make sense of the epistemology of science in the context of the time evolution paradigm, because without such an assumption we would not be able to use memories and records of the past to perform empirical confirmation. But 'Naturalness' assumptions do not only show up in physics in the context of the choice of initial states; as argued by Wallace (2019), we also habitually make a similar Naturalness assumption regarding the values of the fundamental constants. Say we describe the finite precision with which the fundamental constants can be known by some probability distribution over the space of possible values for them; then the relevant Naturalness assumption requires that this distribution is given by '*some not-ridiculously complicated function, relative to the uniform distribution*' (Wallace, 2019, p. 8) over the space. Moreover, this Naturalness assumption appears to play a comparable role to the Naturalness assumption that we make in the initial condition case. In particular, Wallace argues that we need a Naturalness assumption about the constants if we are to be able to derive from microscopic laws of physics the stable regularities that we expect to see in macroscopic physics; for if we use a distribution over constants which is not Natural, then the details of the macroscopic dynamics may depend very sensitively on precise details of the constants, so we will not be able to derive the expected stable regularities.

Evidently then the Naturalness of the fundamental constants in the above sense is related to autonomy$_{ml}$ as discussed in section 11.3, and indeed in the literature the two are often conflated. However, Williams (2015) emphasises that there are two distinct notions in this vicinity: '*one notion of naturalness according to which naturalness problems are failures of an expectation about the autonomy of scales, and a second notion according to which naturalness problems stem from a parameter [or theory] being "unlikely" or "improbable"*'. I will refer to the former as autonomy$_{ml}$ and the latter as 'statistical naturalness.' Using this terminology, the expectation that the fundamental constants should be drawn from a Natural distribution is evidently a kind of statistical naturalness assumption. To show that the two notions are distinct, Williams points out that low-energy supersymmetry obeys autonomy$_{ml}$, but it is an open question whether or not it obeys statistical naturalness, thus proving that these notions are not by definition extensionally equivalent. But nonetheless, although these two conceptions of naturalness track somewhat different properties, I will argue that there is a close connection between them, based on their *epistemic* role.

Recent literature on 'naturalness' has seen a lively debate about whether or not empirically observed failures of naturalness are a problem (Grinbaum, 2012; Rosaler et al., 2019). In section 11.3 we noted some concerns one might have about naturalness in general, but one criticism that has been levelled at statistical naturalness in particular is interestingly similar to an argument I discussed in section 12.1. That is, since the space of possible values for the constants is continuous, no features of the constants can be regarded as 'likely' or otherwise until we specify a measure over the possible values for the constants. And – as in the case of the initial state – we only have one set of fundamental constants, and thus we have only one data point relevant to this measure, meaning that we cannot establish the appropriate measure empirically (Hossenfelder, 2019).

But – as in the case of the initial state – even if there are no empirical reasons to choose some particular measure over the values for the constants, there may be *extra-empirical* reasons. For just as fine-tuned initial conditions could interfere with the robustness of memories and records, so fine-tuned distributions of constants could potentially interfere with the robustness of stable macroscopic regularities, and thus since we need to presuppose our existing knowledge of macroscopic physics in order to use macroscopic instruments to study microscopic physics, it could be the case that we are obliged for epistemic reasons to assume a Natural measure over constants.

At this juncture one might object that the Naturalness of fundamental constants cannot possibly have the same central role in the epistemology of science as Naturalness of initial conditions, for there is an obvious disanalogy: the number of degrees of freedom characterising the initial conditions of the universe is surely much larger than the number of degrees of freedom characterising the fundamental constants. After all, if the initial condition is specified by fixing the values for some number of distinct and independent degrees of freedom at *every* point in spacetime, then the number of degrees of freedom characterising the initial state will be infinite – or, if space is discretised and finite in extent – still enormously large, whereas the number of fundamental constants is usually thought to be relatively small – for example, one way of counting fundamental constants gives a total of 19 in our best current physical theories (Uzan, 2010).

This difference is relevant because it influences the extent to which 'fine-tuning' the initial state and the constants respectively can lead to epistemic weirdness. We have seen that by choosing an un-Natural initial state exhibiting certain kinds of fine correlations, we can arrange for the spontaneous formation of 'memories' and 'records' which include false representations of the past – indeed, by appropriate fine-tuning of the initial state we can produce almost any macroscopic course of history we like, apart from histories violating general constraints that follow directly from the microscopic dynamics, such as conservations principles. By contrast, if we're dealing with a relatively small set of fundamental constants, we won't be able to produce any macroscopic course of history whatsoever just by fine-tuning those constants. Certainly, adjustments to the fundamental constants can have a significant impact on the macroscopic dynamics – for example, in the context of QFT, it is possible to choose un-Natural distributions with constants that cancel out in a precise way such that a term which would otherwise have rapidly grown in size as we move to larger distance scales instead remains relatively small (Wallace, 2019), as in the examples described in section 11.3. However, effects like this still appear to paint with fairly broad strokes, as it were; for the whole idea of the fundamental constants is that they are *universal* and apply *everywhere*, which makes it seem unlikely that we can create highly specific local effects just by fine-tuning some of the constants. For example, it seems implausible to think it would be possible to arrange, purely by fine-tuning the fine structure constant and the cosmological constant against each other, for there to arise a Boltzmann brain with a distinct memory of a global pandemic in the year 2020, even though no such thing actually occurred; and even granting the ability to also fine-tune another seventeen fundamental constants, there still don't seem to be enough degrees of freedom here to arrange the occurrence of arbitrary macroscopic events at will. Thus although a failure of Naturalness for the fundamental constants would certainly have consequences for science, and in particular for our usual way of thinking about reductionism (Wallace, 2019), it would not necessarily produce epistemic weirdness in the same way as fine-tuning of the initial conditions.

However, in a certain sense this response is an oversimplification. For how can we be so sure that there are only nineteen (or thereabouts) fundamental constants? It would not change the argument substantially if there were just a few more or less – but what if there were *many* more? Indeed, what if there were an *infinite* number of fundamental constants? This possibility is not an idle one, since in the formulation of quantum field theory we must go through a process called 'renormalization' in order to express the predictions of the theory in terms of values of the constants at scales where we can measure them; and this process often requires us to add additional terms to the Lagrangian. Moreover, it is known that for certain kinds of QFTs – which are known as 'non-renormalisable' – the number of terms needed is infinite, meaning that we will end up with an infinite number of coupling constants (Butterfield and Bouatta, 2015). Such theories are often ruled out on methodological grounds because it would be impossible to extract empirical content out of them: we would need an infinite number of measurements to fix the values of all of the constants before we could ever predict anything. But from the epistemic point of view, before we can even worry about the usability of such a theory, there is an even more basic issue to be addressed: if we have an infinite number of fundamental constants to fine-tune, then we *could* adjust these constants to produce basically any macroscopic history we would like, modulo certain restrictions arising from conservation principles and the like. Indeed, if we had an infinite or very large number of fundamental constants to adjust, then we could presumably produce almost the same kinds of results by fine-tuning the constants as we could by fine-tuning the initial conditions, so a Naturalness assumption about the constants would be necessary for exactly the same epistemic reasons as the Naturalness assumption about the initial conditions.

Moreover, it should be emphasised that one of the things we can achieve by fine-tuning is precisely to hide the true number of fundamental constants: as noted, terms which would otherwise have a significant impact on the large-distance dynamics can be suppressed to arbitrarily small values. One might imagine, therefore, that with an infinite number of constants to adjust we would be able to arrange for the macrodynamics to look just as if there are only nineteen or thereabouts fundamental constants up until a certain point in history, and then to suddenly and dramatically diverge from that. Or alternatively, we could just fine-tune the constants so as to produce something like a Boltzmann brain with memories which suggest that there are only nineteen fundamental constants, even though there are, in fact, many more. So we cannot straightforwardly invoke empirical evidence to reassure ourselves that the number of fundamental constants is small. It would seem that the only way we can confidently arrive at that conclusion is to make some kind of Naturalness assumption about the constants, meaning that it would be unlikely for these constants to be fine-tuned in such a way that our evidence could significantly deceive us about the true number of fundamental constants. Thus it seems that we may, in fact, need a Naturalness assumption about the values of the constants in order to rule out epistemic weirdness that could arise from fine-tuning of the constants if there were many more constants than we currently believe.

So from this point of view, empirically observed failures of Naturalness for the fundamental constants may be interpreted as pointing to another problem of empirical incoherence, akin to those introduced in chapter 3 and discussed throughout this book. That is, it is essential to assume something like Naturalness for the fundamental constants as a presupposition for the possibility of doing science, in order to rule out certain extreme kinds of epistemic weirdness; but now it seems that the process of doing science has led to the conclusion that our initial assumption about Naturalness was at least partly wrong, so the theory we have arrived at seems in some sense self-undermining.

Returning to the question of the relation between statistical naturalness and autonomy$_{ml}$, it is evident that the specific ways in which failures of statistical naturalness may undermine empirical scientific knowledge that I have just described essentially come down to failures of autonomy$_{ml}$, so one might wonder why we cannot just focus on autonomy$_{ml}$ and get rid of these statistical formulations of naturalness. For example, Williams (2015) offers historical evidence suggesting that the notion of statistical naturalness began as a way of quantifying autonomy$_{ml}$, but has since become somewhat decoupled from its original motivation, in ways that Williams regards as problematic.

However, I think there is a case to be made that statistical naturalness has a meaningful place in its own right in the epistemology of science. To see this, note that although autonomy$_{ml}$ can be empirically tested to a certain extent, we can't straightforwardly use empirical methods to rule out the possibility of failures of autonomy$_{ml}$ so severe as to undermine our existing empirical knowledge of macroscopic reality (much as we can't use straightforward empirical methods to rule out epistemic weirdness resulting from fine-tuned initial states, or branching-world possibilities, or failures of intersubjectivity). We must instead employ the same kind of bootstrapping approach that I have been advocating throughout this book, embarking on empirical enquiries in the hope that we will ultimately arrive at a coherent account of reality which reassures us that severe failures of autonomy$_{ml}$ likely do not occur. And at least in the context of quantum field theory, significant failures of autonomy$_{ml}$ typically require some kind of significant 'fine-tuning' of the fundamental constants, as in the case of precise cancellations of constants in QFT. This means that a set of constants drawn from a Natural distribution is quite unlikely to lead to failures of autonomy of scales, and therefore one way in which our empirical enquiries can serve to affirm our assumption that severe failures of autonomy$_{ml}$ do not occur is by providing us with reasons to think that the distribution of constants is, in fact, Natural. This means that we have reason to care about fine-tunings even if those particular fine-tunings would not themselves have much impact on autonomy$_{ml}$, on the principle that it looks ad hoc and incoherent to allow certain kinds of fine-tunings but still insist that the more serious fine-tunings – the kind of fine-tunings which *would* undermine our existing empirical knowledge – do not occur. This is the same issue we encountered in the case of superdeterminism, where again the worry was that allowing certain kinds of fine-tunings decreases our justification for denying the existence of other, more epistemically problematic fine-tunings. Thus there may be good epistemic reasons to prefer that our theories should obey some form of statistical naturalness, perhaps in the form of a Naturalness assumption about the fundamental constants, because evidence for statistical naturalness also counts as *indirect* evidence for autonomy$_{ml}$.

Moreover, even if we decide that Naturalness for the fundamental constants is not essential from the epistemic point of view, one might still worry that there is a problem with the overall coherence of a viewpoint which allows that Naturalness is violated for the fundamental constants but maintains that it is not violated for the initial condition of the universe. For we have noted that there is no straightforward empirical way to check whether the initial state of the universe contains the kinds of fine correlations which will later result in the spontaneous production of false memories and records, so if we insist on Naturalness for initial states but not constants, this would lead to a system of beliefs in which we accept that Naturalness is violated in a few small empirically observable ways but we continue to insist that it is not violated in any of the ways that we would not be able to observe. Wallace (2019, p. 21) makes a similar point: '*If Naturalness arguments just brutely fail in particle physics and cosmology then there can be no reliable methodological argument for assuming them elsewhere [say, in statistical mechanics].*' Thus certain kinds of violations of statistical naturalness for the fundamental constants do seem liable to undermine some of the assumptions we need to make sense of the epistemology of science.

Thus, although I think Williams is right to identify the need for autonomy$_{ml}$ as the central motivation for the naturalness criterion, there may still be a place for a more general statistical notion of naturalness, as part of the background theoretical structure which reassures us that autonomy$_{ml}$ does hold in our actual world. Of course, this is not to say that all statistical formulations of naturalness are useful – after all, judgements of statistical naturalness depend sensitively on the way in which we choose the measure relative to which constants are considered likely or unlikely, and not all choices will yield judgments which are useful gauges of the extent to which a theory can be expected to obey autonomy$_{ml}$. But perhaps understanding statistical naturalness in terms of its relation to the autonomy$_{ml}$ could help distinguish between better and worse applications of the notion, thus clarifying which failures of statistical naturalness are of concern from the epistemic point of view and which are not. For example, Williams (2019, p. 22, online version) suggests that some modern applications of statistical naturalness require '*entirely divorcing naturalnesss from the effective field theory context and its attendant autonomy of scales-based justification*,' and if this is so, an analysis of statistical naturalness in terms of its epistemic function would likely suggest that these particular applications are illegitimate.

12.3.1 Anthropic Arguments

I have argued that the epistemic difficulties associated with fine-tuning don't arise merely from the fact that a few specific constants (Λ_0, c, and Λ_M,m_H) appear to be fine-tuned; the problem is rather with the consistency of a view which allows these constants to be fine-tuned while denying the existence of any more general fine-tuning.

As a corollary of this argument, it would seem that fine-tuning of certain constants may be acceptable after all if we can come up with some principled reason why this *specific* set of constants should be fine-tuned; for this would give a robust justification for continuing to assume that all of the other initial conditions and constants to which this reason does *not* apply are still likely to conform to a Natural distribution.

For example, one approach which has been suggested to alleviate worries about naturalness involves appealing to anthropic considerations in a multiverse (Carroll, 2006; Giudice, 2013; Friederich, 2021). To see how this might work, let us begin with a further observation about the fine-tuning of the constants Λ_0 and c, and $m_{H,0}$ and v, as discussed in section 11.3: these constants are not just fine-tuned, they appear to be, in some sense, fine-tuned 'for life.' For example, it has been argued that life could not exist if the cosmological constant at macroscopic scales Λ_M were significantly greater than it is, and therefore life could not exist if it were not the case that Λ_0 and c cancel out to give a very small value for the cosmological constant (Weinberg, 1987; Barnes, 2012).

This makes it possible to give what Friederich (2021, p. 53) calls '*the standard fine-tuning argument for the multiverse*.' This argument is based on the strong anthropic principle, which says that the universe in which we find ourselves must be a universe which is compatible with the existence of life (Hacking, 1987). The idea is that if there exists a large and diverse multiverse, then presumably the chances are good that there will be at least one universe in it which is compatible with the existence of life. So, '*if there is a sufficiently diverse multiverse, it is neither surprising that there is at least one universe that is hospitable to life nor that we find ourselves in a life-friendly one*' (Friederich, 2021, p. 53). Thus the existence of a multiverse makes the fine-tuning of the fundamental constants much more probable, and therefore, if we appeal to Bayesian methodology, the fine-tuning of the fundamental constants can be regarded as empirically confirming the hypothesis of a multiverse. Versions of this argument are endorsed by many physicists and philosophers, including Smart (1989), Parfit (2004), Susskind (2008) and Tegmark (2015).

However, the standard fine-tuning argument remains controversial. Many of the difficulties centre around what is known as the inverse gambler's fallacy, which '*consists in inferring from an event with a remarkable outcome that there have likely been many more events of the same type in the past, most with less remarkable outcomes*' (Friederich, 2021, p. 54; Hacking, 1987). White (2000, p. 263) argues that the standard fine-tuning argument commits this fallacy, since it supposes '*that the existence of many other universes makes it more likely that this one – the only one that we have observed – will be life-permitting*' and Draper et al. (2007) and Landsman (2016) make similar arguments.

Essentially, the debate comes down to the question of whether we ought to make predictions by first stipulating that we are in some particular universe and then asking what is likely for that universe, in which case the standard argument does commit the gambler's fallacy, or whether we should first and foremost regard ourselves as being selected from an ensemble of universes, in which case the standard argument does not commit the gambler's fallacy. Thus this debate is linked to the issues discussed in section 7.3 regarding the problem of making predictions for ordinary events in a landscape multiverse. I argued there that for most ordinary scientific predictions in the context of a multiverse, we should be thought of as simply siting ourselves in some a particular universe and then asking what is likely for this universe, rather than imagining a process in which our location is selected from the full multiverse; and one might worry, therefore, that there is some inconsistency involved in the standard fine-tuning argument. For if in most ordinary cases we make predictions by first siting ourselves within a particular universe and making predictions relative to that universe, what justifies suddenly switching to the other point of view when it happens to be convenient for the purpose of making an anthropic argument?

Fundamentally, the issue is that it is very unclear whether it is reasonable to apply our usual confirmation theory – e.g. Bayesian confirmation theory – to the multiverse hypothesis, even though it is very unlike the ordinary kind of situations for which our confirmation theory was originally developed. For example, I argued in chapter 8 that the Everettian multiverse is exactly the kind of hypothesis for which we may not be able to simply do standard Bayesian updating as in ordinary science, and one might think that similar points could apply to other multiverses as well. So we can potentially give a better argument for the multiverse if we stop trying to empirically confirm it in the usual Bayesian observational way, and instead appeal to meta-theoretic issues related to epistemic rationality.

Specifically, the idea would be to use anthropic considerations to explain the fine-tunings of certain constants while still retaining our justification for believing that the remaining constants are likely to have Natural values. Now, just postulating a multiverse by itself will not get the desired result, because if we postulate that the constants in the individual universes belonging to the multiverse are all drawn from a Natural distribution, then it is still very unlikely that we would find ourselves in a universe with fine-tuned constants of the kind that we actually see in our world. So from this point of view it looks like we are still forced towards an un-Natural distribution over the constants. However, it is common practice when reasoning in multiverse contexts to suggest that the existence of life in our universe should be treated as part of our background knowledge when making predictions. So, given that life does exist in our universe, in a multiverse context we can start from a Natural distribution over the constants and immediately predict that the constants in our actual universe must be fine-tuned in the specific ways needed to make life possible, but that we should expect them to be otherwise Natural. This suggests the following alternative argument for the multiverse:

1. In order to avoid undermining our methods of scientific enquiry, we must assume that the fundamental constants are drawn from a Natural distribution.
2. Observational evidence tells us that some of the actual fundamental constants are fine-tuned.

3. The probability of obtaining the observed fine-tunings of the fundamental constants, assuming the constants are drawn from a Natural distribution, is very low.
4. Life would not be possible in our universe if the actual fundamental constants were not fine-tuned in the way that they are.
5. The probability of obtaining the observed fine-tunings of the fundamental constants, assuming the constants are drawn from a Natural distribution and also conditioning on the observation that our universe is compatible with life, is relatively high.
6. If there is just one universe, the fact that our universe is compatible with life is just a coincidence which cannot be used to explain fine-tunings.
7. If there is a sufficiently large multiverse, then it is very likely there will be some universes which are compatible with life, and we will necessarily find ourselves in one of those universes, so if there is a multiverse it is reasonable to condition on the fact that our universe is compatible with life in order to explain fine-tunings.
8. Therefore, in order to explain the observed fine-tunings of certain fundamental constants whilst still maintaining the assumption that the fundamental constants are drawn from a Natural distribution, we should assume there is a sufficiently large multiverse.

Note that although some Bayesian methodology is used here to compare the relative probabilities of fine-tunings with and without the background knowledge that this universe is compatible with life, this argument is not trying to use the observed fine-tunings to perform direct empirical confirmation of the hypothesis of a multiverse. Rather the idea is that postulating a multiverse offers a natural way to defuse the threat that the fine-tunings appear to pose to the epistemology of science, since it leads to a robust, non-ad-hoc way of explaining the fine-tuned constants, whilst still allowing us to assume that the constants are drawn from a Natural distribution. That is, the multiverse hypothesis may help resolve the threat of empirical incoherence as discussed in section 12.3, for it allows us to condition on the existence of life, and once we condition in this way there no longer appears to be any kind of inconsistency in postulating the existence of a certain set of fine-tunings whilst continuing to deny the existence of other kinds of fine-tunings which would undermine our means of enquiry in extreme ways. And in accordance with the arguments of chapter 8, I think this kind of epistemic approach is potentially a valuable alternative way of thinking about the existence of the multiverse, since the more common Bayesian approach requires us to stretch standard Bayesian methodology far beyond the 'ordinary science' context to which it is best suited.

12.4 Superdeterminism Again

At this point, one might naturally wonder if we could not adopt some similar strategy in order to solve the problems for superdeterminism 1.0 discussed in section 12.1. That is, could we perhaps use some kind of anthropic argument to justify the assumption that the initial condition of the universe is fine-tuned in exactly the way required by superdeterminism 1.0, but otherwise Natural, thus defusing the threat of epistemic weirdness in a similar way?

Unfortunately, though, there is an important disanalogy which blocks this style of argument in the superdeterminism case. To see this, let's imagine for the sake of argument that we can come up with some reason to think that violations of the Bell inequalities are necessary for life – for example, there have been various arguments that consciousness has something to do with quantum mechanics (??), and if these ideas were to work out then perhaps a case could be made that consciousness can't exist without Bell violations. We could then argue that in a multiverse picture, if physics is local in all of the universes, we should expect to find ourselves in a universe in which the initial state is fine-tuned in exactly the way needed to give apparent violations of Bell inequalities, but that we should otherwise expect that state to be Natural.

However, violations of the Bell inequalities can only be demonstrated by appeal to relative frequencies across many experiments, and superdeterministic accounts of these experiments typically say that the outcomes in each experiment are determined by a *different* set of fine correlations in the initial state; so in order to predict Bell violations across an ensemble of experiments, we will have to individually fine-tune one set of parameters in the initial state for every Bell experiment in the ensemble. So in practice, we would need to do some individual fine-tuning for almost every Bell experiment that has ever been performed, and most likely also for umpteen examples of natural phenomena which involve some kind of non-locality. And yet an argument that Bell violations are necessary for life would not seem to explain why *all* of these parameters must be fine-tuned. If Bell violations are necessary for life due to their role in the brain, for example, that would explain why the parameters controlling the relevant correlations in the brain must be fine-tuned, but it would not explain why the parameters controlling Bell experiments in the lab must be fine-tuned, since those experiments are not taking place inside anyone's brain. So it would seem that the only way to use anthropic considerations to make superdeterminism 1.0 work would be to come up with some reason why it is necessary for life that Bell violations should *universally* be seen in certain kinds of experiments, even if those experiments are implemented in a way which has nothing to do with life; and it is hard to see how to make that case.

Moreover, this point generalises beyond anthropic arguments. Since there is only one value of the cosmological constant Λ_M at any given macroscopic scale, controlled by just two constants in the relevant bare Lagrangian, it seems not unreasonable to think we could give some kind of explanation for its fine-tuning, either by explaining the value of the cosmological constant at macroscopic scales directly as in the anthropic argument, or explaining how the two constants controlling it come to be related in this specific way. This would allow us to accept this one instance of apparent fine-tuning whilst still allowing us to maintain the Naturalness assumption more generally. But to make something similar work for superdeterminism 1.0, we would have to come up with an explanation not just for the value of one or two parameters, but many thousands of parameters whose values we would normally expect to be independent. So while there are potentially ways to address the epistemic concerns associated with fine-tuning of a few of the fundamental constants, it seems very difficult to imagine any way to address the epistemic concerns associated with fine-tuning of the initial conditions as needed for superdeterminism 1.0, and thus it seems unlikely that superdeterminism of this kind could feature in a satisfactory solution to the epistemic version of the measurement problem.

13
Where to from Here?

Thus far in this book we have seen that many existing primitive ontology approaches to the measurement problem have difficulty accommodating QFT; and meanwhile, we have seen that although unitary-only approaches can typically accommodate QFT, many well-known unitary-only approaches involve some kind of 'epistemic weirdness' which may render them self-undermining. Now, of course, it is still possible that one or more of the primitive ontology approaches might ultimately be made compatible with QFT; and meanwhile, I have noted that there are routes one might take within the various unitary-only interpretations to address the epistemic weirdness, although the results still appear to lead to systems of belief with fairly low coherence. But for the duration of this chapter I will suppose for the sake of argument that none of these possibilities work out, in order to ask – what then? If we rule out all of these unitary-only approaches, and also the existing well-developed primitive ontology approaches, what possibilities remain to solve the epistemic version of the measurement problem?

As in chapter 11, this discussion comes with the caveat that I will not be able to mention every approach which has been proposed. Here I will describe a few ongoing research programs that seem promising to me personally, but, of course, there are many other interesting possibilities which also deserve further development, so this chapter should be taken only as a sampling of the possible routes forward, not an exhaustive account of all possibilities.

13.1 Modified Unitary-Only Approaches

Since it is clear that reproducing the empirical predictions of QFT is easiest within the context of a unitary-only approach, one natural way forward is to take as our starting point one of the unitary-only approaches, and then simply see what is the most minimal change we can make to it if we want to avoid epistemic weirdness.

For example, in the case of the Everett interpretation, we can potentially avoid the probability problem by specifying that exactly one of the Everett histories is selected and realised according to the mod-squared amplitude probability distribution – or alternatively, perhaps we could say that all of the Everettian histories with mod-squared amplitude above a certain value are selected and realised, giving a kind of many-worlds approach with all of the maverick branches removed. Meanwhile, in the case of the observer-relative interpretations, we could overcome the intersubjectivity problem by adding a specific postulate ensuring that perspectives can become connected in certain kinds of physical interactions, or by relativizing outcomes to something larger than an individual observer. All of these routes seem interesting,

Saving Science from Quantum Mechanics. Emily Adlam, Oxford University Press. © Oxford University Press (2025).
DOI: 10.1093/9780197808887.003.0013

although each also has technical and conceptual difficulties, so let us now explore in more detail what they might look like.

13.1.1 Cross-Perspective Links in Relational Quantum Mechanics

Let us begin with the case of the observer-relative interpretations. In this section I will focus on relational quantum mechanics (RQM) because Adlam and Rovelli (2023, p. 7) have proposed an alteration to RQM which is intended to solve exactly the set of epistemic problems raised in chapter 9. In this approach, a new postulate known as Cross-Perspective Links (CPL) is added to RQM: '*In a scenario where some observer Chidi measures a variable V of a system S, then provided that Chidi does not undergo any interactions which destroy the information about V stored in Chidi's physical variables, if Alice subsequently measures Chidi in a basis corresponding to the physical variable representing Chidi's information about the variable V, then Alice's measurement result will match Chidi's measurement result.*' Adlam and Rovelli (2023) specify that the information about *V* is destroyed (relative to Alice) precisely when Alice performs a measurement on Chidi which does not commute with the variable associated with that information – which is to say, once Alice has performed a supermeasurement on Chidi, the information about Chidi's measurement outcome is no longer accessible to her in subsequent measurements.

Clearly this assumption solves the intersubjectivity problem, since it tells us that observers can indeed exchange information during appropriate kinds of physical interactions. Furthermore, since decoherence typically occurs in the coarse-grained position basis, almost all interactions involving well-decohered macroscopic observers will take place in the coarse-grained position basis, which means that in the case where Chidi and Alice are both macroscopic observers, it will almost always be the case that Chidi's information about *V* is stored in his brain in the coarse-grained position basis, and Alice's interaction with him will also take place in the coarse-grained position basis, and thus the CPL postulate will apply and will ensure that Alice and Chidi successfully share information. That is, the CPL postulate, in combination with decoherence, seems adequate to reassure us that the vast majority of our interactions with other observers will work just as we originally imagined in classical physics, with information successfully shared between different observers. Indeed, in order to perform an interaction with another observer which is *not* of the kind covered by CPL, you would have to maintain full coherent control over her, which would require isolating her and her surroundings from all decoherence effects and then exerting fine control over every one of the unimaginably large number of microscopic degrees of freedom that compose her. No human being has ever performed such an interaction and it is very likely that we will never be able to perform such interactions, so the CPL postulate provides very robust reassurance that we can generally rely on measurement results reported by other observers to provide meaningful information.

But what is the justification for the CPL assumption? One might be tempted to offer a transcendental argument to the effect that a condition like CPL is necessary for the possibility of objective scientific knowledge, so we are entitled to simply assume CPL without any particular empirical or theoretical support for it. And indeed, I think we must certainly *begin* the process of scientific enquiry by assuming something like CPL. However, as discussed throughout this book, in general we expect that our scientific enquiries will provide empirical and/or theoretical support for the basic assumptions that we need to make in order to get the process of enquiry started, so our initially unjustified assumptions about the reliability of our means of enquiry can be retrospectively justified in a progressive coherentist sense. I have argued that approaches like the Everettian picture and the observer-relative picture do not really achieve this retrospective justification in a satisfying way, but perhaps the RQM+CPL approach is better off in this regard?

Well, one point that can be made in favour of the CPL assumption is that arguably it is not entirely ad hoc. Adlam and Rovelli (2023) argue that it can be motivated by the physicalist idea that the internal state of an observer must supervene on ordinary physical properties. For the physical properties of physical systems are usually relevant to their interactions with other systems, so from this point of view it is quite natural to think that the internal state of an observer must be accessible to other observers by means of some kind of physical interaction, and the interactions referenced in CPL seem like the most obvious candidate. This idea is quite consistent with the general philosophy behind RQM, since one of the founding principles of the view is the idea that observers are just physical systems like any other – this principle is the reason why RQM, unlike other perspectival interpretations, postulates that any physical system whatsoever can have facts relativized to it. So there is a sense in which CPL does cohere quite well with the rest of the system of beliefs associated with RQM – and indeed, arguably the original formulation of RQM deserves criticism precisely because it tells us that observers are just physical systems but it also makes the internal state of each observer into a kind of extra-physical fact which can never be accessed by any other observer.

However, even if we accept CPL as a legitimate assumption, there are still a number of details of RQM+CPL that remain to be fleshed out. Most importantly for the present discussion, it's not yet obvious that RQM+CPL can accommodate QFT, so it's not yet clear that it is actually any better off than the PO approaches. The difficulty is that both RQM itself, and the CPL assumption as formulated above, seem to assume that we are dealing with well-defined individual systems to which facts can be relativised, but in many regimes of quantum field theory there *are* no well-defined systems – as we saw in chapter 11, even particles are emergent in QFT. So it's not clear what an RQM formulation of QFT would look like – what would the relative facts be relativised to? Perhaps we might imagine relativizing facts to regions of spacetime rather than individual physical systems[1], but this would lead to many new technical questions – for example, would we then want to say that there exists some kind of composition structure such that a region contained inside another region typically shares some relative facts with its containing region? And if not, how do we identify the relative facts associated with a given conscious observer? After all, there doesn't

[1] Thanks to Carlo Rovelli for this suggestion.

seem to be any precise way of defining exactly which region of spacetime I myself currently occupy, and we presumably wouldn't want to have to say that the 'relative facts' that RQM ascribes to me look completely different depending on whether you identify my region with my whole body, or just my brain, or just a part of my brain, or with various different collections of cells. Also, although regions of spacetime are precisely defined within QFT, there are strong indications that within quantum gravity that the structure of spacetime itself will also be only approximate and emergent (Huggett and Wüthrich, 2013), so one might worry that this solution is only kicking the can down the road, since the same problem may ultimately re-emerge in quantum gravity.

Another concern is that CPL as defined above applies to *every* observer in RQM, and RQM tells us that any physical system whatsoever can play the role of an observer. So postulating CPL seems to forced us to say that every time any two systems interact – even if it is just a pair of electrons briefly passing by each other – a cross-perspective link is created. Furthermore, in cases where the 'observers' are relatively small, Mucino et al. (2022) and Brukner (2021) argue that there won't typically be any well-defined basis in which the interaction is naturally described as taking place, which leads Adlam and Rovelli (2023) to suggest that CPL may ultimately need to be formulated in such a way that for microscopic interactions, cross-perspective links form in not just one basis but some range of bases, with a unique cross-perspective link arising only at larger scales due to the fact that decoherence favours certain bases of interaction. There are some technical difficulties with this response because it is not clear what exactly we are to say about the quantum state of one system relative to another after such an interaction – we might have to accept that systems can have more than one quantum state relative to one another, with unique quantum states also arising only at larger scales due to the role of decoherence. But in any case, even if this proposal can be made to work, it comes with the downside that RQM+CPL is then committed to an unimaginably vast network of cross-perspective links joining up all the individual microscopic particles in the world in many different ways; and furthermore most of this structure is superfluous, given that the point of CPL was merely to ensure that there is a relationship between the perspectives of various macroscopic, conscious observers who belong to some epistemic community.

Therefore it would seem preferable if something like the CPL postulate could somehow be made to emerge naturally in the process of decoherence, so we would end up with cross-perspective links forming most commonly between the perspectives of large systems like macroscopic observers, in the basis favoured by decoherence, but not between the 'perspectives' of individual microscopic particles. This might even help with the QFT problem as described above, since then we could potentially accept that 'observers' in QFT do indeed emerge only in some appropriate limit. But it's unclear how to make this work because both RQM itself and the CPL postulate seem to rely on everything being precisely defined; and thus if we say that CPL applies only to macroscopic observers and not to fundamental particles, we seem to be lapsing back into a version of the Heisenberg cut, with all the associated problems – where do we put the cutoff, and how do we achieve this without reverting to treating observers as unanalysed primitives? So there is definitely still work to be done on finding a version of CPL which handles this issue better, and it remains to be seen if a fully realised version of the approach can be made to work for QFT.

13.1.2 More General Relativisations

Another way of modifying the postulates of the observer-relative interpretations would involve postulating that measurement outcomes are relativised not to individual observers but to some more general 'context.' For example, Healey's desert pragmatism (Healey, 2012) relativises states and outcomes to the situation defined by a 'decoherence environment,' i.e. a region of spacetime in which environmental decoherence stabilises the value of some variable such that it can meaningfully be assigned a value. Similarly, Ormrod and Barrett (2024) envision relativising states and outcomes to sets of systems, rather than individual systems.

Neither of these approaches are observer-relative interpretations in the sense in which I have used that term, since they relativise measurement outcomes to something other than individual observers or systems. And this is potentially a good way of avoiding the intersubjectivity problems discussed in chapter 9 because a decoherence environment can contain many different observers who will all share the same information. Indeed, according to Healey (2022a, p. 15), the ubiquity of environmental decoherence means that all of the human race belong to the same decoherence environment, and thus in desert pragmatism, '*scientists and other observers ... share a single context of assessment, all their measurement outcomes are immanently objective facts.*'

Now, the way in which Healey (2012) presents desert pragmatism seems best suited to answer the question of how the quantum formalism should be interpreted, as discussed in section 5.2, rather than to resolve the epistemic version of the measurement problem. For he regards desert pragmatism as maintaining that quantum mechanics '*further[s] the goals of physics without itself offering novel representations or descriptions of physical reality*, (p. 1)' and in particular, that '*while not itself issuing descriptive claims about physical reality, quantum theory does advise an agent on the scope and limitations of descriptive claims it may make in a given situation* (p. 2)'. And the descriptive claims licensed by quantum mechanics according to desert pragmatism are typically 'Non-Quantum Magnitude Claims,' specifying that a certain classical or quasi-classical variable takes a value in a certain range. These claims may involve making claims about measurement *outcomes*, but it's not clear that such claims would include any representation or description of the physical process of measurement itself. So we might well agree that desert pragmatism is the right way of interpreting the quantum formalism, and yet still consider that from an epistemic point of view there is a further need to give a physical description of what is going on during a quantum measurement, in order to fulfil the mandate for an account of reality which affirms the reliability of our means of enquiry. That is, in order to use the notion of 'decoherence environments' to solve the epistemic version of the measurement problem, we would ideally want to show how to give a description of the physical nature of the measurement process which has the consequence that measurement outcomes are indeed relativized to decoherence environments, as suggested by desert pragmatism.

If such an account could be given, it would presumably not be a unitary-only approach to quantum mechanics. For although the contexts to which Healey relativizes measurement outcomes are defined by decoherence, the mere existence of decoherence is not enough to ensure that all the observers in the same decoherence environment share the same outcomes, as desert pragmatism requires. In pure unitary quantum mechanics, decoherence has the consequence that for any measurement taking place within a decoherence environment, the post-measurement state will look like a classical mixture of well-defined, stable measurement outcomes. But the pure unitary description does not contain any representation of the specific measurement outcomes obtained by individual observers, and thus it also has no mechanism to ensure that all observers within the decoherence environment will agree on the outcome of a certain measurement – they will all see that measurement as having a stable, quasi-classical outcome, but nothing in the standard formalism guarantees that they will all see the *same* quasi-classical outcome. So although desert pragmatism itself can avoid postulating additional structure, since it is not seeking to describe or represent reality, any descriptive account realising the vision of measurement outcomes relativized to decoherence environments *would* have to add something to unitary quantum mechanics, and thus it would be a package deal rather than merely an account of reality for the standard unitary formalism.

One challenge for such an approach would be to define 'decoherence environments' in a precise way. For decoherence is a continuous process, and thus there is no exact boundary between 'decohered' and 'not decohered,' and hence also no exact boundary to any region of spacetime dominated by decoherence processes. This seems acceptable if we are working in the original desert pragmatist picture where no attempt is made to represent or describe physical reality, but if we are hoping to give an account of the physical nature of the measurement process which could underpin the desert pragmatist approach, then we would presumably have to find a way of precisifying the definition of the decoherence environments, unless we are willing to accept irreducible vagueness in reality itself.

One way to precisify the decoherence environments might involve adopting an approach similar to RQM+CPL. Indeed, Healey (2022b) notes that RQM+CPL would likely end up looking quite similar to desert pragmatism because once cross-perspective links are added to RQM, then essentially all macroscopic observers who are in some kind of causal contact will end up sharing the same measurement outcomes. Thus CPL plus decoherence plausibly can be expected to lead to composites made up of observers sharing information which roughly coincide with the decoherence environments envisioned by Healey.

Alternatively, one could perhaps try to define decoherence environments by directly forming a composite higher-level entity to which outcomes can be relativized, rather than building the composites up dynamically by means of individual interactions as in RQM+CPL. There are certainly reasons to find this top-down approach appealing – in particular, such an approach would avoid postulating the large network of superfluous microscopic cross-perspective-links that we end up with RQM+CPL.

However, arguably the approach taken by RQM+CPL has some advantages from the epistemic point of view. For if our solution to the intersubjectivity problem is simply to form high-level 'contexts' such that one context can contain the whole epistemic community, then there is still no mechanism by which observers can obtain information about the contents of any context other than their own, and thus in some sense the epistemic problem recurs – since we only have empirical access to one context, we will have little reason to think that the same regularities still apply in other such contexts. This is a concern because, as noted in chapter 9, the usual motivation for adopting a relational approach to quantum mechanics is to allow us to maintain that unitary quantum mechanics is universal – and in particular, as discussed in section 3.2, the goal is to allow us to say that quantum mechanics can be correct for two distinct contexts in an Extended Wigner's Friend case, even though this may initially appear to lead to a contradiction. But if we don't have any good reason to think that quantum mechanics applies in any contexts other than our own, then there is no cause to think a contradiction would arise here in any case, and so the motivation for relativizing outcomes to contexts in the first place evaporates. Thus the approach of directly forming higher-level composites may still face some kind of self-undermining problem, since the legitimacy of extrapolating from the contents of one context to the universality of unitary quantum mechanics would then be questionable.

By contrast, in an approach based on a dynamical postulate like cross-perspective links, the dynamics are specifically tailored so as to allow observers to gain access to measurement outcomes obtained by other observers, thus ensuring that we can indeed obtain direct empirical evidence about regularities which hold in contexts other than our own. Now, the distinction is somewhat subtle – in practice, for any given observer the set of relative facts to which she will have access in RQM+CPL will be virtually identical to the relative facts ascribed to her decoherence environment by a version of desert pragmatism based on directly forming high-level contexts, so it is not the case that the observer in RQM+CPL has a larger *quantity* of empirical evidence. However, in RQM+CPL that empirical evidence is interpreted as being collected from multiple different 'contexts,' whereas in the version of desert pragmatism based on forming high-level contexts, that very same empirical data is understood as being associated with just a single large context. Thus arguably the observer in RQM+CPL has better grounds to perform induction to the conclusion that quantum mechanics applies in all contexts, since she has chosen to interpret her data as providing her with direct insight into multiple contexts. The difference between the epistemic virtues of these two approaches helps demonstrate why one might think it is important to add some descriptive account of reality to the desert pragmatist view, since without such a thing we cannot fully assess the epistemic status of the empirical evidence available to us.

To conclude this discussion of modified observer-relative approaches, the point I'd like to emphasize is that solving the epistemic problems of the observer-relative interpretations is not a simple matter of just postulating that intersubjective information-sharing between perspectives is possible; there are a number of complex technical questions to be resolved about how this can be achieved and what it would ultimately look like in the context of QFT. Intersubjectivity therefore cannot be simply a minor addition to the theory, but rather requires the addition of substantive new structures. Adding cross-perspective-links to RQM involves postulating models mathematically different to the models of standard quantum mechanics, since no such links appear in unitary quantum mechanics; and relativising observations to entities larger than individual observers will likely also require mathematically distinct models, if we demand that such an approach offers a model of the physical process of measurement itself. So the epistemic difficulties I have discussed throughout this book are not minor issues which can easily be defined away – resolving them in a way which satisfactorily solves the epistemic version of the measurement problem will likely require real technical work, which may take putative solutions to the measurement problem in interesting new directions.

13.1.3 SWRCH Consistent Histories

In chapter 6 we saw that the central epistemic difficulty for the Everett interpretation is the difficulty of giving a probabilistic interpretation of mod-squared amplitudes in a context in which all measurement outcomes always occur. It seems natural, therefore, to ask if it is possible to formulate an approach akin to the Everett interpretation in which we probabilistically select and actualise just one history out of the set, thus settling questions about the interpretation of probability. Evidently in order to do this we must first extract from the Everettian multiverse a well-defined set of quasi-classical histories and a probability distribution over them, which is not trivial, particularly if one is working within the version of the Everett approach in which 'worlds' and hence histories are typically regarded as being emergent and approximate.

However, in section 3.3 I described a formalism which could potentially work here – the consistent histories approach. Thus in this section I will discuss in more detail the epistemic difficulties faced by the SWRCH version of the consistent histories formalism, in which we first select just one consistent set, and then probabilistically select and actualise exactly one history from it.

We saw in section 3.3 that a SWRCH version of the consistent histories picture encounters epistemic weirdness connected to the use of records in empirical confirmation, since in a generic consistent set of histories it will not be the case that each of the constituent histories contains macroscopic records providing accurate information about the actual past of that history. Now, there is an obvious way of addressing this issue: since the consistent histories framework doesn't tell us anything about how we should choose the consistent set from which we will probabilistically select our history, we can simply select a consistent set with the property that all of the histories in it are made up of quasi-classical events containing mostly accurate records of the past. Gell-Mann and Hartle (1995, p. 8) have formulated a stronger version of the decoherence condition which achieves exactly this. Their strong decoherence condition singles out those sets which have '*records that guarantee the permanence of the past.*' Specifically, they require that at each step in the chain of projections, the next projection is chosen from the set of projections that commute with generalised records of history up to that moment. Here, 'generalised records' are sets of orthogonal projection operators which distinguish between the different possible histories in a consistent set – i.e. they define measurements such that if you performed that measurement, the outcome would tell you which history from the set you are currently in. Thus the important consequence of the strong decoherence condition is that the projections are chosen in such a way that they always leave untouched a set of records which faithfully record the actual past. So we can simply stipulate that the consistent set from which our world is actually selected obeys the strong decoherence condition rather than merely the weak decoherence condition, and then we are somewhat safer from epistemic weirdness connected with records.

However, one might worry that there is something problematic about simply stipulating that our history is selected from a consistent set which has exactly the properties we would like it to have in order to assure us of the epistemic status of scientific knowledge. After all, Gell-Mann and Hartle offer no *reason* to think that our own history is, in fact, chosen from a consistent set obeying the decoherence condition, so this must simply be added as an additional assumption. Yet shouldn't the epistemic security of science rest on more robust foundations? Indeed, is this assumption really any better than the Everettian assumption ESA, or various other ad hoc assumptions covered in this book?

In fact, I do think the strong decoherence assumption seems a little better than the ESA, since it can be clearly formulated in mathematical terms, and it doesn't require us to make any essential reference to perspectives, consciousnesses, or subjective experiences. But nonetheless the assumption does seem somewhat ad hoc. In particular, note that as with a number of the ad hoc assumptions discussed in this book, we can't straightforwardly get empirical evidence for it or refine it in response to empirical evidence – for after all, we have no way to check if our records of the past are accurate except by comparing them to other records of the past, but as noted in section 3.3, in the consistent histories picture, records will typically agree even if they are not accurate, since all records of a given event will typically all be entangled with each other. Nor is the strong decoherence assumption implied or even suggested by any ontological commitments or structural feature of the consistent history formalism because all sets of consistent histories are equivalent from the point of view of the formalism.

Now, at this point the proponent of consistent histories might try to make a No Worse Off argument similar to those discussed in chapter 7. For example, they might suggest that the assumption that our actual history is selected from a set of strongly decoherent histories looks very similar to the induction assumption, i.e. the assumption that the history of our actual universe is one in which the future resembles the past, and thus they might contend that if it is acceptable to make the induction assumption, then surely the strong decoherence assumption is equally reasonable.

However, I think this analogy is not so strong. First, as noted in section 7.4, the induction assumption is responsive to empirical enquiry, whereas the assumption of strong decoherence cannot so easily be adjusted in response to empirical evidence: if our history is not strongly decoherent, nonetheless it may seem strongly decoherent to us, since the entanglement between records may keep them mutually consistent and thus prevent us from noticing that records have been updated or replaced. So the strong decoherence assumption ultimately looks more ad hoc than the induction assumption, since it is not so susceptible to progressive refinement as our theoretical and empirical knowledge develops. Second, the two assumptions also work at very different levels of generality. The induction assumption can be thought of as simply an assumption that there *exist* laws of nature – i.e. that there exist some universal regularities which can at least sometimes be codified as scientific laws accessible to human observers. The assumption does not say anything about what those laws are, other than that they are at least somewhat universal, so it is not an assumption about the workings of one particular scientific theory but rather a general precondition which must to some extent be satisfied if we can have any hope of formulating any theories in the first place. The strong decoherence assumption is much more specific: it is formulated within the framework and language of one specific scientific theory, so it doesn't make any sense unless we first assume that a) there are universal regularities which can be codified as laws, and b) that these laws are (at least roughly) the laws of quantum mechanics, as encoded in the consistent histories formalism. The point is that we might reasonably be more tolerant towards very general assumptions about the preconditions required to make science possible at all versus assumptions made to plug holes in some specific scientific theory – for the whole point of formulating a scientific theory is to accommodate the empirical evidence, so if the theory we come up with can't do that in a natural way by making use of the concepts we have specifically designed to accommodate the empirical evidence, then surely something has gone wrong, and we should try to do better at formulating a theory which does in a natural way what it sets out to do.

So, in fact, a better analogy for the strong decoherence assumption might be the assumption that we are not Boltzmann brains, as discussed in section 7.2. The similarities with the strong decoherence assumption are immediately apparent – both assumptions are necessary in order to secure the reliability of memories and records, and both are in similar ways not so responsive to empirical evidence. And just like the strong decoherence assumption, the assumption that you are not a Boltzmann brain needs to be formulated within a specific scientific theory – in this case, statistical mechanics. In fact, as discussed in section 7.2, the assumption that we are not Boltzmann brains is usually derived from the Past Hypothesis, which likewise can only be formulated within the framework of a particular theory, since we must appeal to entropy or some other feature of some region of the state space before we can assume that the initial state belongs to this region.

These similarities suggest that it may be useful to invoke some of the literature on the Past Hypothesis as we try to understand the status of the strong decoherence assumption in the consistent histories setting. For example, it has often been argued that the Past Hypothesis needs to be at least somewhat lawlike if it is to be generalisable in the way we need it to be in order to ensure that entropy is increasing everywhere and at all times. Price (2004, pp. 20–21) writes, '*As actually used in contemporary cosmology, the hypothesis of the smooth early universe is ... taken to be projectible, in the sense that everyone expects it and its consequences [the existence of galaxies, for example] to continue to hold, as we look further and further out into space. The hypothesis is thus being accorded a lawlike status, rather than treated as something that "just happens".*' And clearly much the same argument can be made about the strong decoherence assumption: if it is just a contingent fact that our actual history is selected from a strongly decoherent consistent set, then even if it is true that the set of histories we inhabit has been strongly decoherent up until now, it seems we have little reason to expect that the set will *continue* being strongly decoherent. So there are good reasons to think that a successful version of the SWRCH approach would require us to elevate the strong decoherence requirement to a law, thus guaranteeing that the set will be strongly decoherent across all of time.

In addition, I argued in section 7.2 that the Past Hypothesis is less ad hoc than the Everettian assumption ESA, since the ontology of a time-evolution theory demands that some arbitrary choice of initial state is made, whereas the ontology of the Everett interpretation does not demand that we make any stipulation about what observers will see. So the relevant question for us here is whether the strong decoherence assumption is more similar to the Past Hypothesis or to the ESA in this regard. And, in fact, I think it is probably closer to the Past Hypothesis because the ontology of a SWRCH approach is based on a history being selected from some consistent set, and if we stay entirely within the standard consistent histories formalism then there is no real reason to select one particular set rather than another. Thus the ontology of the SWRCH approach does indeed demand that some arbitrary stipulation about the choice of consistent set is made, so it is arguably less of a stretch to simply stipulate that the set in question is, in fact, a strongly decoherent one.

That said, this argument depends quite sensitively on how one thinks of the ontology of the consistent histories formalism. One might worry that idea that just one history gets 'actualised' is itself somewhat ad hoc, given the picture of reality provided by the consistent histories formalism – after all, all of the other histories are there within the formalism and they look equally valid at the mathematical level. And if one takes the view that all of the histories are in some sense real, then actually the strong decoherence assumption looks more like the assumption ESA than the Past Hypothesis – it amounts to the stipulation that we should assume we are in inside a particular kind of history, even though we know all the other histories exist and most have conscious agents within them. Indeed, on this construal the consistent histories picture seems even worse off than the Everettian one because the Everettian at least has mod-squared amplitudes which at least at a formal level provide a well-defined classical probability distribution across all decoherent histories, whereas the consistent histories approach doesn't assign mod-squared amplitudes or probabilities to sets of histories – the probabilities are assigned only inside consistent sets, providing us with no guidance at all about how to choose one set rather than the other. So the Everettian may have work to do to convince us that mod-squared amplitudes can have a probabilistic interpretation, but the consistent histories approach simply has no hope of convincing us that there is some kind of probabilistic selection of a consistent set, because there is nothing in the formalism which could possibly play this role.

To conclude, the strong decoherence assumption looks somewhat epistemically problematic, but there may be ways to make it less so. In particular, what seems essential is to have a more robust understanding of the ontology of a SWRCH approach, in order to resist the intuition that all of the histories are equally real. That is, rather than using the somewhat vague term 'actualised,' we would like to know what kind of physical ontology is associated with this picture and how the configuration of that ontology is determined by the particular consistent history which is selected and 'actualised.' I will now discuss a proposal due to Kent demonstrating one possible way to achieve this.

13.1.4 Kent's Solution to the Lorentzian Quantum Reality Problem

Kent's proposed solution to the measurement problem is based on a simple idea: there is no collapse of the wavefunction, so we just allow the wavefunction to undergo its standard unitary evolution over the whole course of history, and then, at the end of time, we imagine a single measurement being performed on the final state to determine the actual content of reality. As Kent (2014, p. 5, online version) puts it, '*An event occurs if and only if it leaves effective records in the final time ... measurement.*' So, for example, Kent (2017) imagines a measurement at the end of time on the positions of photons which have been reflected off matter at various points, and then defines the actual content of reality by stipulating that the expectation value of the components of the stress-energy tensor at a spacetime point x are given by the quantum expectation value at that point, conditional on some of the results from that final measurement. The specific details of the stress-energy tensor formulation will not be important for us here; the key point is simply that the actual course of history is, in effect, determined by whatever is recorded in the outcome of the measurement performed on the final state. This model is mathematically different from any models appearing in standard unitary quantum mechanics, and thus Kent's proposal is a package deal rather than merely an account of reality.

Now, Kent's model is not expressed within the consistent histories formalism. This is clear from the fact that Kent explicitly sets out to create a Lorentz-covariant model, meaning that the model respects the symmetries encoded in the theory of special relativity, whereas the consistent histories formalism is not Lorentz-covariant in its standard formulation. But nonetheless Kent's model is, in a sense, aiming to offer a principled, lawlike way of defining a single set of histories from which we can probabilistically select and actualise just one history. Specifically, Kent singles out the set of histories which are recorded in the final state in the position basis for photons (or something like this – the precise details will depend on the specific definition of the final measurement, which varies in different versions of Kent's approach).

From the point of view of the epistemic version of the measurement problem, one key advantage of Kent's view is that it provides an illuminating account of what exactly a measurement *is*, and thus it explains, without needing to include observers or measurements in its fundamental ontology, why quantum measurements are a particularly special kind of physical interaction. For when we make a measurement, we create a record of the measurement outcome in our brains (and measuring devices, lab books, and so on). Moreover, we know that decoherence will very quickly create many records of the states of these macroscopic systems in their environments, meaning that some record of these states is almost guaranteed to be preserved until the end of time, so definite values for those states will be actualised by the final measurement. Now, when the system we are measuring is itself a macroscopic system, there is nothing particularly significant about this, because there would already have been decoherence-induced records of the state of the measured system which were almost guaranteed to be preserved until the end of time, so a definite value of that state would have been actualised by the final measurement with or without our measurement. But if we are measuring a microscopic quantum system, typically there will *not* yet be any records of the state of the system in its environment, so if we had not performed a measurement there would most likely *not* have been a record of that state preserved until the end of time, so no definite value of the state would have been actualised by the final measurement. Thus in this picture we can see clearly what is so special about the process of performing a quantum measurement: a quantum measurement is a physical interaction which magnifies a microscopic state up to macroscopic scales, and thus creates stable persisting records of it which are likely to endure until the end of time. Thus in Kent's picture it is no longer surprising that quantum measurements appear to involve a different kind of dynamics from ordinary quantum evolution: measurements do, in a real sense, *cause* quantum states to become definite, because the magnification of those states up to macroscopic scales is what ensures that they will subsequently be actualised by the final measurement.

Note also that in Kent's approach, it follows from the nature of the measurement process that measurement outcomes are unique and intersubjectively accessible, and that there is always a stable ongoing record of the measurement outcome, since otherwise the measurement would not be actualised by the final measurement and therefore could not be observed by anyone. This is similar to what is achieved by the idea that measurement involves a wavefunction collapse, thus creating a unique, intersubjectively accessible event which can written in stable classical records. Indeed, from an epistemic point of view Kent's approach is even better than a wavefunction collapse, since the result of a wavefunction collapse isn't guaranteed to have an ongoing record, whereas in Kent's approach any actual physical event such as a measurement outcome must have a record which persists until the end of time.

Now, I don't necessarily want to argue that Kent's approach is definitely the right solution to the measurement problem. There are certainly still issues to be addressed here – in particular most of Kent's discussions are sited in the context of ordinary quantum mechanics rather than quantum field theory per se, so a natural next step would be to explore its implementation within QFT. It's also very unclear how Kent's approach would work in the context of quantum gravity, since in its current formulation it requires a pre-existing fixed spacetime background. However, this approach does provide a nice example of what we should ideally be seeking in a solution to the epistemic version of the measurement problem. Specifically, we want an account of reality in which measurement outcomes are typically intersubjectively accessible, typically have stable records, and typically are an accurate reflection of the underlying probabilities or mod-squared amplitudes; and we want all of this to follow in a natural way from *what the account has to say about the physical nature of the measurement process*, rather than as the result of further ad hoc stipulations.

And this, ultimately, is where many of the unitary-only approaches discussed in this book fall short – there may or may not be some way to cobble together various assumptions to get to a broadly reasonable account of the measurement process in these approaches, but none of them really offer a picture in which it is a *natural* feature of the measurement process that measurements yield intersubjectively accessible, stable, and unique outcomes. Of course, given the holistic nature of scientific knowledge, we cannot expect that quantum mechanics *on its own* will certify the reliability of the entire measurement process; as Curiel (2020) emphasizes, the analysis of a quantum measurement requires us to rely on many other theories, such as our knowledge of the electronics of our detectors, or our knowledge of the biology of the visual processes we use to read our detectors. But the point is that a precondition for any of these theories to be meaningful in the first place is that classical events like measurement outcomes must be intersubjectively accessible, stable, and unique, and thus we would like our solution to the measurement problem to deliver at least these basic features in a natural way, with the remaining details to be filled in by higher-level theories that we can justifiably make use of if we have some reasonable solution to the epistemic version of the measurement problem.

Kent's model also serves as a valuable counterexample to refute claims that are sometimes made about the impossibility of achieving an epistemically natural account of measurement in the quantum context. For example, quantum contextuality (see section 2.4.1) tells us that quantum measurements cannot in general be understood as simply revealing the value of the measured observable, and one might initially take this to mean that it cannot possibly be the case that the quantum measurements reliably yield meaningful information about the structures they are probing. But Kent's approach shows that this is not right. For it is indeed true in Kent's model that quantum measurements don't simply reveal pre-existing values of quantities, since the observed values are determined by the final measurement and thus don't typically exist at all before they are measured. But nonetheless, because the distribution of measurement outcomes across history is determined altogether in a single final measurement obeying quantum Born rule probabilities, it remains true that individual measurement outcomes reliably provide information about the fundamental structures of quantum theory, in virtue of the role played by those structures in determining the distribution of outcomes. So although in this model quantum measurements don't provide information about pre-existing values of variables, they do provide information about the theoretical structures of quantum mechanics that we are aiming to probe, and this feature emerges from the model in a natural way rather than being added as an ad hoc assumption.

Likewise, one might think that since some unitary-only approaches deny that quantum mechanical states are a representation of any physical system, it would not be reasonable to expect such approaches to have the consequence that measurements yield reliable information about the structures involved in quantum theory. But Kent's approach is a counterexample to this claim as well. For it is true in his model that quantum mechanical states do not play a direct representational role: the actual states and properties of physical systems are given by the distribution of values for the stress-energy tensor as defined by the final measurement, not the quantum states. But nonetheless, the quantum mechanical wavefunction has an important part to play in defining the distribution of those values, and thus measurement outcomes realised by that distribution of values serve to give us information about quantum mechanical structures, even though such structures do not themselves play a descriptive or representational role. Thus again, Kent's model shows that it is possible to imagine a scenario in which measurements naturally provide information about the structures they are supposed to be probing, even if we don't think of those structures as playing a straightforwardly representational role.

Of course, the ingredients Kent uses to achieve all this are not unique to his approach. Really the key insight of his picture is that measurement is a special kind of physical process which magnifies microscopic variables and thereby encodes them in decoherence-stabilised states of macroscopic systems. And this is true within most unitary-only interpretations as well – the only real difference is that Kent adds a stipulation which converts macroscopic states stabilised by decoherence into a well-defined and unique quasi-classical history, thus eliminating the most obvious kinds of epistemic weirdness. This seems like a direction which could potentially be pursued in other ways; so even if Kent's approach is not the right answer in the end, it certainly offers some useful ideas about what a viable solution to the measurement problem might look like.

13.2 Anti-Reductionism

It is notable that our search for solutions to the measurement problem which work for QFT whilst avoiding epistemic weirdness has led us to approaches which have something in common – a kind of anti-reductionism. For example, we noted that in RQM+CPL there are good reasons to think that 'cross-perspective links' should only appear at some non-microscopic level; we saw that desert pragmatism seems to involve defining 'contexts' to which outcomes are relativised at some non-microscopic level; we noted in section 3.3 that in a SWRCH approach the histories featuring in a consistent set are typically made up of largely macroscopic events; and we saw that Kent's approach also has a somewhat non-reductionist character, since it is largely the *macroscopic* events which get actualised by the final measurement.

Now, the term 'reductionism' is used in a variety of different ways in philosophy – for example, sometimes it refers to the thesis that all mental events can be reduced to the physical (Velmans, 1998), or indeed that *everything* can be reduced to the physical. But here I will focus on a specific form sometimes known as 'microphysicalism' (Hüttemann, 2003) or 'petty reductionism' (Weinberg, 2001) – the idea that all large-scale phenomena are ultimately reducible to smaller-scale phenomena, and hence we reach more 'fundamental' levels of reality as we go to smaller and smaller scales. Sometimes petty reductionism is coupled with the view that there exists a smallest, most fundamental level from which everything else emerges, but I will not focus on that assumption here: my concern is specifically with the view that given two scales S_1, S_2 with $S_1 > S_2$, everything happening at scale S_1 supervenes on what is happening at scale S_2.

We should also distinguish between *methodological* reductionism(Franklin and Robertson, 2023) and what I will call ontological reductionism. The former refers to the view that the appropriate way to study physical systems is to derive their macroscopic dynamics from some microscopic dynamics, together with appropriate auxiliary assumptions or Nagelian bridge laws. The latter amounts to the claim that all goings-on at large scales entirely supervene on goings-on at smaller scales, regardless of whether or not these relations can be written down by human scientists. The truth of ontological reductionism does not necessarily entail the truth of methodological reductionism, since even if supervenience relations exist, they may not take a form which is methodologically useful for scientific purposes – for example, in some cases the relevant bridge laws or auxiliary assumptions could be too complicated for scientists to practically formulate. Philosophical discussion of anti-reductionism has often taken the form of arguing against methodological reductionism – for example, Batterman (2023) identifies many examples drawn from many-body physics in which deriving macroscopic dynamics from microscopic dynamics does not appear to be the most methodologically fruitful approach – or arguing against the idea that the best explanations are always the ones formulated at the smallest scales (Strevens, 2008; Craver and Kaplan, 2020). However both of these kinds of arguments are to some extent relativized to pragmatic human interests, whereas in this section I want to focus on the idea that not only methodological reductionism but also *ontological* reductionism could fail in some regimes, which I understand to be a claim about physical matters of fact, independent of human interests. Obviously the failure of ontological reductionism would also have consequences for methodological reductionism, since if ontological reductionism fails in some regime then it is likely methodological reductionism would also be false in that regime, but methodological reductionism is not my main focus here.

Now, as argued by Adlam (2024a), there are a number of indications within quantum field theory that moving away from petty reductionism could be a useful way to understand the theory. But rather than focusing on quantum mechanics itself, let us begin by taking a broader historical perspective. It seems fair to say that up until the advent of quantum mechanics, petty reductionism has always involved reducing large things to things which are smaller but of a somewhat similar type – which is to say, we were always reducing classical things to smaller classical things, like particles, classical waves, classical fields, and so on. The objects used in classical reductions may have been unfamiliar in some ways, but they could all be understood as inhabiting a well-defined, objective, intersubjectively accessible reality in which things happened in just one way, giving rise to relatively simple reduction relations between the macroscopic and the microscopic: as Wallace (2018, p. 21) puts it, in such reductions it is fairly straightforward to provide '*a description of a theory's empirical content directly in its microphysical vocabulary*.' But things are no longer so simple in quantum mechanics: we are now trying to reduce our essentially classical macroscopic reality to smaller things which are very non-classical – and in particular, they appear to be non-classical in the sense that they cannot always be understood as inhabiting a single shared objective reality. As a result, it is no longer possible to express all the empirical content directly in the microphysical vocabulary.

Thus it is in some ways very unsurprising to find that continuing to insist on petty reductionism has led us into epistemic weirdness. For petty reductionism entails that all of the properties of the macroscopic world are ultimately inherited from the microscopic world, and thus as soon as you make your 'fundamental' level somewhat non-classical, it is difficult to stop the non-classicality from leaking upwards to be replicated at larger distance scales. Of course, our past experience with reduction does indicate that somewhat novel phenomena may emerge as one goes to larger scales, but in a Nagelian reduction, as in any logically rigorous deduction, you can really only get out what you put in, so the large-distance phenomena will still tend to have the same essential character as the fundamental level. In particular, if you have a lot of non-classical features in your microscopic physics, you are likely to get non-classical features in the macroscopic reality that supervenes on it, such as the existence of multiple macroscopic worlds as in the Everett picture, or multiple incommensurable macroscopic perspectives as in the observer-relative approaches. And conversely, if you want to *suppress* all such features in the macroscopic limit, you will probably have to engage in some fairly complex fine-tuning to make sure the relevant effects all exactly cancel out, in which case your emergent classical reality will be lacking in robustness, since any small perturbation at the microscopic level could make the quantum features reappear at the macroscopic level.

This, presumably, is the instinct behind the primitive ontology approaches: if we want the macroscopic world to supervene on the microscopic world, and we want the macroscopic world to be quasi-classical (and *robustly* so), the microscopic world had better be pretty close to classical as well. But the problem is that quantum mechanics itself is, obviously, non-classical, and quantum field theory is even more non-classical, and the prospects of somehow getting all of this to emerge out of an even more fundamental level which is somehow quasi-classical seem fairly remote. So as long as we insist on petty reductionism, it seems hard to stop the quantumness from infecting the larger distance scales, leading to epistemic weirdness.

But if we are willing to reconsider petty reductionism, there is a natural alternative: rather than starting from something quasi-classical at the smallest, most fundamental level and building upwards, hoping to derive first quantum mechanics and then the classical world in the appropriate limit, we could instead start with some quasi-classical variables defined at a larger scale where we already expect some quasi-classical behaviour. Evidently if this could be made to work it would be quite straightforward to arrive at a unique, shared, stable quasi-classical macroscopic world supervening on these larger-scale quasi-classical variables, and as we have seen throughout this book, an account of reality which provides us with a unique, shared, stable macroscopic reality tends to avoid most kinds of epistemic weirdness.

That said, any approach of this kind must proceed with caution, for petty reductionism is not just a dogma – there are a number of compelling reasons why we expect scientific theories to obey this feature, and thus it cannot be abandoned too lightly. The first reason is simply the fact that petty reductionism has had spectacular empirical success. A great many important discoveries in the history of human science have involved reducing the behaviour of some large things to the behaviour of a smaller number of small things – 118 elements reduced to arrangements of electrons, protons, and nuclei; the inheritance of thousands of different characteristics reduced to arrangements of four DNA molecules; a variety of phenomena involving heat reduced to molecular motion; and so on. Clearly these incredible accomplishments are a compelling argument in favour of petty reductionism and provide a strong presumption in favour of continuing to search for scientific explanations of this kind.

However, the fact that petty reductionism is successful on a large range of scales does not necessarily mean that it will be successful at *all* scales. It is true that in the practice of science we commonly assume that a type of explanation that is known to work well in a range of times, places, or regimes will likely also work in other times, places, or regimes – the principle of induction being the most obvious example. However, the case of petty reductionism is somewhat unusual because inherent in the formulation of this principle is the idea that not all scales are equal; petty reductionism specifically tells us that there is apparently some kind of direction of fundamentality which has consequences for the relationships between different scales. Given this, it would not be reasonable to invoke a principle of homogeneity of scales to insist that petty reductionism must hold in the same way at every possible scale, since the whole point of petty reductionism is that scales are not always homogeneous! Thus although it makes sense to continue searching for explanations in the tradition of petty reductionism, we should also be open to the possibility that this mode of description breaks down at certain scales – though certainly the evidence would have to be quite compelling in order to convince us to overthrow many centuries of scientific tradition.

The second reason to favour petty reductionism is a practical one. Physics is a quantitative subject which seeks to find systematic descriptions of the universe which are precise, mathematical, and relatively simple, and therefore physics needs to deal with objects which can be defined to a high level of precision. Certain kinds of macroscopic objects may appear quite well-defined if we are only considering effects on macroscopic scales – a billiard ball, for example, or a pendulum. But once we move to effects on smaller scales it becomes apparent that these things are actually defined quite vaguely, and we encounter a range of problems which will be familiar to mereologists (Unger, 1980; van Inwagen, 1990; Lewis, 1993; Donnelly, 2009) – for example, there are always some number of atoms near the surface of the billiard ball such that it is ambiguous whether or not these atoms are a part of the billiard ball or not, and thus it would be hard to arrive at a sensible way of doing physics in which the billiard ball were treated as fundamental across both larger and smaller scales, because things would potentially turn out quite differently depending on apparently arbitrary choices about how exactly we define the boundaries of the billiard ball.

For example, this kind of issue is likely part of the reason why an interpretation of quantum mechanics which takes measuring instruments as primitive seems unappealing to most physicists, as instanced by the comments of Albert and Bell quoted in section 5.2. There is no well-defined way of specifying the exact physical boundaries of a measuring instrument, so an approach in which 'measuring instruments' play a fundamental role would likely make indeterminate predictions in many scenarios, since we would not be able to say definitively whether a measuring instrument is present or not. For example, suppose I come into your laboratory and start taking particles away from your interferometer one by one: at which point does it cease to be an interferometer, and thus cease to have the effects we expect an interferometer to produce in the corresponding quantum domain? It's hard to see how there could be a well-defined answer to that question if we insist on treating the interferometer rather than its constituent parts as primitive. So it is quite easy to postulate a 'non-reductionist' approach in an approximate, hand-wavy way, but quite difficult to make that into a precisely defined framework which provides unambiguous predictions.

In addition, the systematisations provided by physical laws are typically supposed to offer some kind of simplification relative to the un-systematised data – it is, after all, the fact that we are able to effect such notable simplifications which gives us reason to believe that there *are* physical laws or organising principles in the first place. And often this kind of simplification is achieved by explaining the behaviour of a range of large objects in terms of the behaviour of a smaller number of small objects – e.g. explaining the properties of all 118 elements on the periodic table just in terms of arrangements of electrons, protons, and neutrons. So another problem with treating macroscopic objects as fundamental is that there are just far too many different types of macroscopic objects, and therefore it is difficult to see how we could achieve any significant simplifications from a systematisation based on these macroscopic objects.

Now, these obstacles are not necessarily insuperable – petty reductionism has been the dominant scientific paradigm for a very long time, so comparatively little serious thought has gone into considering how to make a non-reductionist picture work – but they do indicate that abandoning petty reductionism is not an easy fix for all of our problems. There is still be a lot of work to be done to show how we could actually formulate a non-reductionist theory powerful and unifying enough to be a serious competitor against existing reductionist theories.

However, some interesting possibilities in this direction are exemplified in the approaches we have discussed in this chapter. For example, in Kent's picture our reality ultimately supervenes on values for the stress-energy tensor which are defined at individual spacetime points, so in that sense Kent's picture obeys petty reductionism; however, Kent defines values for the stress-energy tensor in a way that can largely be expected to represent what we would describe as 'macroscopic' events, rather than microscopic ones, for it is largely macroscopic events which reliably produce records that persist until the end of time, so they can be actualised by the final measurement. Thus by combining an initially microscopic description with an extended temporal development, Kent arrives at a precise way of defining what are in some sense *macroscopic* variables entirely within the microphysical language. Making use of the interplay of spatial and temporal dimensions as a way of implementing a non-reductionist program is a suggestive idea which could certainly be explored further.

An alternative and somewhat more radical approach involves simply denying that a non-reductionist picture must find a precise way of defining macroscopic objects – perhaps by invoking something like Chen's fundamental nomic vagueness (Chen, 2022). However, clearly such an approach would have to be adopted cautiously; for as I have emphasised throughout the book, one key reason why it is important to solve the epistemically construed measurement problem is that arriving at a description of the physical nature of measurement which explains how those measurements give us access to the relevant theoretical structures is an important part of the justification we have for believing that our means of enquiry are actually providing meaningful information. And in order for this justification to be compelling, we should not make things too easy for ourselves: we should demand a high standard of precision and clarity, in order that achieving a coherent account of reality which accounts properly for its own epistemology can actually be seen as a meaningful achievement. So if we really want to allow vagueness in the definition of fundamental objects, we must do that in a way which is not too trivial to achieve: we must demand precision, if not in the world itself, then in the formulation and application of the laws. So this route may be promising, but it still does not offer an easy way out of the problem of defining macroscopic ontology.

13.3 Quantum Gravity

In this book I have largely focused on the epistemic issues that arise in our existing theories of quantum mechanics and QFT. Both of these theories are already well-developed and have been empirically verified to a high degree of accuracy, so at least prima facie it seems quite unlikely that solving the measurement problem will lead to any major changes in our empirical knowledge of the regimes relevant to these two theories – the goal is not to make new predictions in these regimes, but to arrive at a coherent account which fills out the details of the knowledge we already have.

However, work on a theory which unifies quantum mechanics with General Relativity is still ongoing, and as yet no theory of quantum gravity has made substantive novel testable predictions, so it is entirely possible that a satisfactory solution to the measurement problem could have a meaningful impact on the empirical predictions for regimes relevant to quantum gravity. Thus this is the area in which thinking about the measurement problem is most likely to have a direct impact on scientific progress; so I will now close by briefly discussing the way in which epistemic considerations of the kind we have encountered throughout this book could potentially be relevant for ongoing work on quantum gravity.

First, epistemic concerns are very relevant to a key conceptual issue in quantum gravity, known as the 'problem of time' (Kuchar, 2011). This issue arises in the process of Dirac quantisation, which is the standard methodology used to create a quantum theory for a system such that certain possible transformations on that system are deemed to be physically meaningless (Dirac, 1930). For example, we typically think that translations of the whole universe are meaningless, since there is presumably no external space in which the whole universe could move. To quantise such a system, we start out with a very general 'kinematical Hilbert space,' and then we impose constraints to arrive at a 'physical Hilbert space' containing all the physically possible states, i.e. the states which are invariant under the transformations we consider meaningless. So, for example, we expect the state space of quantum gravity to be invariant under translations of the whole universe, and this is enforced by means of the 'Hamiltonian constraint' $H||\Psi\rangle = 0$, where H is the Hamiltonian of the whole universe and $||\Psi\rangle$ is the universal quantum state.

This equation is known as the 'Wheeler-DeWitt equation' (Rovelli and Vidotto, 2015), and it entails that all observables of the theory must commute with the Hamiltonian H: that is, applying an operator for an observable followed by the Hamiltonian H produces the same result as applying the Hamiltonian H followed by the operator for that observable. However, the operator H is also the generator of time evolution in this theory, and thus anything which commutes with H necessarily has the same expectation value both before and after any period of time evolution – which to say, any observable which commutes with the Hamiltonian cannot undergo changes in time. So it seems to follow from the Dirac quantisation approach that no observable in quantum gravity can ever change – so change, and even time itself, seems to have vanished altogether.

Many different answers to this dilemma have been proposed, but one possibility is to say we should simply accept that quantum gravity is showing us that time and change are merely illusory (Barbour, 1994a,b). In the terminology I have used in this book, this amounts to saying that the absence of time is merely a form of benign weirdness, so we need not resist it. However, Healey (2002) raises concerns about the problem of time which have a similar epistemic character to the concerns regarding quantum measurement that I have been discussing in this book. Healey points out that measurement is a process which takes place *in time*, and which by definition involves a *change* in the state of knowledge of the observer performing the measurement. Thus he argues that the absence of time leads to empirical incoherence: '*There can be no reason whatever to accept any theory of gravity – quantum or classical – which entails that there can be no observers, or that observers can have no experiences, some occurring later than others, or that there can be no change in the mental states of observers, or that observers cannot perform different acts at different times. It follows that there can be no reason to accept any theory of gravity – quantum or classical – which entails that there is no time, or that there is no change*' (p. 300). In the terminology I have used in this book, Healey is arguing that the unreality of time is a form of epistemic weirdness rather than merely benign weirdness, meaning that it is not something we can just choose to accept. Indeed, he describes the problem in terms very similar to those I used for the measurement problem back in section 5.3: '*It is important to note that it does not follow that no such theory can be true. But any such theory would have the peculiar feature that, if true, there could be no reason to accept it*' (Healey, 2002, p. 300).

And thus Healey's comments are useful to clarify what exactly is needed to satisfactorily solve the 'problem of time' – either we must deny that time is 'unreal' in quantum gravity, or if we are forced to accept that it is 'unreal,' in some sense, we are going to have to refine and clarify our understanding of the physical process of measurement in order to understand how scientific enquiry is possible without time being 'real.' A number of research programmes, including the Page-Wootters approach (Page and Wootters, 1983; Baumann et al., 2021), the quantum reference frame formalism (Cramer, 1986; Castro-Ruiz et al., 2020), and the evolving constants approach (Rovelli, 2002), have arisen with the aim of solving the problem of time by extracting quasi-temporal structures out of the equations of quantum gravity. And this is certainly important – observers can't have access to quasi-temporal structures if there *are* no such structures, so extracting such structures is a necessary condition for addressing the epistemically construed problem of time. But it is not a sufficient condition because we still need to be able to show how these quasi-temporal structures connect up with observation in a way that allows us to make sense of the scientific practice of performing experiments (apparently) *in time*, in order to understand how such experiments could give us meaningful information, and furthermore to understand what they are actually giving us information about. This does not necessarily mean we have to accept the reality of time – perhaps there is some other way of understanding the epistemology of the measurement process in a 'timeless' universe. But the problem cannot simply be ignored.

Moreover, it should also be noted that our difficulties in connecting observers up with temporal structures in quantum gravity are potentially exacerbated by the fact that we don't even have a fully satisfactory description of observation even in standard textbook quantum mechanics. The Page-Wootters formalism (Page and Wootters,

1983; Baumann et al., 2021) and quantum reference frame formalism (Giacomini et al., 2019; Vanrietvelde et al., 2020; Giacomini and Brukner, 2022) largely focus on extracting unitary evolution out of the timeless setting of quantum gravity, rather than describing measurements and observations, and thus they neatly sidestep the measurement problem. But clearly we are never going to fully resolve the *epistemic* aspects of the problem of time unless we have something to say about how measurements and observations are realised in this context. This raises questions about the common idea that it is possible to formulate a theory of quantum gravity in an interpretation-neutral way: there is a very real possibility that lacunae in our understanding of quantum mechanics will carry over to quantum gravity, and one might even wonder if these limitations are to some degree responsible for the difficulties that we have faced in endeavouring to unify quantum mechanics and gravity.

Similar points can also be made with regard to another major problem faced by a number of approaches to quantum gravity: it's typically difficult to derive local observables out of a theory of quantum gravity. Local observables are observable quantities associated with specific regions of spacetime, an idea which is formalised by the algebraic approach to quantum field theory, which formalises quantum field theories as assignations of algebras to local regions of spacetime (Fredenhagen, 2015). But many physicists believe that a viable theory of quantum gravity must be background-independent (Colosi et al., 2005), which means such a theory will not have pre-existing regions of spacetime to which we can assign observables: somehow we must get the regions and observables to emerge from the theory together, so none of our usual methods of defining local observables will be applicable.

Furthermore, we really do need to be able to define some local observables in a theory of quantum gravity, for without them we will never be able to make sense of the physical content of the theory: if quantum gravity is ever to be connected to the real world, then eventually we have to get out of it something which can be identified as an observation being made, or an experience being had, or even just an event occurring, and all of those things are most naturally formalised as local observables. So the local observables problem is, again, a problem of epistemology: no matter how good the theory is, it can't be fully satisfactory if we don't properly understand how to connect it with the observations which have led us to formulate it in the first place.

Now, one successful approach to defining local observables in the context of quantum gravity involves starting from a classical spacetime background and then describing quantum gravity effects as occurring inside some bounded region. For example, in covariant Loop Quantum Gravity one typically extracts observables by specifying some temporally extended surface and then calculating transition amplitudes from one side of the surface to the other (Rovelli and Vidotto, 2014). We can imagine the boundary as being embedded in a classical spacetime, so states defined on it are indeed local observables. And, in fact, this kind of approach is presumably adequate to predict anything we can actually observe – after all, we always experience ourselves as being located in a classical spacetime, so we will always observe quantum gravity effects through a classical boundary. But this kind of description is adequate only if we are content to give a purely first-person description of quantum gravity: as soon as we try to incorporate other observers and see the relations between them, we are going to have to be able to describe quantum gravity without presupposing a classical boundary, in order that we can show how classical regions of spacetime and various different observers emerge out of it. And as we saw in chapter 10, there are good reasons to think that a satisfactory scientific epistemology requires both first-person and third-person pictures, which suggests that we probably cannot get away with presupposing classical surfaces forever – ultimately we will have to figure out how to get local observables out of a fully background independent theory without starting from a classical boundary.

But in light of the measurement problem, it is not really surprising that we are facing this difficulty in quantum gravity; for it exactly parallels the problems we have encountered in interpreting standard quantum mechanics. As we saw in chapter 2, one way of understanding the measurement problem is to say that we don't know how to extract physical content out of quantum mechanics except by simply postulating a classical observer external to the theory who makes these observations; that is to say, without a solution to the measurement problem we don't know how to incorporate observers into the theory. And this is exactly the problem that we are facing in quantum gravity: we have a number of options for how to write down quantum states of spacetime, but in order to extract physical content we seemingly need to postulate a classical observer who makes observations on a classical boundary. Again, the problem is that we don't know how to incorporate observers into the theory. Moreover, the epistemic construal of the measurement problem seems particularly relevant here. For it tells us that the heart of the measurement problem is the need to offer a coherent account of the physical nature of measurement and observation; and that is exactly the thing we are struggling with in quantum gravity. So focusing on the epistemic aspects of the measurement problem is not just relevant for those of us who are particularly interested in epistemology – it is also plausibly important for further scientific progress, and could give new insight into the obstacles that stand in the way of us arriving at a fully realised theory of quantum gravity.

14
Conclusion

Throughout this book, we have seen that by taking seriously the need to have a robust epistemology for science, we can gain new insight into the path to a viable solution to the measurement problem. To close, I'd like to reflect again on the ways in which quantum mechanics, and the measurement problem in particular, emphasises certain interesting features of the epistemology of science.

One conclusion we can draw is that scientific knowledge is in some ways more fragile than we might at first have imagined. Of course, we already knew that there are many ways the world could be which would significantly undermine our ability to draw conclusions using the methods of science – this is true of many traditional sceptical hypotheses, like the brain-in-vat hypothesis, the Boltzmann brain hypothesis, and so on. But the measurement problem has introduced us to many more such possibilities – an Everettian branching world, an observer-relative world, a consistent histories world drawn from a set which is not strongly decoherent, a world with a finely-tuned initial state or fundamental constants, and so on. And moreover, these are not merely abstract possibilities but rather hypotheses which, some would argue, are quite strongly supported by the scientific evidence. Indeed, as we have seen throughout this book, there are significant technical difficulties standing in the way of writing down even *one* solution to the measurement problem which accommodates all of the scientific evidence but which does not undermine our ability to draw conclusions using the methods of science. So even when the measurement problem is eventually solved in a satisfactory way (as I optimistically believe it will be!) this episode in the history of science will forever stand as a reminder that the fact that we live in a world in which it is possible to gain reliable scientific information by means of observations and measurements is highly non-trivial and should not be taken for granted.

Now, in a foundationalist picture of scientific knowledge it seems we would have to simply assume from the start that none of these kinds of epistemic weirdness hold in the actual world, and thus one might worry that the mere existence of these possibilities undermines all of scientific knowledge. How could we possibly justify the claim that we really know something about the contents of reality if our knowledge rests on a teetering stack of assumptions which are not only unverifiable, but such that we cannot even say anything meaningful about how likely they are to be true? Answering this question inevitably pushes us towards a holistic, coherentist conception of scientific knowledge: our confidence in the deliverances of science is grounded on our success in arriving at a coherent account of reality which offers a model of the physical nature of the measurement process and thus affirms that the measurements on which it is based are indeed providing meaningful information. Indeed, from this point of view, the possibility of various kinds of epistemic weirdness may actually

Saving Science from Quantum Mechanics. Emily Adlam, Oxford University Press. © Oxford University Press (2025).
DOI: 10.1093/9780197808887.003.0014

reassure us that such a bootstrapping approach to scientific epistemology is reasonable – for when we do find a coherent account, we will have better reason to think that our account gets something right about reality, since we can now see more clearly that it was not inevitable that we should be able to arrive at such an account! Moreover, this is precisely why the measurement problem demands a solution: as increasingly emphasized in the modern epistemology of measurement, the possibility of scientific knowledge and the validity of the scientific process rests crucially on our ability to provide models of the measurement process which show how measurements are related to the reality they are supposed to be informing us about, and although we have some understanding of how to model measurements in quantum mechanics, that understanding cannot be fully satisfactory as long as it fails to explain how the specific outcome we actually observe comes about.

One implication of this approach to the measurement problem is that a complete epistemology of science must include more than just a process of incremental Bayesian belief-updating based on individual observations. Of course, there is no doubt that in many circumstances this kind of belief-updating is appropriate and useful, but the holistic character of scientific knowledge is still there in the wings and must be taken into account when our ordinary belief-updating leads us to beliefs which are in tension with the background knowledge relative to which these beliefs were originally updated. And, of course, the holistic aspects of scientific knowledge can be expected to be particularly relevant when we are dealing with fundamental questions about the physical nature of measurement, because assumptions about the meaning of measurement outcomes are involved in virtually every established scientific result.

Focusing on the epistemic aspects of the measurement problem has also underlined the importance of the social and community-driven aspects of scientific knowledge. This point has often been emphasized in feminist philosophy of science, but the point has perhaps not permeated fully to the philosophy of physics, possibly because the relevance of the social aspects of science may seem less obvious in physics than it is in the social and biological sciences. But nonetheless, in physics, as in any part of science, scientific knowledge is built and shared by the entire community: our epistemic peers provide us with the language and concepts that we use to formulate and interpret our physical theories, with confirmation that their experiments have shown results similar to our own, and with a community-driven notion of objectivity and epistemic rationality that helps weed out at least some of the pragmatic and aesthetic biases inherent in our own view of the world. So on this view, observer-relative accounts of quantum mechanics, and even accounts like the Everett approach which rely crucially on irreducibly indexical first-person information, don't really do justice to the nature of scientific enquiry – the knowledge we get from science is just not the kind of thing that could ever be private or first-person in nature, since it is from its very inception a product of a whole community.

These issues also potentially highlight a novel aspect of the debate about 'realism' in science. Of course, this debate is expansive and wide-ranging, and there are many different reasons why one might be an empiricist, operationalist, or some other flavour of anti-realist. But one common route to this kind of view involves the argument that we can be more or less completely certain about the empirical regularities that we have observed, but any beliefs we may form about theoretical entities or structures on the basis of empirical evidence are necessarily extrapolating beyond that evidence and therefore come with a significantly elevated risk of being wrong. Thus it is often suggested that it is *epistemically rational* to be epistemically committed only to the predictions made by science about observable things, or that limiting our epistemic commitments to empirical phenomena is more epistemically safe than having epistemic commitments to theoretical entities or structures.

However, we have seen throughout this book that seeking to come up with a consistent account of reality which reproduces all the empirical evidence whilst also making sense of scientific epistemology is an important part of scientific enquiry which can lead us to refine, update, and precisify our scientific knowledge – and not only the theoretical scientific knowledge, but also the empirical scientific knowledge! From this point of view, the epistemically cautious thing to do is *not* to refrain from formulating a picture of unobservable parts of physical reality. Quite the reverse; by declining to offer a complete account of reality, including unobservable parts, the empiricists, operationalists, anti-realists, and so on are missing a valuable opportunity to further test and refine their set of scientific beliefs. The idea that empiricist and operationalist views are more epistemically cautious than realism seems to assume that 'realism' amounts to nothing more than taking the same empirical knowledge espoused by the empiricists and then adding some additional, less well-founded beliefs about theoretical entities, but we have seen throughout this book that realism is not just additive in this sense: rather the goal is to figure out how to accommodate empirical observations in a coherent picture of the whole of reality, and making progress towards this goal actually makes our empirical beliefs *more* secure. Empiricists, operationalists, and anti-realists want to limit their commitments in order to minimise epistemic risk, but counterintuitively, it appears that seeking to minimise epistemic risk in this way can actually make one's belief system less epistemically safe!

To finish on an optimistic note – I really believe that taking these epistemic considerations seriously may offer a route out of the long impasse over the measurement problem because these considerations offer a new and in some ways more concrete metric for judging the success of various proposed interpretations, as opposed to appealing to subjective notions like intuition and understanding. Once the need for a coherent epistemology is taken into consideration, the space of possible solutions to the measurement problem becomes considerably narrower, and thus the shape of a viable solution becomes considerably clearer. Moreover, it seems entirely possible that some of our difficulties in understanding quantum mechanics arise from the fact that this is a setting where epistemic considerations and epistemic restrictions become particularly relevant – in the context of quantum mechanics it is no longer possible for us to simply ignore the way in which the theories we come up with are inextricably bound up with the limitations placed on us by our situation as physically embodied observers. So it is my hope that studying the epistemology of the measurement problem could ultimately help us understand what it is we are currently missing and get us closer to a satisfactory solution. Indeed, although the intractability of the measurement problem is inconvenient for us right now, from a scientific point of view it is a gift, because the more difficult it is to find a coherent account of our empirical experience, the more the possibilities will be narrowed down, and the more confidence we can have that we have achieved meaningful scientific progress when we eventually find a way through the bottleneck.

Bibliography

S. Abramsky and C. Heunen. Operational Theories and Categorical Quantum Mechanics, 2013.

E. Adlam. Do We Have Any Viable Solution to the Measurement Problem? *Foundations of Physics*, 53(44), 2023a. URL https://doi.org/10.1007/s10701-023-00686-x.

E. Adlam. The Problem of Confirmation in the Everett Interpretation. *Studies in History and Philosophy of Science Part B: Studies in History and Philosophy of Modern Physics*, 47:21–32, 2014. ISSN 1355-2198. doi: http://dx.doi.org/10.1016/j.shpsb.2014.03.004. URL http://www.sciencedirect.com/science/article/pii/S1355219814000276.

E. Adlam. Spooky Action at a Temporal Distance. *Entropy*, 20(1):41, 2018.

E. Adlam. Laws of Nature as Constraints. *Foundations of Physics*, 52, 2022a. URL https://doi.org/10.1007/s10701-022-00546-0.

E. Adlam. Two Roads to Retrocausality. *Synthese*, 200(422), 2022b.

E. Adlam. Does Science need Intersubjectivity? The Problem of Confirmation in Orthodox Interpretations of Quantum Mechanics. *Synthese*, 2022c.

E. Adlam. Is There Causation in Fundamental Physics? New Insights From Process Matrices and Quantum Causal Modelling. *Synthese*, 201(5):1–40, 2023b. doi: 10.1007/s11229-023-04160-z.

E. Adlam. Disappearing without a Trace: The Arrows of Time in Kent's Solution to the Lorentzian Quantum Reality Problem. *The British Journal for the Philosophy of Science*, 0(ja):null, 2023c. doi: 10.1086/726085.

E. Adlam. Quantum Field Theory and the Limits of Reductionism. *Synthese*, 204(3):1–37, 2024a. doi: 10.1007/s11229-024-04734-5.

E. Adlam. The Combination Problem for Relational Quantum Mechanics, 2024b.

E. Adlam. What Does '(Non)-absoluteness of Observed Events' Mean? *Foundations of Physics*, 54(13), 2024c.

E. Adlam and C. Rovelli. Information is Physical: Cross-Perspective Links in Relational Quantum Mechanics. *Philosophy of Physics*, Nov 2023. doi: 10.31389/pop.8.

D. Albert. The Sharpness of the Distinction between the Past and the Future. In A. Wilson, editor, *Chance and Temporal Asymmetry*. Oxford University Press, 2014. ISBN 9780199673421. URL https://books.google.com/books?id=5kZYBAAAQBAJ.

D. Z. Albert. *Time and Chance*. Harvard University Press, 2000.

D.Z. Albert. *Quantum Mechanics and Experience*. Harvard University Press, 1994. ISBN 9780674741133. URL https://books.google.ca/books?id=HYEZD0Mh8JEC.

V. Allori. Primitive Ontology and the Structure of Fundamental Physical Theories. In *The Wave Function: Essays on the Metaphysics of Quantum Mechanics*. Oxford University Press, 04 2013. ISBN 9780199790807. doi: 10.1093/acprof:oso/9780199790807.003.0002. URL https://doi.org/10.1093/acprof:oso/9780199790807.003.0002.

V. Allori. Primitive Ontology in a Nutshell. *International Journal of Quantum Foundations*, 1(3): 107–122, June 2015. URL http://philsci-archive.pitt.edu/11651/.

C. Anastopoulos. *Particle or Wave: The Evolution of the Concept of Matter in Modern Physics*. Princeton University Press, 2008.

L. Antony. Quine as Feminist: The Radical Import of Naturalized Epistemology. In *Only Natural: Gender, Knowledge, and Humankind*. Oxford University Press, 10 2022. ISBN 9780190934361. doi: 10.1093/oso/9780190934361.003.0002. URL https://doi.org/10.1093/oso/9780190934361.003.0002.

D. M. Armstrong. *What is a Law of Nature?* Cambridge University Press, 1983.

A. Aspect, P. Grangier, and G. Roger. Experimental Tests of Realistic Local Theories via Bell's Theorem. *Phys. Rev. Lett.*, 47:460–463, Aug 1981. doi: 10.1103/PhysRevLett.47.460. URL http://link.aps.org/doi/10.1103/PhysRevLett.47.460.

A. J. Ayer, editor. *Logical Positivism*. Greenwood Press, Westport, Conn., 1961.

G. Bacciagaluppi. *Remarks on Space-Time and Locality in Everett's Interpretation*, pages 105–122. Springer Netherlands, Dordrecht, 2002. ISBN 978-94-010-0385-8.

G. Bacciagaluppi. Probability, Arrow of Time and Decoherence. *Studies in History and Philosophy of Modern Physics*, 38:439–456, 2007.

G. Bacciagaluppi. The Role of Decoherence in Quantum Mechanics. In Edward N. Zalta, editor, *The Stanford Encyclopedia of Philosophy*. Metaphysics Research Lab, Stanford University, Fall 2020 edition, 2020.

Y. Balashov and M. Janssen. Presentism and Relativity. *British Journal for the Philosophy of Science*, 54(2):327–346, 2003. doi: 10.1093/bjps/54.2.327.

J. Barbour. The Timelessness of Quantum Gravity: I. The Evidence From the Classical Theory. *Classical and Quantum Gravity*, 11:2853–73, 1994a.

J. B. Barbour. The Timelessness of Quantum Gravity: Ii. The Appearance of Dynamics in Static Configurations. *Classical and Quantum Gravity*, 11:2875–97, 1994b.

L.A. Barnes. The Fine-Tuning of the Universe for Intelligent Life. *Publications of the Astronomical Society of Australia*, 29(4):529–564, 2012. doi:10.1071/AS12015

J.A. Barrett. *The Quantum Mechanics of Minds and Worlds*. OUP Oxford, 1999. ISBN 9780191583254. URL https://books.google.com/books?id=intcadNS4soC.

J.A. Barrett and E.K. Chen. Algorithmic Randomness and Probabilistic Laws, 2023.

A. Barzegar. QBism Is Not So Simply Dismissed. *Foundations of Physics*, 50(7):693–707, May 2020. ISSN 1572-9516. doi: 10.1007/s10701-020-00347-3. URL http://dx.doi.org/10.1007/s10701-020-00347-3.

R. W. Batterman. *Studies in History and Philosophy of Science*, 97:130–132, 2023. ISSN 0039-3681. doi: https://doi.org/10.1016/j.shpsa.2022.08.017. URL https://www.sciencedirect.com/science/article/pii/S0039368122001315.

V. Baumann, F. Del Santo, A. R. H. Smith, F. Giacomini, E. Castro-Ruiz, and C. Brukner. Generalized probability rules from a timeless formulation of Wigner's friend scenarios. *Quantum*, 5:524, Aug 2021. ISSN 2521-327X. doi: 10.22331/q-2021-08-16-524. URL http://dx.doi.org/10.22331/q-2021-08-16-524.

L. Becker. That von Neumann Did Not Believe in a Physical Collapse. *British Journal for the Philosophy of Science*, 55(1):121–135, 2004. doi: 10.1093/bjps/55.1.121.

J. Bell. Against 'Measurement'. *Physics World*, August 1990.

J. S. Bell. Bertlmann's Socks and the Nature of Reality. *J. Phys. Colloq.*, 42:41–62, 1981. doi: 10.1051/jphyscol:1981202.

J. S. Bell. *Speakable and Unspeakable in Quantum Mechanics*. Cambridge University Press, second edition, 2004. ISBN 9780511815676. URL http://dx.doi.org/10.1017/CBO9780511815676. Cambridge Books Online.

N. Berenstain. Privileged-perspective realism in the quantum multiverse. In David Glick, George Darby, and Anna Marmodoro, editors, *The Foundation of Reality: Fundamentality, Space, and Time*. Oxford University Press, 2020.

V.B. Berestetskii, L.P. Pitaevskii, and E.M. Lifshitz. *Quantum Electrodynamics*. Elsevier Science, 2012. ISBN 9780080503462. URL https://books.google.com/books?id=Tpk-lqyr3GoC.

P. Berghofer, and H.A. Wiltsche. Phenomenological Approaches to Physics: Mapping the Field. In H.A. Wiltsche, and P. Berghofer, editors, *Phenomenological Approaches to Physics*. Synthese Library, vol 429. Springer, Cham, 2020.

M. Bild, M. Fadel, Y. Yang, U. von Lüpke, P. Martin, A. Bruno, and Y. Chu. Schrödinger Cat States of a 16-microgram Mechanical Oscillator. *Science*, 380(6642):274–278, 2023. doi: 10.1126/science.adf7553. URL https://www.science.org/doi/abs/10.1126/science.adf7553.

M. Bitbol. Is the Life-World Reduction Sufficient in Quantum Physics? *Continental Philosophy Review*, (4):1–18, 2021. doi: 10.1007/s11007-020-09515-8.

M. Black. The Gap Between "Is" and "Should". *The Philosophical Review*, 73(2):165–181, 1964. ISSN 00318108, 15581470. URL http://www.jstor.org/stable/2183334.

D. Blackwell and L. Dubins. Merging of Opinions with Increasing Information. *The Annals of Mathematical Statistics*, 33(3):882–886, 1962. ISSN 00034851. URL http://www.jstor.org/stable/2237864.

N. Bohr. The Quantum Postulate and the Recent Development of Atomic Theory. *Nature*, 121: 580–590, 1928.

N. Bohr. Natural Philosophy and Human Cultures. *Nature*, (143):268–272, 1939.

N. Bohr. Unity of Knowledge. In Niels Bohr, editor, *Atomic physics and human knowledge*, pages 67–82. Wiley, 1958.

K.-W. Bong, A. Utreras-Alarcón, F. Ghafari, Y.-C. Liang, N. Tischler, E.G. Cavalcanti, G.J. Pryde, and H.M. Wiseman. A strong no-go theorem on the Wigner's friend paradox. *Nature Physics*, 16(12):1199–1205, aug 2020. doi: 10.1038/s41567-020-0990-x.

L. BonJour. Can Empirical Knowledge Have a Foundation? *American Philosophical Quarterly*, 15(1):1–13, 1978. ISSN 00030481.

L. BonJour. *In Defense of Pure Reason: A Rationalist Account of a Priori Justification*. Cambridge University Press, Cambridge, 1998.

M. Born. Quantenmechanik der Stoßvorgänge. *Physik*, page 803–827, 1926.

N. Bostrom. *Anthropic Bias: Observation Selection Effects in Science and Philosophy*. Routledge, 2002.

R. Bousso and B. Freivogel. A Paradox in the Global Description of the Multiverse. *Journal of High Energy Physics*, 2007(06):018–018, June 2007. ISSN 1029-8479. doi: 10.1088/1126-6708/2007/06/018. URL http://dx.doi.org/10.1088/1126-6708/2007/06/018.

P.W. Bridgman. *The Logic of Modern Physics*. Macmillan paperbacks edition. Macmillan, 1980. ISBN 9780405125942. URL https://books.google.ca/books?id=i5PvAAAAMAAJ.

H. Brown and D. Wallace. Solving the Measurement Problem: De broglie–bohm Loses out to Everett. *Foundations of Physics*, 35(4):517–540, Apr 2005. ISSN 1572-9516. doi: 10.1007/s10701-004-2009-3. URL http://dx.doi.org/10.1007/s10701-004-2009-3.

S.C. Brown. Count Rumford and the Caloric Theory of Heat. *Proceedings of the American Philosophical Society*, 93(4):316–325, 1949. ISSN 0003049X. URL http://www.jstor.org/stable/3143157.

Č. Brukner. *On the Quantum Measurement Problem*, pages 95–117. Springer International Publishing, Cham, 2017.

C. Brukner. Qubits are not observers – a no-go theorem. *arXiv:2107.03513 [quant-ph]*, 2021. doi: 10.48550/ARXIV.2107.03513. URL https://arxiv.org/abs/2107.03513.

J. Bub. Bananaworld: Quantum Mechanics for Primates. 11 2012.

J. Bub and I. Pitowsky. Two dogmas about quantum mechanics, 2008.

C.P. Burgess and G.D. Moore. *The Standard Model: A Primer*. Cambridge University Press, 2006. ISBN 9780511256523. URL https://books.google.com/books?id=o32xAQAACAAJ.

J. Butterfield, and N. Bouatta. *Renormalization for Philosophers*. Poznan Studies in the Philosophy of the Sciences & the Humanities, 2015

T.Y. Cao and S.S. Schweber. The Conceptual Foundations and the Philosophical Aspects of Renormalization Theory. *Synthese*, 97(1):33–108, 1993. ISSN 00397857, 15730964. URL http://www.jstor.org/stable/20117830.

R. Carnap and R.A. George. *The Logical Structure of the World: And, Pseudoproblems in Philosophy*. Open Court classics. Open Court, 2003. ISBN 9780812695236. URL https://books.google.co.uk/books?id=WgY2ZMsJtQgC.

M. Carrier. Constructing or Completing Physical Geometry? On the Relation between Theory and Evidence in Accounts of Space-Time Structure. *Philosophy of science, 57, pp. 369-394*, 57, 09 1990. doi: 10.1086/289564.

S. Carroll. What Scientific Idea is Ready for Retirement? *edge.org*. URL https://edge.org/response-detail/25322.

S.M. Carroll. Is our Universe natural? *Nature*, 440 (7088):1132–1136, 2006 Apr 27. doi: 10.1038/nature04804. PMID: 16641984.

S.M. Carroll. Why Boltzmann Brains Are Bad, 2017. URL https://arxiv.org/abs/1702.00850.

S.M. Carroll. Beyond Falsifiability: Normal Science in a Multiverse, 2018.

S.M. Carroll and A. Singh. Mad-Dog Everettianism: Quantum Mechanics at its Most Minimal. In Anthony Aguirre, Brendan Foster, and Zeeya Merali, editors, *What is Fundamental?*, pages 95–104. Springer Verlag, 2019.

E. Castro-Ruiz, F. Giacomini, A. Belenchia, and Č. Brukner. Quantum Clocks and the Temporal Localisability of Events in the Presence of Gravitating Quantum Systems. *Nature Communications*, 11(1), May 2020. ISSN 2041-1723. doi: 10.1038/s41467-020-16013-1. URL http://dx.doi.org/10.1038/s41467-020-16013-1.

L. Catani, M. Leifer, G. Scala, D. Schmid, and R.W. Spekkens. Aspects of the Phenomenology of Interference that are Genuinely Nonclassical. *Physical Review A*, 108(2), August 2023a. ISSN 2469-9934. doi: 10.1103/physreva.108.022207. URL http://dx.doi.org/10.1103/PhysRevA.108.022207.

L. Catani, M. Leifer, D. Schmid, and R.W. Spekkens. Why Interference Phenomena do not Capture the Essence of Quantum Theory. *Quantum*, 7:1119, September 2023b. ISSN 2521-327X. doi: 10.22331/q-2023-09-25-1119. URL http://dx.doi.org/10.22331/q-2023-09-25-1119.

C.M. Caves, C.A. Fuchs, and R. Schack. Quantum Probabilities as Bayesian Probabilities. *Physical Review A*, 65(2), January 2002. ISSN 1094-1622. doi: 10.1103/physreva.65.022305. URL http://dx.doi.org/10.1103/PhysRevA.65.022305.

D. Chalmers. The combination problem for panpsychism. In G. Brüntrup and L. Jaskolla, editors, *Panpsychism*. Oxford University Press, 2016.

V. Chandrasekaran, R. Longo, G. Penington, and E. Witten. An algebra of observables for de Sitter space. *Journal of High Energy Physics*, 2023(2), February 2023. ISSN 1029-8479. doi: 10.1007/jhep02(2023)082. URL http://dx.doi.org/10.1007/JHEP02(2023)082.

H. Chang. *Inventing Temperature: Measurement and Scientific Progress*. OUP Usa, New York, US, 2004.

H. Chang. Scientific Progress: Beyond Foundationalism and Coherentism. *Royal Institute of Philosophy Supplements*, 61:1–20, 2007. doi: 10.1017/S1358246100009723.

H. Chang. Operationalism. In E.N. Zalta, editor, *The Stanford Encyclopedia of Philosophy*. Metaphysics Research Lab, Stanford University, Fall 2021 edition, 2021.

E.K. Chen. Bell's Theorem, Quantum Probabilities, and Superdeterminism, July 2020. URL http://philsci-archive.pitt.edu/18307/. Forthcoming in E. Knox and A. Wilson (eds.), The Routledge Companion to the Philosophy of Physics. (This new version of the paper contains a new section that discusses the recent model of superdeterminism by G. S. Ciepielewski, E. Okon, and D. Sudarsky (2020).).

E.K. Chen. Fundamental Nomic Vagueness. *The Philosophical Review*, 131(1):1–49, jan 2022. doi: 10.1215/00318108-9415127.

T. Chiang. *Stories of Your Life and Others*. Knopf Doubleday Publishing Group, 2010. ISBN 9781931520898. URL https://books.google.com/books?id=WTBA-wzX7cwC.

C. Chihara. Some Problems for Bayesian Confirmation Theory. *The British Journal for the Philosophy of Science*, 38(4):551–560, 1987. doi: 10.1093/bjps/38.4.551. URL https://doi.org/10.1093/bjps/38.4.551.

P. M. Churchland. The Ontological Status of Observables: In Praise of the Superempirical Virtues. In *A Neurocomputational Perspective: The Nature of Mind and the Structure of Science*. The MIT Press, 04 1992. ISBN 9780262270328. doi: 10.7551/mitpress/4940.003.0011. URL https://doi.org/10.7551/mitpress/4940.003.0011.

D. Colosi. On Unitary Evolution in Quantum Field Theory in Curved Spacetime. *The Open Nuclear &; Particle Physics Journal*, 4(1):13–20, August 2011. ISSN 1874-415X. doi: 10.2174/1874415x01104010013. URL http://dx.doi.org/10.2174/1874415X01104010013.

D. Colosi, L. Doplicher, W. Fairbairn, L. Modesto, K. Noui, and C. Rovelli. Background Independence in a Nutshell: The Dynamics of a Tetrahedron. *Classical and Quantum Gravity*, 22(14):2971–2989, June 2005. ISSN 1361-6382. doi: 10.1088/0264-9381/22/14/008. URL http://dx.doi.org/10.1088/0264-9381/22/14/008.

A.H. Compton and W. Heisenberg. *The Physical Principles of the Quantum Theory*, pages 117–166. Springer Berlin Heidelberg, Berlin, Heidelberg, 1984.

O.C. de Beauregard. Méchanique quantique. *Comptes Rendus Académie des Sciences*, 236(1632), 1953.

D. Craig and P. Singh. Consistent Histories in Quantum Cosmology. *Foundations of Physics*, 41(3):371–379, feb 2010. doi: 10.1007/s10701-010-9422-6.

J. Cramer. The Transactional Interpretation of Quantum Mechanics. *Reviews of Modern Physics*, 58:647–687, 01 1986. doi: 10.1103/RevModPhys.58.647.

C. F. Craver and D.M. Kaplan. Are More Details Better? On the Norms of Completeness for Mechanistic Explanations. *The British Journal for the Philosophy of Science*, 71(1):287–319, 2020. doi: 10.1093/bjps/axy015. URL https://doi.org/10.1093/bjps/axy015.

P. Cruse and D. Papineau. Scientific realism without reference. In M. Marsonet, editor, *The Problem of Realism*, pages 174–189. Ashgate, 2002.

M.E. Cuffaro. The Measurement Problem Is a Feature, Not a Bug; Schematising the Observer and the Concept of an Open System on an Informational, or (Neo-)Bohrian, Approach. *Entropy*, 25(10):1410, 2023. URL https://www.mdpi.com/1099-4300/25/10/1410.

M.E. Cuffaro and S. Hartmann. The Open Systems View. *Philosophy of Physics*, 2(1), 2024.

E. Curiel. Schematizing the observer and the epistemic content of theories, 2020.

O.C. de Beauregard. Méchanique Quantique. *Comptes Rendus Académie des Sciences*, 236(1632), 1953.

R. David, and K. Thébault. Against the empirical viability of the Deutsch-Wallace-Everett approach to quantum mechanics. *Studies in History and Philosophy of Modern Physics*, 47: 55–61, 2014.

M. Dascal. What's left for the neo-Copenhagen theorist. *Studies in History and Philosophy of Science Part B: Studies in History and Philosophy of Modern Physics*, 72, 310–321.

B. de Finetti. *Theory of Probability: A Critical Introductory Treatment.* Wiley Series in Probability and Statistics. Wiley, 2017. ISBN 9781119286370. URL https://books.google.com/books?id=FzjoDQAAQBAJ.

W.M. de Muynck. Towards a Neo-Copenhagen Interpretation of Quantum Mechanics. *Foundations of Physics*, 34(5):717–770, May 2004. ISSN 0015-9018. doi: 10.1023/b:foop.0000022186.59877.0c. URL http://dx.doi.org/10.1023/B:FOOP.0000022186.59877.0c.

J.B. DeBrota, C.A. Fuchs, and R. Schack. Respecting One's Fellow: QBism's Analysis of Wigner's Friend. *Foundations of Physics*, 50(12):1859–1874, August 2020. ISSN 1572-9516. doi: 10.1007/s10701-020-00369-x. URL http://dx.doi.org/10.1007/s10701-020-00369-x.

W. Demopoulos. Generalized probability measures and the framework of effects. In *Probability in Physics*, pages 201–217. Springer, 2012.

D.C. Dennett. Real Patterns. *Journal of Philosophy*, 88(1):27–51, 1991. doi: 10.2307/2027085.

D. Deutsch. Quantum Theory as a Universal Physical Theory. *International Journal of Theoretical Physics*, 24:1–41, 1985. URL https://api.semanticscholar.org/CorpusID:17530632.

D. Deutsch. Quantum Theory of Probability and Decisions. *Proceedings of the Royal Society of London. Series A: Mathematical, Physical and Engineering Sciences*, 455(1988):3129–3137, August 1999. ISSN 1471-2946. doi: 10.1098/rspa.1999.0443. URL http://dx.doi.org/10.1098/rspa.1999.0443.

A. Di Biagio and C. Rovelli. Relational Quantum Mechanics is about Facts, not States: A Reply to Pienaar and Brukner. *Foundations of Physics*, 62(52), 2022. URL https://doi.org/10.1007/s10701-022-00579-5.

A. Di Biagio, P. Donà, and C. Rovelli. The Arrow of Time in Operational Formulations of Quantum Theory. *Quantum*, 5:520, Aug 2021. ISSN 2521-327X. doi: 10.22331/q-2021-08-09-520. URL http://dx.doi.org/10.22331/q-2021-08-09-520.

M. Dickson. Is Measurement a Black Box? On the Importance of Understanding Measurement Even in Quantum Information and Computation. *Philosophy of Science*, 74(5):1019–1032, 2007. ISSN 00318248, 1539767X. URL http://www.jstor.org/stable/10.1086/525641.

D. Dieks. Quantum Mechanics and Perspectivalism, 2018. URL https://arxiv.org/abs/1801.09307.

L. Diósi. Models for Universal Reduction of Macroscopic Quantum Fluctuations. *Phys. Rev. A*, 40:1165–1174, Aug 1989. doi: 10.1103/PhysRevA.40.1165. URL https://link.aps.org/doi/10.1103/PhysRevA.40.1165.

P.A.M. Dirac. *The Principles of Quantum Mechanics.* Oxford University Press, 1930. ISBN 9781794767119. URL https://books.google.ca/books?id=IvrDDwAAQBAJ.

M. Donnelly. Mereological Vagueness and Existential Vagueness. *Synthese*, 168(1):53–79, 2009. doi: 10.1007/s11229-008-9312-z.

I. Douven and W. Meijs. Bootstrap Confirmation Made Quantitative. *Synthese*, 149(1):97–132, 2006. ISSN 00397857, 15730964. URL http://www.jstor.org/stable/20118726.

F. Dowker and A. Kent. On the Consistent Histories Approach to Quantum Mechanics. *Journal of Statistical Physics*, 82:1575–1646, March 1996. doi: 10.1007/BF02183396.

K. Draper, P. Draper, and J. Pust. Probabilistic arguments for multiple universes. *Pacific Philosophical Quarterly*, 88 (3):288–307, 2007.

F. Dretske. Epistemic Operators. The Journal of Philosophy, 67 (24):1007–1023, Dec. 24, 1970.

F.I. Dretske. *Perception, Knowledge and Belief: Selected Essays.* Cambridge Studies in Philosophy. Cambridge University Press, 2000. ISBN 9780521777421. URL https://books.google.com/books?id=JMJhcBYWAWQC.

P. Duhem. *The Aim and Structure of Physical Theory.* Princeton: Princeton University Press, 1954.

D. Dürr, S. Goldstein, R. Tumulka, and N. Zanghì. Bohmian Mechanics and Quantum Field Theory. *Physical Review Letters*, 93(9), Aug 2004. ISSN 1079-7114. doi: 10.1103/physrevlett.93.090402. URL http://dx.doi.org/10.1103/PhysRevLett.93.090402.

D. Dürr, S. Goldstein, T. Norsen, W. Struyve, and N. Zanghì. Can Bohmian Mechanics be made Relativistic? *Proceedings of the Royal Society A: Mathematical, Physical and Engineering Sciences*, 470(2162):20130699, feb 2014. doi: 10.1098/rspa.2013.0699.

J. Earman. *Bayes or Bust?: A Critical Examination of Bayesian Confirmation Theory.* MIT Press, 1992. ISBN 9780262519007. URL https://books.google.com/books?id=4LhNEAAAQBAJ.

M. Egg and M. Esfeld. Primitive Ontology and Quantum State in the Grw Matter Density Theory. *Synthese*, 192(10):3229–3245, 2015. ISSN 00397857, 15730964. URL http://www.jstor.org/stable/24704651.

A. Einstein. On the Electrodynamics of Moving Bodies. *Annalen der Physik*, 17:891 – 921, 1905.

A. Einstein. Quantum Mechanics and Reality. *Dialectica 2*, 3-4:320 – 324, 1948.

A. Einstein, M. Born, and H. Born. *The Born-Einstein letters: correspondence between Albert Einstein and Max and Hedwig Born from 1916-1955, with commentaries by Max Born.* Macmillan, 1971. URL https://books.google.co.uk/books?id=HvZAAQAAIAAJ.

A. Elga. Self-Locating Belief and the Sleeping Beauty Problem. *Analysis*, 60(2):143–147, 2000. doi: 10.1111/1467-8284.00215.

M. Esfeld. From the Measurement Problem to the Primitive Ontology Programme. In *Do Wave Functions Jump?: Perspectives of the Work of GianCarlo Ghirardi. Volume 198 of Fundamental Theories of Physics.* Valia Allori, Angelo Bassi, Detlef Durr, and Nino Zanghi, editors. Springer Nature, 2020.

M. Esfeld, D. Lazarovici, M. Hubert, and D. Dürr. The Ontology of Bohmian Mechanics. *The British Journal for the Philosophy of Science*, 65(4):773–796, 2014. ISSN 00070882, 14643537. URL http://www.jstor.org/stable/24562842.

H. Everett. 'Relative State' Formulation of Quantum Mechanics. *Reviews of Modern Physics*, 29, 1957.

A. Fedrizzi and Massimiliano Proietti. Quantum physics: our study suggests objective reality doesn't exist, 2023. URL https://theconversation.com/quantum-physics-our-study-suggests-objective-reality-doesnt-exist-126805.

H. Feigl. *The Origin and Spirit of Logical Positivism*, pages 21–37. Springer Netherlands, Dordrecht, 1981.

B. Fitelson. A Probabilistic Theory of Coherence. *Analysis*, 63(3):194–199, 2003. ISSN 00032638, 14678284. URL http://www.jstor.org/stable/3329309.

A. Franklin. Whence the Effectiveness of Effective Field Theories? *The British Journal for the Philosophy of Science*, 71(4):1235–1259, 2020. doi: 10.1093/bjps/axy050. URL https://doi.org/10.1093/bjps/axy050.

A. Franklin. A Challenge for Humean Everettians, Nov 2023. URL https://philsci-archive.pitt.edu/22716/.

A. Franklin and K. Robertson. A Middle Way: A Non-Fundamental Approach to Many-Body Physics by Robert Batterman: Autonomy and Varieties of Reduction. *Studies in History and Philosophy of Science*, 97:123–125, 2023. ISSN 0039-3681. doi: https://doi.org/10.1016/j.shpsa.2022.08.014. URL https://www.sciencedirect.com/science/article/pii/S0039368122001285.

D. Frauchiger and R. Renner. Quantum Theory Cannot Consistently Describe the Use of Itself. *Nature Communications*, 9(1), Sep 2018. ISSN 2041-1723. doi: 10.1038/s41467-018-05739-8. URL http://dx.doi.org/10.1038/s41467-018-05739-8.

K. Fredenhagen. *An Introduction to Algebraic Quantum Field Theory*, pages 1–30. Springer International Publishing, Cham, 2015.

S. French. *A Phenomenological Approach to Quantum Mechanics: Cutting the Chain of Correlations.* Oxford University Press, 11 2023. ISBN 9780198897958. doi: 10.1093/oso/9780198897958.001.0001. URL https://doi.org/10.1093/oso/9780198897958.001.0001.

S. Friederich. *Multiverse Theories: A Philosophical Perspective.* Cambridge University Press, 2021. ISBN 9781108487122. URL https://books.google.com/books?id=cioNEAAAQBAJ.

R. Frigg. *GRW Theory (Ghirardi, Rimini, Weber Model of Quantum Mechanics)*, pages 266–270. Springer Berlin Heidelberg, Berlin, Heidelberg, 2009.

C. Fuchs. In M. Schlosshauer, editor, *Elegance and Enigma: The Quantum Interviews*, The Frontiers Collection. Springer Berlin Heidelberg, 2011. ISBN 9783642208805. URL https://books.google.com/books?id=YV1CIauzQ6YC.

C. A. Fuchs. QBism, the Perimeter of Quantum Bayesianism. *ArXiv e-prints*, March 2010.

C. A. Fuchs, N. D. Mermin, and R. Schack. An Introduction to QBism with an Application to the Locality of Quantum Mechanics. *American Journal of Physics*, 82(8):749–754, August 2014. doi: 10.1119/1.4874855.

C.A. Fuchs. Interview with a Quantum Bayesian, 2012. https://arxiv.org/pdf/1207.2141

C.A. Fuchs. QBism, Where Next?, 2023. https://arxiv.org/abs/2303.01446

C.A Fuchs and R. Schack. QBism and the Greeks: Why a Quantum State does not Represent an Element of Physical Reality. *Physica Scripta*, 90(1):015104, December 2014. ISSN 1402-4896. doi: 10.1088/0031-8949/90/1/015104. URL http://dx.doi.org/10.1088/0031-8949/90/1/015104.

C.A. Fuchs and B.C. Stacey. QBism: Quantum Theory as a Hero's Handbook, 2019. https://arxiv.org/abs/1612.07308

H. Gaifman and M. Snir. Probabilities over Rich Languages, Testing and Randomness. *Journal of Symbolic Logic*, 47(3):495–548, 1982. doi: 10.2307/2273587.

M. Gell-Mann and J.B. Hartle. Strong decoherence. 1995. doi: 10.48550/ARXIV.GR-QC/9509054. URL https://arxiv.org/abs/gr-qc/9509054.

M. Gell-Mann and J.B. Hartle. Equivalent Sets of Histories and Multiple Quasiclassical Realms, 1996.

G. C. Ghirardi, A. Rimini, and T. Weber. Unified Dynamics for Microscopic and Macroscopic Systems. *Phys. Rev. D*, 34:470–491, Jul 1986. doi: 10.1103/PhysRevD.34.470. URL http://link.aps.org/doi/10.1103/PhysRevD.34.470.

F. Giacomini and Č. Brukner. Quantum Superposition of Spacetimes Obeys Einstein's Equivalence Principle. *AVS Quantum Science*, 4(1):015601, mar 2022. doi: 10.1116/5.0070018.

F. Giacomini, E. Castro-Ruiz, and C. Brukner. Quantum Mechanics and the Covariance of Physical Laws in Quantum Reference Frames. *Nature Communications*, 10, 01 2019. doi: 10.1038/s41467-018-08155-0.

G.F. Giudice. Naturalness after LHC8. In 2013 EPS Conference Proceedings, 2013.

D. Glick. QBism and the Limits of Scientific Realism, August 2019. URL http://philsci-archive.pitt.edu/16382/.

C. Glymour. Theory and Evidence. *British Journal for the Philosophy of Science*, 32(3):314–318, 1981.

P. Goff. Why Panpsychism Doesn't Help Us Explain Consciousness. *Dialectica*, 63(3):289–311, 2009. doi: 10.1111/j.1746-8361.2009.01196.x.

P. Goff. There is No Combination Problem. In M. Blamauer, editor, *The Mental as Fundamental: New Perspectives on Panpsychism*, pages 131–140. Ontos Verlag, 2011.

H. Greaves. On the Everettian Epistemic Problem. *Studies in History and Philosophy of Modern Physics*, 38(1):120–152, 2006. doi: 10.1016/j.shpsb.2006.05.004.

H. Greaves and W. Myrvold. Everett and Evidence. In S. Saunders, J. Barrett, A. Kent, and D. Wallace, editors, *Many Worlds?: Everett, Quantum Theory & Reality*. Oxford University Press, 2010.

H. Greaves and D. Wallace. Justifying Conditionalization: Conditionalization Maximizes Expected Epistemic Utility. *Mind*, 115(459):607–632, 2006. doi: 10.1093/mind/fzl607.

R. Griffiths. Consistent Histories and the Interpretation of Quantum Mechanics. *Journal of Statistical Physics*, 36:219–272, 07 1984. doi: 10.1007/BF01015734.

A. Grinbaum. Which Fine-Tuning Arguments Are Fine? *Foundations of Physics*, 42(5):615–631, 2012. doi: 10.1007/s10701-012-9629-9.

A. Grünbaum. The Duhemian Argument. *Philosophy of Science*, 27(1):75–87, 1960. ISSN 00318248, 1539767X. URL http://www.jstor.org/stable/185307.

A.H Guth. Eternal Inflation and its Implications. *Journal of Physics A: Mathematical and Theoretical*, 40(25):6811–6826, June 2007. ISSN 1751-8121. doi: 10.1088/1751-8113/40/25/s25. URL http://dx.doi.org/10.1088/1751-8113/40/25/S25.

I. Hacking. Do We See Through a Microscope? *Pacific Philosophical Quarterly*, 62(4):305–322, 1981. doi: 10.1111/j.1468-0114.1981.tb00070.x.

I. Hacking. The Inverse Gambler's Fallacy: The Argument from Design. The Anthropic Principle Applied to Wheeler Universes. *Mind*, 96(383): 331–340, 1987. http://www.jstor.org/stable/2254310

J. Halpern. Sleeping Beauty Reconsidered: Conditioning and Reflection in Asynchronous Systems. *Oxford Studies in Epistemology*, 1, 2006.

J.Y. Halpern and M.R. Tuttle. Knowledge, Probability, and adversaries. *Proceedings of the eighth annual ACM Symposium on Principles of distributed computing*, 1989. URL https://api.semanticscholar.org/CorpusID:1453307.

D. Haraway. Situated Knowledges: The Science Question in Feminism and the Privilege of Partial Perspective. *Feminist Studies*, 14(3):575–599, 1988. ISSN 00463663. URL http://www.jstor.org/stable/3178066.

L. Hardy. Operational general relativity: Possibilistic, probabilistic, and quantum, 2016. URL https://arxiv.org/abs/1608.06940.

R. Healey. Can Physics Coherently Deny the Reality of Time? *Royal Institute of Philosophy Supplement*, 50:293–, 2002. doi: 10.1017/s1358246100010614.

R. Healey. Quantum Theory: A Pragmatist Approach. *British Journal for the Philosophy of Science*, 63(4):729–771, 2012. doi: 10.1093/bjps/axr054.

R. Healey. Quantum relativity without relativism, March 2022a. URL https://philsci-archive.pitt.edu/20397/. For publication in a special issue of Foundations of Physics on Relational Quantum Mechanics.

R. Healey. Securing the Objectivity of Relative Facts in the Quantum World, July 2022b. URL https://philsci-archive.pitt.edu/20846/. For publication in a special issue of Foundations of Physics on Relational Quantum Mechanics.

R. Healey. Scientific Objectivity and Its Limits. *The British Journal for the Philosophy of Science*, 75(3): 639–662, 2024.

B. Hedden. Time-Slice Rationality. *Mind*, 124(494):449–491, 2015. doi: 10.1093/mind/fzu181.

W. Heisenberg. The development of the interpretation of the quantum theory. 1955. URL https://api.semanticscholar.org/CorpusID:118361861.

C.G. Hempel. Problems and Changes in the Empiricist Criterion of Meaning. *Revue Internationale de Philosophie*, 4(11):41–63, 1950. ISSN 00488143, ISSN 00488143, http://www.jstor.org/stable/23932368.

C.G. Hempel. The Concept of Cognitive Significance: A Reconsideration. *Proceedings of the American Academy of Arts and Sciences*, 80(1):61–77, 1951. ISSN 01999818. URL http://www.jstor.org/stable/20023635.

C.G. Hempel. *Philosophy of Natural Science*. Prentice-Hall, 1966.

L. Henderson. The Problem of Induction. In Edward N. Zalta and Uri Nodelman, editors, *The Stanford Encyclopedia of Philosophy*. Metaphysics Research Lab, Stanford University, Winter 2022 edition, 2022.

B. Hensen, H. Bernien, A.E Dréau, A. Reiserer, N. Kalb, M.S Blok, J. Ruitenberg, R.FL Vermeulen, R.N Schouten, C. Abellán, et al. Experimental loophole-free violation of a bell inequality using entangled electron spins separated by 1.3 km. *arXiv preprint arXiv:1508.05949*, 2015.

J. Hey, T. Neugebauer, and C. Pasca. *Georges-Louis Leclerc de Buffon's 'Essays on Moral Arithmetic'*, pages 245–282. 08 2010. ISBN 978-3-642-13982-6. doi: 10.1007/978-3-642-13983-3_15.

D.R. Hofstadter and D.C. Dennett. *The Mind's I: Fantasies and Reflections on Self and Soul*. Bantam New Age Books: Nonfiction. Bantam Books, 1982. ISBN 9780553345841. URL https://books.google.com/books?id=hz22QgAACAAJ.

P.R. Holland. *The Quantum Theory of Motion: An Account of the de Broglie-Bohm Causal Interpretation of Quantum Mechanics*. Cambridge University Press, 1995. ISBN 9780521485432. URL https://books.google.co.uk/books?id=BsEfVBzToRMC.

S. Hossenfelder. Screams for Explanation: Finetuning and Naturalness in the Foundations of Physics. *Synthese*, Sep 2019. ISSN 1573-0964. doi: 10.1007/s11229-019-02377-5. URL http://dx.doi.org/10.1007/s11229-019-02377-5.

S. Hossenfelder. Superdeterminism: A guide for the perplexed, 2020. URL https://arxiv.org/abs/2010.01324.

S. Hossenfelder and T. Palmer. Rethinking Superdeterminism. *Frontiers in Physics*, 8:139, 2020. ISSN 2296-424X. doi: 10.3389/fphy.2020.00139. URL https://www.frontiersin.org/article/10.3389/fphy.2020.00139.

D. Howard. Who Invented the "Copenhagen Interpretation"'? A Study in Mythology. *Philosophy of Science*, 71(5):669–682, 2004. doi: 10.1086/425941.

C. Howson. The 'old evidence' problem. *British Journal for the Philosophy of Science*, 42(4):547–555, 1991. doi: 10.1093/bjps/42.4.547.

N. Huggett and C. Wüthrich. Emergent Spacetime and Empirical (in)coherence. *Studies in History and Philosophy of Science Part B: Studies in History and Philosophy of Modern Physics*, 44(3):276–285, aug 2013. doi: 10.1016/j.shpsb.2012.11.003.

D. Hume. *An Enquiry Concerning Human Understanding: A Critical Edition*, volume 3. Oxford University Press, 2000.

E. Husserl. *Edmund Husserl's Ideas, Volume II*, pages 15–39. Springer Netherlands, Dordrecht, 1970.

A. Hüttemann. *What's Wrong with Microphysicalism?* Routledge, New York, 2003.

P. van Inwagen. *Material Beings*. Cornell paperbacks. Cornell University Press, 1990. ISBN 9780801483066. URL https://books.google.com/books?id=mFaBDQAAQBAJ.

P. Van Inwagen. Changing the Past. *Oxford Studies in Metaphysics*, 5:3–40, 2010.

J. Ismael. Closed Causal Loops and the Bilking Argument. *Synthese*, 136(3):305–320, 2003. doi: 10.1023/a:1025170026539.

F. Jackson. What Mary Didn't Know. *The Journal of Philosophy*, 83(5):291–295, 1986. ISSN 0022362X. URL http://www.jstor.org/stable/2026143.

M. Janas, M.E. Cuffaro, and M. Janssen. *Understanding Quantum Raffles: Quantum Mechanics on an Informational Approach - Structure and Interpretation (Foreword by Jeffrey Bub)*. Springer, 2021.

E.T. Jaynes and G.L. Bretthorst. *Probability Theory: The Logic of Science*. Cambridge University Press, 2003. ISBN 9780521592710. URL https://books.google.com/books?id=tTN4HuUNXjgC.

R.C. Jeffrey. *The Logic of Decision*. University of Chicago Press, New York, NY, USA, 1965.

H. Jeffreys. *The Theory of Probability*. Oxford Classic Texts in the Physical Sciences. OUP Oxford, 1998. ISBN 9780191589676. URL https://books.google.com/books?id=vh9Act9rtzQC.

J.-L. Jiang, C. Ecker, and L. Rezzolla. Bayesian Analysis of Neutron-star Properties with Parameterized Equations of State: The Role of the Likelihood Functions. *The Astrophysical Journal*, 949(1):11, May 2023. ISSN 1538-4357. doi: 10.3847/1538-4357/acc4be. URL http://dx.doi.org/10.3847/1538-4357/acc4be.

A.N. Jordan and I.A. Siddiqi. *Quantum Measurement: Theory and Practice*. Cambridge University Press, 2024. ISBN 9781009117548. URL https://books.google.com/books?id=7ETzEAAAQBAJ.

J.M. Joyce. The Development of Subjective Bayesianism. In Dov M. Gabbay, John Woods, and Akihiro Kanamori, editors, *Handbook of the History of Logic*, pages 10–415. Elsevier, 2004.

I. Kant. *Critique of Pure Reason*. Bohn's philosophical library. G. Bell and sons, 1890. URL https://books.google.co.uk/books?id=yflBAAAAYAAJ.

I. Kant. Concerning the Ultimate Ground of the Differentiation of Directions in Space. In David Walford and Ralf Meerbote, editors, *The Cambridge Edition of the Works of Immanuel Kant. Theoretical Philosophy, 1755–1770*, pages 365–72. Cambridge University Press, 1768.

R. E. Kastner. On the Status of the Measurement Problem: Recalling the Relativistic Transactional Interpretation, 2017. URL https://arxiv.org/abs/1709.09367.

R. E. Kastner and John G. Cramer. Quantifying Absorption in the Transactional Interpretation, 2018. URL https://arxiv.org/abs/1711.04501.

T. Kelly. Epistemic Rationality as Instrumental Rationality: A Critique. *Philosophy and Phenomenological Research*, 66(3):612–640, 2003. ISSN 00318205. URL http://www.jstor.org/stable/20140564.

A. Kent. Solution to the Lorentzian Quantum Reality Problem. *Phys Rev A*, 90(1):012107, July 2014. doi: 10.1103/PhysRevA.90.012107.

A. Kent. Quantum Reality via Late-time Photodetection. *Physical Review A*, 96(6), Dec 2017. ISSN 2469-9934. doi: 10.1103/physreva.96.062121. URL http://dx.doi.org/10.1103/PhysRevA.96.062121.

S. Kochen and E.P. Specker. The Problem of Hidden Variables in Quantum Mechanics. In C.A. Hooker, editor, *The Logico-Algebraic Approach to Quantum Mechanics*, The University of Western Ontario Series in Philosophy of Science, pages 293–328. Springer Netherlands, 1975.

H. Kragh. Max Planck: The Reluctant Revolutionary. *Physics World*, 13(12):31, dec 2000. doi: 10.1088/2058-7058/13/12/34.

K. V. Kuchar. Time and Interpretations of Quantum Gravity. *International Journal of Modern Physics D*, 20(supp01):3–86, 2011. doi: 10.1142/S0218271811019347. URL https://doi.org/10.1142/S0218271811019347.

K. Landsman. The Fine-Tuning Argument: Exploring the Improbability of Our Existence. In Landsman, K., and van Wolde, E., editors, *The Challenge of Chance. The Frontiers Collection*. Springer, Cham, 2016. https://doi.org/10.1007/978-3-319-26300-7_6

K. Landsman. *Foundations of Quantum Theory: From Classical Concepts to Operator Algebras*. Imprint: Springer, Cham, 2017.

K. Landsman. Bohmian Mechanics is Not Deterministic. *Foundations of Physics*, 52(4), July 2022. ISSN 1572-9516. doi: 10.1007/s10701-022-00591-9. URL http://dx.doi.org/10.1007/s10701-022-00591-9.

M. Lange. Hume and the Problem of Induction. In D.M. Gabbay, S. Hartmann, and J. Woods, editors, *Handbook of the History of Logic. Volume 10: Inductive Logic*, pages 43–91. Elsevier, 2011.

L. Laudan. Normative Naturalism. *Philosophy of Science*, 57(1):44–59, 1990. ISSN 00318248, 1539767X. URL http://www.jstor.org/stable/187620.

A.Y. Lee. Degrees of consciousness. *Noûs*, 57(3):553–575, 2023. doi: https://doi.org/10.1111/nous.12421. URL https://onlinelibrary.wiley.com/doi/abs/10.1111/nous.12421.

G. Leegwater. When Greenberger, Horne and Zeilinger meet Wigner's Friend, 2018.

D. Lewis. *Counterfactuals*. Wiley, 2013. ISBN 9781118696415. URL https://books.google.co.uk/books?id=bCvnk3JMvfAC.

D. Lewis. Attitudes de dicto and de se. *Philosophical Review*, 88(4):513–543, 1979. doi: 10.2307/2184843.

D. Lewis. A Subjectivist's Guide to Objective Chance. In R.C. Jeffrey, editor, *Studies in Inductive Logic and Probability*, pages 83–132. University of California Press, 1980.

D. Lewis. Humean Supervenience Debugged. *Mind*, 103(412):473–490, 1994. doi: 10.1093/mind/103.412.473.

D. Lewis. Why Conditionalize. In A. Eagle, editor, *Philosophy of Probability: Contemporary Readings*, pages 403–407. Routledge, 1999.

D.K. Lewis. Many, but Almost One. In K. Cambell, J. Bacon, and L. Reinhardt, editors, *Ontology, Causality and Mind: Essays on the Philosophy of D. M. Armstrong*, pages 23–38. Cambridge University Press, 1993.

E.H. Lieb. The Stability of Matter and Quantum Electrodynamics, 2002.

A. Linde. Sinks in the landscape, Boltzmann brains and the cosmological constant problem. *Journal of Cosmology and Astroparticle Physics*, 2007(01):022–022, January 2007. ISSN 1475-7516. doi: 10.1088/1475-7516/2007/01/022. URL http://dx.doi.org/10.1088/1475-7516/2007/01/022.

P. Lipton. *Inference to the Best Explanation*. Routledge/Taylor and Francis Group, London and New York, 1991.

M. Lockwood. The Grain Problem. In H.M. Robinson, editor, *Objections to Physicalism*, pages 271–291. Oxford University Press, 1993.

Barry Loewer. The Emergence of Time's Arrows and Special Science Laws from Physics. *Interface Focus*, 2(1):13–19, 2012. doi: 10.1098/rsfs.2011.0072. URL https://royalsocietypublishing.org/doi/abs/10.1098/rsfs.2011.0072.

F. London and E. Bauer. The Theory of Observation in Quantum Mechanics. In Wheeler and Zurek (pp. 217–259). Princeton, NJ: Princeton University Press, 1939.

H.E. Longino. *Science as Social Knowledge: Values and Objectivity in Scientific Inquiry*. Philosophy of science. Princeton University Press, 1990. ISBN 9780691020518. URL https://books.google.com/books?id=S8fIbD19BisC.

R. Loudon. *The Quantum Theory of Light*. OUP Oxford, 2000. ISBN 9780191589782. URL https://books.google.com/books?id=AEkfajgqldoC.

E. Mach. *The Science of Mechanics*. Creative Media Partners, LLC, 1893. ISBN 9781297399848. URL https://books.google.com/books?id=Two3rgEACAAJ.

T. Maudlin. Bell's other assumption(s). (talk, quoted in E. Chen, 'Bell's Theorem, Quantum Probabilities, and Superdeterminism,' *The Routledge Companion to the Philosophy of Physics*, 2020), 2019.

T. Maudlin. Three Measurement Problems. *Topoi*, 14(1):7–15, 1995. doi: 10.1007/bf00763473.

G. Maxwell. Theories, perception and structural realism. In R.G. Colodny, editor, *The Nature and Function of Scientific Theories: Essays in Contemporary Science and Philosophy*, pages 3–34. University of Pittsburgh Press, 1970.

K.J. McQueen and L. Vaidman. In defence of the self-location uncertainty account of probability in the many-worlds interpretation, February 2018. URL http://philsci-archive.pitt.edu/15195/.

C.J. G. Meacham. Sleeping Beauty and the Dynamics of de Se Beliefs. *Philosophical Studies*, 138(2):245–269, 2008. doi: 10.1007/s11098-006-9036-1.

N. D. Mermin. There is no Quantum Measurement Problem. *Physics Today*, 75(6):62–63, 06 2022. ISSN 0031-9228. doi: 10.1063/PT.3.5027. URL https://doi.org/10.1063/PT.3.5027.

R.W. Miller. *Fact and Method: Explanation, Confirmation and Reality in the Natural and the Social Sciences*. Princeton University Press, 1987.

B. Monton. Life is Evidence for an Infinite Universe, December 2004. URL http://philsci-archive.pitt.edu/3507/.

R. Mucino, E. Okon, and D. Sudarsky. Assessing relational quantum mechanics. *Synthese*, 200(5):1–26, 2022. doi: 10.1007/s11229-022-03886-6.

W. Myrvold. Review: The Quantum Mechanics of Minds and Worlds. *Philosophy of Science*, 69(3):536–538, 2002. ISSN 00318248, 1539767X. URL https://www.jstor.org/stable/10.1086/342458.

T. Nagel. What is It Like to be a Bat? *Philosophical Review*, 83(October):435–50, 1974. doi: 10.2307/2183914.

A. Navarro-Boullosa, A. Bernal, and J. Alberto Vazquez. Bayesian Analysis for Rotational Curves with ℓ-boson Stars as a Dark Matter Component. *Journal of Cosmology and Astroparticle Physics*, 2023(09):031, September 2023. ISSN 1475-7516. doi: 10.1088/1475-7516/2023/09/031. URL http://dx.doi.org/10.1088/1475-7516/2023/09/031.

E. Nelson. Stochastic Mechanics of Relativistic Fields. *Journal of Physics: Conference Series*, 504, 03 2014. doi: 10.1088/1742-6596/504/1/012013.

M.A. Nielsen and I.L. Chuang. *Quantum Computation and Quantum Information*. Cambridge University Press, New York, NY, USA, 10th edition, 2011. ISBN 1107002176, 9781107002173.

J.D. Norton. Cosmic Confusions: Not Supporting Versus Supporting Not. *Philosophy of Science*, 77(4):501–523, 2010. doi: 10.1086/656006.

R. Nozick. *Invariances: The Structure of the Objective World*. Belknap Press of Harvard University Press, Cambridge, Mass., 2001.

M. Ohnesorge. The Epistemic Privilege of Measurement: Motivating a Functionalist Account. *Philosophy of Science*, page 1–11, 2023. doi: 10.1017/psa.2023.47.

A. Oldofredi. The Bundle Theory Approach to Relational Quantum Mechanics. *Foundations of Physics*, 51(1), February 2021. ISSN 1572-9516. doi: 10.1007/s10701-021-00407-2. URL http://dx.doi.org/10.1007/s10701-021-00407-2.

E.T. Olson. Personal identity. In S.P. Stich and T.A. Warfield, editors, *The Blackwell Guide to Philosophy of Mind*, pages 352–368. Blackwell, 2003.

E.T. Olson. Personal Identity. In E.N. Zalta and U. Nodelman, editors, *The Stanford Encyclopedia of Philosophy*. Metaphysics Research Lab, Stanford University, Fall 2023 edition, 2023.

N. Ormrod and J. Barrett. A no-go theorem for absolute observed events without inequalities or modal logic, 2022.

N. Ormrod and J. Barrett. Quantum influences and event relativity, 2024.

D.N. Page and W.K. Wootters. Evolution without evolution: Dynamics described by stationary observables. *Phys. Rev. D*, 27:2885–2892, Jun 1983. doi: 10.1103/PhysRevD.27.2885. URL https://link.aps.org/doi/10.1103/PhysRevD.27.2885.

T. N. Palmer. Invariant set theory, 2016. URL https://arxiv.org/abs/1605.01051.

D. Parfit. *Why Anything? Why This? In Tim Crane & Katalin Farkas, Metaphysics: a guide and anthology*. New York: Oxford University Press, 2004.

D. Papineau. A Fair Deal for Everettians. In *Many Worlds?: Everett, Quantum Theory, and Reality*. Oxford University Press, 06 2010. ISBN 9780199560561. doi: 10.1093/acprof:oso/9780199560561.003.0009. URL https://doi.org/10.1093/acprof:oso/9780199560561.003.0009.

J. Pearl. *Causality*. Cambridge University Press, 2009.

R. Penrose. On Gravity's Role in Quantum State Reduction. *Gen. Rel. Grav.*, 28:581–600, 1996. doi: 10.1007/BF02105068.

M.E. Peskin and D.V. Schroeder. *An Introduction To Quantum Field Theory*. Frontiers in physics. Westview Press, 1995. ISBN 9780813345437. URL https://books.google.co.uk/books?id=EVeNNcslvX0C.

J. Pienaar. A Quintet of Quandaries: Five No-Go Theorems for Relational Quantum Mechanics. *Foundations of Physics*, 51(5), Oct 2021. ISSN 1572-9516. doi: 10.1007/s10701-021-00500-6. URL http://dx.doi.org/10.1007/s10701-021-00500-6.

I. Pitowsky. Quantum mechanics as a theory of probability, 2005.

U.T. Place. The Causal Potency of Qualia: Its Nature and its Source. *Brain and Mind*, 1(2):183–192, 2000. doi: 10.1023/a:1010023129393.

H. Poincaré. *Science and Hypothesis*. Walter Scott publishing Company, 1907. URL https://books.google.com/books?id=gbbaAAAAMAAJ.

H. Price. *Time's Arrow & Archimedes' Point: New Directions for the Physics of Time*. Oxford Paperbacks: Philosophy. Oxford University Press, 1996. ISBN 9780195117981. URL https://books.google.co.uk/books?id=wJXmCwAAQBAJ.

H. Price. On the origins of the arrow of time: Why there is still a puzzle about the low entropy past. In C. Hitchcock, editor, *Contemporary Debates in Philosophy of Science*, pages 219–239. Blackwell, 2004.

H. Price. Decisions, Decisions, Decisions: Can Savage Salvage Everettian Probability? 06 2010. doi: 10.1093/acprof:oso/9780199560561.003.0014. URL https://doi.org/10.1093/acprof:oso/9780199560561.003.0014.

S. Psillos. A Philosophical Study of the Transition From the Caloric Theory of Heat to Thermodynamics: Resisting the Pessimistic Meta-Induction. *Studies in History and Philosophy of Science Part A*, 25(2):159–190, 1994. doi: 10.1016/0039-3681(94)90026-4.

W. V. Quine. Epistemology Naturalized. In *Ontological Relativity and Other Essays*. Columbia University Press, 1968.

W. Quine. Two dogmas of empiricism. In *Can Theories be Refuted?*, pages 41–64. Springer, 1976.

F. Ramsey. Truth and Probability. In Antony Eagle, editor, *Philosophy of Probability: Contemporary Readings*, pages 52–94. Routledge, 1926.

J. Read and B. Le Bihan. The Landscape and the Multiverse: What's the Problem?, 2021. URL http://philsci-archive.pitt.edu/18846/. Forthcoming in Synthese.

H. Reichenbach. *The Theory of Relativity and a Priori Knowledge*. University of California Press, Berkeley, 1965.

M. Ridley. Quantum Probability from Temporal Structure. *Quantum Reports*, 5(2):496–509, 2023. ISSN 2624-960X. doi: 10.3390/quantum5020033. URL https://www.mdpi.com/2624-960X/5/2/33.

S. Rivat and A. Grinbaum. Philosophical Foundations of Effective Field Theories. *European Physical Journal A*, 56(3), 2020.

J.T. Roberts. Laws About Frequencies. 2009. URL http://philsci-archive.pitt.edu/4785/.

J. Rosaler, R. Harlander, G. Schiemann, and M. Á. Carretero Sahuquillo. Naturalness, Hierarchy, and Fine-Tuning. *Foundations of Physics*, 49(9):855–859, 2019. doi: 10.1007/s10701-019-00288-6.

G.H. Rosenberg. The Boundary Problem for Phenomenal Individuals. In R. Stuart Hameroff, Kaszniak, W. Alfred, and A. Scott, editors, *Toward a Science of Consciousness Ii : The Second Tucson Discussions and Debates*. MIT Press, 1998.

R.D. Rosenkrantz. *Foundations and Applications of Inductive Probability*. Ridgeview Publishing Company, 1981. ISBN 9780917930034. URL https://books.google.com/books?id=VCTvAAAAMAAJ.

A. Roth. Practical Intersubjectivity. In F. Schmitt, editor, *Socializing Metaphysics : The Nature of Social Reality*, pages 65–91. Rowman & Littlefield, 65–91, 2003.

C. Rovelli. Relational Quantum Mechanics. *International Journal of Theoretical Physics*, 35(8):1637–1678, Aug 1996. ISSN 1572-9575. doi: 10.1007/bf02302261. URL http://dx.doi.org/10.1007/BF02302261.

C. Rovelli. Partial Observables. *Physical Review D*, 65(12), June 2002. ISSN 1089-4918. doi: 10.1103/physrevd.65.124013. URL http://dx.doi.org/10.1103/PhysRevD.65.124013.

C. Rovelli and F. Vidotto. *Covariant Loop Quantum Gravity: An Elementary Introduction to Quantum Gravity and Spinfoam Theory*. Cambridge University Press, 2014.

C. Rovelli and F. Vidotto. Compact Phase Space, Cosmological Constant, and Discrete Time. *Physical Review D*, 91(8), Apr 2015. ISSN 1550-2368. doi: 10.1103/physrevd.91.084037. URL http://dx.doi.org/10.1103/PhysRevD.91.084037.

T. Ryckman. *The Reign of Relativity: Philosophy in Physics 1915–1925*. Oxford University Press, New York, 2005.

G. Ryle. *The Concept of Mind: 60th Anniversary Edition*. Taylor & Francis, 2009. ISBN 9781134012220. URL https://books.google.com/books?id=YXN4AgAAQBAJ.

S. Saunders. The quantum block universe. *Harvard Department of Philosophy preprint*, 1992.

S. Saunders. Branch-counting in the Everett interpretation of quantum mechanics. *Proceedings of the Royal Society A: Mathematical, Physical and Engineering Sciences*, 477(2255), November 2021a. ISSN 1471-2946. doi: 10.1098/rspa.2021.0600. URL http://dx.doi.org/10.1098/rspa.2021.0600.

S. Saunders. The Everett Interpretation: Probability, June 2021b. URL http://philsci-archive.pitt.edu/19362/.

S. Saunders and D. Wallace. Branching and uncertainty, January 2008. URL http://philsci-archive.pitt.edu/3811/. This is a revised version of the paper of the same name deposited on 12 June 2007, as it will appear in the British Journal for the Philosophy of Science.

S. Saunders, J. Barrett, A. Kent, and D. Wallace, editors. *Many Worlds?: Everett, Quantum Theory & Reality*. Oxford University Press, 2010.

L.J. Savage. *The Foundations of Statistics*. Wiley Publications in Statistics, 1954.

L.J. Savage. *The Foundations of Statistics*. Dover Books on Mathematics. Dover Publications, 2012. ISBN 9780486137100. URL https://books.google.com/books?id=N_bBAgAAQBAJ.

S. Schindler. A coherentist conception of ad hoc hypotheses. *Studies in History and Philosophy of Science Part A*, 67:54–64, 2018. ISSN 0039-3681. doi: https://doi.org/10.1016/j.shpsa.2017.11.011. URL https://www.sciencedirect.com/science/article/pii/S0039368117301036.

M. Schlick. *Moritz Schlick Philosophical Papers: Volume 1: (1909–1922)*. Vienna Circle Collection. Springer Netherlands, 1978. ISBN 9789027703149. URL https://books.google.com/books?id=_Jk8MqimEX8C.

D. Schmid, Yílé Ying, and Matthew Leifer. A review and analysis of six extended Wigner's friend arguments, 2023.

E. Schrödinger. Die gegenwärtige situation in der quantenmechanik. *Naturwissenschaften*, 23:807–812, 1935. URL https://api.semanticscholar.org/CorpusID:206795705.

J. Schupbach. *Bayesianism and Scientific Reasoning*. Elements in the Philosophy of Science. Cambridge University Press, 2022.

C.T. Sebens and S.M. Carroll. Self-locating Uncertainty and the Origin of Probability in Everettian Quantum Mechanics. *The British Journal for the Philosophy of Science*, 69(1):25–74, March 2018. ISSN 1464-3537. doi: 10.1093/bjps/axw004. URL http://dx.doi.org/10.1093/bjps/axw004.

H.I. Sharlin and T. Sharlin. *Lord Kelvin, the Dynamic Victorian*. Pennsylvania State University Press, 1979. ISBN 9780271002033. URL https://books.google.com/books?id=2pbvAAAAMAAJ.

A. Shimony. Integral Epistemology in "Naturalistic Epistemology: A Symposium of Two Decades". *Boston Studies in the Philosophy of Science*, 100:299–318, 1987.

A. Shimony. Reality, Causality and Closing the Circle. In *Search for a Naturalistic Worldview Volume I*. Cambridge University Press, 1993a.

A. Shimony, M. Horne, and J. Clauser. Comment on 'The theory of local beables'. *Epistemological Letters*, 13(1), 1976.

A. Shimony. Role of the Observer in Quantum Theory. *American Journal of Physics*, 31:755–773, 1963. URL https://api.semanticscholar.org/CorpusID:120026478.

A. Shimony. Search for a worldview which can accommodate our knowledge of microphysics. 1993b. URL https://api.semanticscholar.org/CorpusID:123620504.

T. Shogenji. Is Coherence Truth Conducive? *Analysis*, 59(4):338–345, 1999. ISSN 00032638, 14678284. URL http://www.jstor.org/stable/3328607.

J. Smart. *Our place in the universe: a metaphysical discussion*. New York, NY, USA: Blackwell, 1989.

C. Smeenk. *Testing Inflation*, page 206–227. Cambridge University Press, 2017.

E. Sober. Likelihood, Model Selection, and the Duhem-Quine Problem. *The Journal of Philosophy*, 101(5):221–241, 2004. ISSN 0022362X. URL http://www.jstor.org/stable/3655662.

M. Solomon. Social Empiricism. *Noûs*, 28(3):325–343, 1994. ISSN 00294624, 14680068. URL http://www.jstor.org/stable/2216062.

R. W. Spekkens. Contextuality for Preparations, Transformations, and Unsharp Measurements. *Physical Review A*, 71(5):052108, May 2005. doi: 10.1103/PhysRevA.71.052108.

P. Spirtes, R. Scheines, and C. Glymour. *Causation, Prediction, and Search*. Adaptive Computation and Machine Learning. MIT Press, 2000. ISBN 9780262194402. URL https://books.google.ca/books?id=vV-U09kCdRwC.

J. Sprenger. The Objectivity of Subjective Bayesianism. *European Journal for Philosophy of Science*, 8(3):539–558, 2018. doi: 10.1007/s13194-018-0200-1.

H. Stein. personal communication, 1963. Cited in Shimony, A. 'Role of the Observer in Quantum Theory,' *American Journal of Physics*.

H. Stein. Some Reflections on the Structure of our Knowledge in Physics. In B. Skyrms D. Prawitz and D. Westerstøahl, editors, *Logic, Methodology and Philosophy of Science*, pages 633–655. Elsevier Science B.V., 1994.

M. Strevens. *Depth: An Account of Scientific Explanation*. Harvard University Press, 2008. ISBN 9780674031838. URL http://www.jstor.org/stable/j.ctv1dv0tnw.

W. Struyve and H. Westman. A minimalist pilot-wave model for quantum electrodynamics. *Proceedings of the Royal Society A: Mathematical, Physical and Engineering Sciences*, 463(2088):3115–3129, sep 2007. doi: 10.1098/rspa.2007.0144.

L. Susskind. *The Cosmic Landscape: String Theory and the Illusion of Intelligent Design*. Little, Brown, 2008.

P. Suppes. A comparison of the meaning and uses of models in mathematics and the empirical sciences. In *The Concept and the Role of the Model in Mathematics and Natural and Social Sciences: Proceedings of the Colloquium sponsored by the Division of Philosophy of Sciences of the International Union of History and Philosophy of Sciences organized at Utrecht, January 1960, by Hans Freudenthal*, pages 163–177. Springer, 1961.

R. I. Sutherland. Bell's Theorem and Backwards in Time Causality. *Int. J. Theor. Phys.*, 22:377–384, 1983. doi: 10.1007/BF02082904.

G. Hooft. The Cellular Automaton Interpretation of Quantum Mechanics, 2015.

E. Tal. *The Epistemology of Measurement: A Model-Based Account*. PhD thesis, University of Toronto, 2012.

E. Tal. Old and New Problems in Philosophy of Measurement. *Philosophy Compass*, 8(12):1159–1173, 2013. doi: 10.1111/phc3.12089.

E. Tal. Individuating Quantities. *Philosophical Studies: An International Journal for Philosophy in the Analytic Tradition*, 176(4):853–878, 2019. ISSN 00318116, 15730883. URL http://www.jstor.org/stable/45147348.

G. Tastevin and F. Laloë. The outcomes of measurements in the de Broglie-Bohm theory, 2021.

C. Tavris. *The Mismeasure of Women*. Touchstone, 1992. URL https://books.google.com/books?id=PcipzwEACAAJ.

M. Tegmark. *Our Mathematical Universe: My Quest for the Ultimate Nature of Reality*. Knopf Doubleday Publishing Group, 2015.

P. Teller. *Conditionalization, Observation, and Change of Preference*. Springer Netherlands, 1976.

P. Teller. Whither Constructive Empiricism? *Philosophical Studies: An International Journal for Philosophy in the Analytic Tradition*, 106(1/2):123–150, 2001. ISSN 00318116, 15730883. URL http://www.jstor.org/stable/4321194.

K.P. Y. Thébault and Richard Dawid. Many Worlds: Decoherent or Incoherent? *Synthese*, 192(5):1559–1580, 2015. doi: 10.1007/s11229-014-0650-8.

C. G. Timpson. On a Supposed Conceptual Inadequacy of the Shannon Information in Quantum Mechanics. *Studies in History and Philosophy of Science Part B: Studies in History and Philosophy of Modern Physics*, 34(3):441–468, 2003. doi: 10.1016/s1355-2198(03)00037-6.

M.G. Titelbaum. The Relevance of Self-Locating Beliefs. *The Philosophical Review*, 117(4):555–605, 2008. ISSN 00318108, 15581470. URL http://www.jstor.org/stable/40606047.

M.G. Titelbaum. Tell Me You Love Me: Bootstrapping, Externalism, and No-Lose Epistemology. *Philosophical Studies*, 149(1):119–134, 2010. doi: 10.1007/s11098-010-9541-0.

R. Tumulka. A Relativistic Version of the Ghirardi Rimini Weber Model. *Journal of Statistical Physics*, 125:821–840, November 2006. doi: 10.1007/s10955-006-9227-3.

R. Tumulka. *A Relativistic GRW Flash Process with Interaction*, pages 321–347. Springer International Publishing, Cham, 2021.

P. Unger. The Problem of the Many. *Midwest Studies in Philosophy*, 5(1):411–468, 1980. doi: 10.1111/j.1475-4975.1980.tb00416.x.

P. Urbach and C. Howson. *Scientific Reasoning: The Bayesian Approach*. Open Court, Chicago, 1993.

J.-P. Uzan. Varying Constants, Gravitation and Cosmology. Living Reviews in Relativity. 14 (2011): 2, Sep. 2010.

A. Valentini. *On the pilot-wave theory of classical, quantum and subquantum physics*. PhD thesis, SISSA, Trieste, 1992.

A. Valentini. Hidden variables, statistical mechanics and the early universe. In J. Bricmont, G. Ghirardi, D. Dürr, F. Petruccione, M. C. Galavotti, and N. Zanghi, editors, *Chance in Physics. Lecture Notes in Physics*, vol 574. Springer, Berlin, Heidelberg, 2001. https://doi.org/10.1007/3-540-44966-3_12

A. Valentini. De Broglie–Bohm Pilot-Wave Theory: Many Worlds in Denial? In *Many Worlds?: Everett, Quantum Theory, and Reality*. Oxford University Press, 06 2010. ISBN 9780199560561. doi: 10.1093/acprof:oso/9780199560561.003.0019. URL https://doi.org/10.1093/acprof:oso/9780199560561.003.0019.

B.C Van Fraassen. Laws and symmetry. Oxford University Press, 1989.

B.C. van Fraassen. Relational Quantum Mechanics: Rovelli's World. *Discusiones Filosóficas*, 11(17):13–51, 2010.

B.C. van Fraassen. *The Scientific Image*. Clarendon Library of Logic and Philosophy. Clarendon Press, 1980. ISBN 9780198244271. URL https://books.google.co.uk/books?id=VLz2F1zMr9QC.

B.C. van Fraassen. *Scientific Representation: Paradoxes of Perspective*. OUP Oxford, 2008. ISBN 9780191613746. URL https://books.google.com/books?id=C8-k2adtM2UC.

B. Van Fraassen. Explanation Through Representation, and its Limits. *Epistemologia*, 1: 30–46, 2012. doi: 10.3280/epis2012-001003.

A. Vanrietvelde, P.A. Hoehn, F. Giacomini, and E. Castro-Ruiz. A change of perspective: switching quantum reference frames via a perspective-neutral framework. *Quantum*, 4:225, Jan 2020. ISSN 2521-327X. doi: 10.22331/q-2020-01-27-225. URL http://dx.doi.org/10.22331/q-2020-01-27-225.

M. Velmans. Goodbye to Reductionism: Complementary First and Third-Person Approaches to Consciousness. In S.R. Hameroff, A.W. Kaszniak, and A. Scott, editors, *Toward a Science of Consciousness II: The Second Tucson Discussions and Debates*, pages 45–52. MIT Press, 1998.

P.E. Vermaas. *A Philosopher's Understanding of Quantum Mechanics: Possibilities and Impossibilities of a Modal Interpretation*. Cambridge University Press, 1999.

R. Viertl. Fuzzy Bayesian Inference. In D. Dubois, M. Asunción Lubiano, H. Prade, M. Ángeles Gil, P. Grzegorzewski, and O. Hryniewicz, editors, *Soft Methods for Handling Variability and Imprecision*, pages 10–15, Berlin, Heidelberg, 2008. Springer Berlin Heidelberg.

A. Vilenkin. Predictions from Quantum Cosmology. *Phys. Rev. Lett.*, 74:846–849, Feb 1995. doi: 10.1103/PhysRevLett.74.846. URL https://link.aps.org/doi/10.1103/PhysRevLett.74.846.

S. Vineberg. Dutch Book Arguments. In E.N. Zalta and U. Nodelman, editors, *The Stanford Encyclopedia of Philosophy*. Metaphysics Research Lab, Stanford University, Fall 2022 edition, 2022.

J. von Neumann. *Mathematical Foundations of Quantum Mechanics: New Edition*. Princeton University Press, 1932. ISBN 9780691178561. URL https://books.google.co.uk/books?id=B3OYDwAAQBAJ.

D. Wallace. *The Emergent Multiverse: Quantum Theory according to the Everett Interpretation*. OUP Oxford, 2012a. ISBN 9780191057397. URL https://books.google.co.uk/books?id=jNaSBAAAQBAJ.

D. Wallace. In defence of naivete: The conceptual status of Lagrangian QFT, 2001.

D. Wallace. Everett and Structure. *Studies in History and Philosophy of Science Part B: Studies in History and Philosophy of Modern Physics*, 34(1):87–105, 2003.

D. Wallace. Epistemology Quantized: Circumstances in which We should come to Believe in the Everett Interpretation. *British Journal for the Philosophy of Science*, 57(4):655–689, 2006. doi: 10.1093/bjps/axl023.

D. Wallace. A formal proof of the Born rule from decision-theoretic assumptions, 2009. URL https://arxiv.org/abs/0906.2718.

D. Wallace. How to Prove the Born Rule. In S. Saunders, J. Barrett, A. Kent, and D. Wallace, editors, *Many Worlds?: Everett, Quantum Theory & Reality*. Oxford University Press, 2010.

D. Wallace. The Everett Interpretation. In R. Batterman, editor, *The Oxford Handbook of Philosophy of Physics, Oxford Handbooks*, Oxford Academic, online edn, 5 Sept. 2013. https://doi.org/10.1093/oxfordhb/9780195392043.013.0014, accessed 21 Mar. 2025.

D. Wallace. Decoherence and its role in the modern measurement problem. *Philosophical Transactions of the Royal Society A: Mathematical, Physical and Engineering Sciences*, 370(1975):4576–4593, Sep 2012b. ISSN 1471-2962. doi: 10.1098/rsta.2011.0490. URL http://dx.doi.org/10.1098/rsta.2011.0490.

D. Wallace. On the Plurality of Quantum Theories: Quantum theory as a framework, and its implications for the quantum measurement problem, 2018. URL http://philsci-archive.pitt.edu/15292/.

D. Wallace. Naturalness and Emergence, February 2019. URL http://philsci-archive.pitt.edu/15757/.

D. Wallace. The sky is blue, and other reasons quantum mechanics is not underdetermined by evidence, May 2022. URL http://philsci-archive.pitt.edu/20537/.

C. Wallmann and J. Williamson. The Principal Principle and Subjective Bayesianism. *European Journal for Philosophy of Science*, 10(1):1–14, 2019. doi: 10.1007/s13194-019-0266-4.

S.A. Walter. Ether and electrons in relativity theory. In J. Navarro, editor, *Ether and Modernity*, pages 67–87. Oxford University Press, 2018.

S. Weinberg. Anthropic bound on the cosmological constant. *Physical Review Letters*, 59(22): 2607–2610, 1987.

S. Weinberg. *10. Reductionism Redux*, pages 107–122. Harvard University Press, Cambridge, MA and London, England, 2001. ISBN 9780674066403. doi: doi:10.4159/9780674066403-011. URL https://doi.org/10.4159/9780674066403-011.

H. Weyl. *Symmetry*. Princeton Science Library. Princeton University Press, 1952. ISBN 9780691023748. URL https://books.google.com/books?id=T43Cmu_EaZAC.

K. Wharton. The Universe is not a Computer. In B. Aguirre, A. Foster and G. Merali, editors, *Questioning the Foundations of Physics*, pages 177–190. Springer, November 2015.

K. Wharton. A New Class of Retrocausal Models. *Entropy*, 20(6):410, May 2018. ISSN 1099-4300. doi: 10.3390/e20060410. URL http://dx.doi.org/10.3390/e20060410.

R. White. Fine-tuning and multiple universes. *Noûs*, 34 (2):260–276, 2000.

E. P. Wigner. *Remarks on the Mind-Body Question*, pages 247–260. Springer Berlin Heidelberg, Berlin, Heidelberg, 1961.

P. Williams. Naturalness, the Autonomy of Scales, and the 125Gev Higgs. *Studies in History and Philosophy of Science Part B: Studies in History and Philosophy of Modern Physics*, 51:82–96, 2015. doi: 10.1016/j.shpsb.2015.05.003.

P. Williams. Two Notions of Naturalness. *Foundations of Physics*, 49(9):1022–1050, 2019. doi: 10.1007/s10701-018-0229-1.

A. Wilson. Everettian Quantum Mechanics without Branching time. *Synthese*, 188(1):67–84, 2012. ISSN 00397857, 15730964. URL http://www.jstor.org/stable/41681629.

Y. Ying, M.M. Ansanelli, A. Di Biagio, E. Wolfe, and E. Gama Cavalcanti. Relating Wigner's Friend scenarios to Nonclassical Causal Compatibility, Monogamy Relations, and Fine Tuning, 2023.

A. Zeilinger. A Foundational Principle for Quantum Mechanics. *Foundations of Physics*, 29(4):631–643, 1999. doi: 10.1023/A:1018820410908.

A. Zeilinger. *Bell's Theorem, Information and Quantum Physics*, pages 241–254. Springer Berlin Heidelberg, Berlin, Heidelberg, 2002. ISBN 978-3-662-05032-3. doi: 10.1007/978-3-662-05032-3_17. URL https://doi.org/10.1007/978-3-662-05032-3_17.

Index